Lecture Notes in Mathematics

Edited by A. Dold and B. Eckmann

674

Zbigniew Fiedorowicz
Stewart Priddy

Homology of Classical Groups Over Finite Fields and Their Associated Infinite Loop Spaces

Springer-Verlag
Berlin Heidelberg New York 1978

Authors
Zbigniew Fiedorowicz
Department of Mathematics
University of Michigan
Ann Arbor, Michigan 48104/USA

Stewart Priddy
Department of Mathematics
Northwestern University
Evanston, Illinois 60201/USA

AMS Subject Classifications (1970): 18H10, 20G40, 55D35, 55F40

ISBN 3-540-08932-2 Springer-Verlag Berlin Heidelberg New York
ISBN 0-387-08932-2 Springer-Verlag New York Heidelberg Berlin

Printing and binding: Beltz Offsetdruck, Hemsbach/Bergstr.
2141/3140-543210

Table of Contents

Introduction

One of the hallmarks of significant advances in mathematics is the discovery of unexpected relationships between diverse areas. Perhaps the most notable recent example of such an advance is D. Quillen's development of higher algebraic K-theory in which algebra and topology are related in a new and fundamental way. On the one hand higher algebraic K-theory introduces topological methods for defining algebraic invariants, such as the higher K-groups of rings. On the other it provides a machine for translating algebraic concepts into topological concepts.

In this monograph we will study one concrete aspect of the subject. From the viewpoint of algebra we look at classical groups π over finite fields $\mathbf{F}_q$ of characteristic p, including among others $\pi = GL(n,\mathbf{F}_q)$, $O(n,\mathbf{F}_q)$, $Sp(2n,\mathbf{F}_q)$, $U(n,\mathbf{F}_{q^2})$. From this viewpoint one is interested in computing for instance, the homology and cohomology of these groups as well as the higher K-groups associated to these groups.

From the viewpoint of topology we look at the infinite loop spaces X associated with the homotopy fiber of

$$\psi^q - 1\colon BG \to BG$$

where ψ^q is the Adams operation in KG-theory and G is one of the continuous classical groups, U, O, Sp. One of the reasons for our interest in these spaces is that they are intimately associated with the J-homomorphism $J\colon SO \to SF$ of G. Whitehead. Indeed the affirmative solution of the Adams conjecture [24] implies the existence of commutative diagrams (localized at a prime ℓ)

$$\begin{array}{ccc} SO & \xrightarrow{J} & SF \\ & {\scriptstyle\tau}\searrow \quad \nearrow{\scriptstyle\sigma} & \\ & JO(q) & \end{array}$$

where $JO(q)$ is the homotopy fiber of

$$BO \xrightarrow{\psi^q - 1} BO$$

where q is a primitive root of unity mod ℓ^2 if ℓ is odd and $q \equiv \pm 3 \pmod 8$ if $\ell = 2$. Under σ, $\pi_* JO(q)$ is then essentially the ℓ-primary component of $J_*: \pi_*(SO) \to \pi_*(SF)$ (cf. May [17] for further details).

These two apparently unrelated topics are connected by homology equivalences of the form

$$K(\pi,1) \to X$$

which we use to compute the homology of the groups and to study the loop structure of the spaces. It is interesting to note that Kan and Thurston [14] have recently shown that to any connected X there corresponds a group π and a homology equivalence of this type.

A glossary of the groups and spaces studied in this paper is listed at the end of this introduction.

Before describing the contents of this paper we shall first

discuss a specific representative example: let $O(n,\mathbf{F}_q)$ be the group of $n \times n$ matrices M over $\mathbf{F}_q$ which leave invariant the quadratic form $x_1^2 + \cdots + x_n^2$ (i.e. $MM^t = I_n$). One of our main results (Theorem III 3.1d) establishes an equivalence of infinite loop spaces

$$\Gamma_0 B\mathcal{O}(\mathbf{F}_q) \xrightarrow{\approx} JO(q)$$

for q odd, where $\Gamma_0 B\mathcal{O}(\mathbf{F}_q)$ is the zero component of the infinite loop space constructed from the classifying spaces $BO(n,\mathbf{F}_q)$. (As a space $\Gamma_0 B\mathcal{O}(\mathbf{F}_q)$ is homotopic to Quillen's plus construction on the classifying space $BO(\infty,\mathbf{F}_q)$.)

Using this equivalence we compute the mod ℓ cohomology ring of $O(n,\mathbf{F}_q)$ $(0 \leq n \leq \infty)$ and for $\ell = 2$ derive explicit formulas for the Dyer-Lashof operations of $JO(q)$.

The principal method of this paper is Quillen's technique of "Brauer lifting" used in [25] to study the general linear groups $GL(n,\mathbf{F}_q)$ via a diagram of the form

$$\begin{array}{ccccccc} U & \longrightarrow & JU(q) & \longrightarrow & BU & \xrightarrow{\psi^q - 1} & BU \\ & & & \nwarrow^{\lambda} & \uparrow{\scriptstyle\beta} & & \\ & & & & \Gamma_0 B\mathcal{GL}(\mathbf{F}_q) & & \end{array}$$

where $\Gamma_0 B\mathcal{GL}(\mathbf{F}_q)$ is the infinite loop space associated with the classifying spaces $BGL(n,\mathbf{F}_q)$ and where the horizontal maps form a fiber sequence. The map β is derived from Brauer lifting. The lift λ is unique since

$$[\Gamma_0 B\mathcal{GL}(\mathbf{F}_q),U] = KU^{-1}(\Gamma_0 B\mathcal{GL}(\mathbf{F}_q)) \cong KU^{-1}(BGL(\infty,\mathbf{F}_q)) = 0$$

The main result of [25] states that λ is a homotopy equivalence.

In applying Quillen's method to the orthogonal case one constructs an analogous diagram

$$\begin{array}{ccccccc} SO & \longrightarrow & JO(q) & \longrightarrow & BO & \xrightarrow{\psi^q - 1} & BSO \\ & & & \nwarrow^{\lambda} & \uparrow \beta & & \\ & & & & \Gamma_0 B\mathcal{O}(\mathbf{F}_q) & & \end{array}$$

At this point however, one encounters the problem that $KO^{-1}(\Gamma_0 B\mathcal{O}(\mathbf{F}_q)) \neq 0$. Hence λ is not uniquely determined by the diagram and if chosen incorrectly will fail to be a homotopy equivalence. In Theorems IV 2.4 and V 2.3, this problem is overcome by showing that if λ is chosen to be an H-map then it is automatically a homotopy equivalence at ℓ.

We should point out that in a recent paper [13], E. M. Friedlander using the methods of etale homotopy theory has established that there is an equivalence of spaces

$$\Gamma_0 B\mathcal{O}(\mathbf{F}_q) \xrightarrow{\cong} JO(q)$$

However due to the above difficulties there seems to be no apparent way of showing that his map is an infinite loop.

This monograph consists of nine chapters and one appendix. In Chapter I we study the spaces X defined by fibrations

$$X \to BG \xrightarrow{\psi^q - 1} BG$$

and compute their mod-ℓ and integral homology. Permutative categories associated with classical groups over a finite field are introduced in Chapter II; these categories give rise to infinite loop spaces which are later compared with the spaces of Chapter I.

In Chapter III we construct the Brauer lift maps relating the spaces of Chapters I and II. Chapter III also contains a statement of our main results and a detailed outline of their proof. The reader who wishes to gain a general impression of this monograph is urged to read this chapter first, occasionally refering back to Chapters I and II for the necessary definitions.

The actual proof of these results is broken down in the following way. In Chapter IV (resp. V) we present our calculations at the prime $\ell = 2$ (resp. ℓ odd). These calculations are based on Chapter VI, where the homology of various small finite groups is computed, and on Chapter VII (resp. VIII), where products of these small groups are used to detect the homology of the classical groups over finite fields.

Finally in Chapter IX we compute mod-2 Dyer-Lashof homology operations for all the spaces associated with the various finite classical groups. The Appendix contains corresponding computations of the multiplicative homology operations associated with the tensor product in the orthogonal groups.

More detailed introductions appear at the beginning of each chapter. The main logical interrelationships between the various chapters are shown in the following diagram

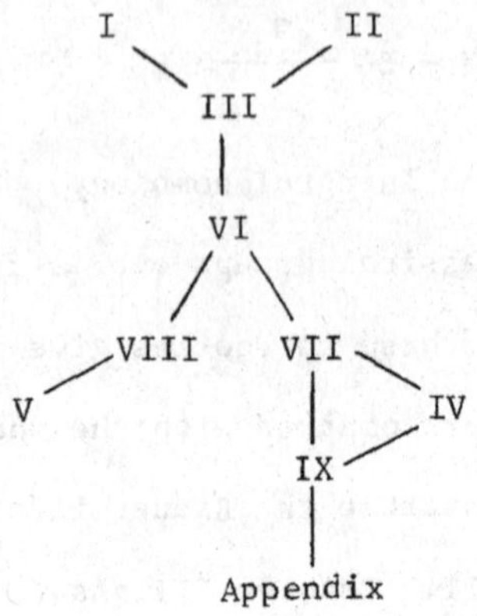

It is a pleasure to acknowledge useful conversations with Frank Adams, Ib Madsen, and Dan Quillen. We are especially grateful to Peter May for his continued interest during the course of this research and for many suggested improvements to an earlier draft.

Our thanks also go to Mrs. Vicki Davis for her fast and efficient typing of this monograph.

Conventions: Throughout this paper all homology and cohomology groups are taken with simple coefficients in $\mathbb{Z}/\ell$, the integers modulo ℓ, unless explicitly indicated to the contrary. The symbols O, Sp, U, Spin will denote the infinite classical groups over $\mathbb{R}$.

Glossary

For the convenience of the reader we give a complete list of the classical groups π and their corresponding infinite loop spaces X studied in this paper

a) $GL(n,\mathbb{F}_q)$ - general linear group.

b) $SL(n,\mathbb{F}_q)$ - special linear group: matrices with determinant = 1.

c) $O(n,\mathbb{F}_q)$ - orthogonal group: matrices preserving the form

$x_1^2 + \cdots + x_n^2$, also denoted $O_+(n,\mathbb{F}_q)$.

d) $O_-(n,\mathbb{F}_q)$ - extraordinary orthogonal group: matrices preserving the form $\mu x_1^2 + x_2^2 + \cdots + x_n^2$, μ a non-square.

e) $SO(n,\mathbb{F}_q)$ - special orthogonal group: orthogonal matrices with determinant = 1.

f) $N(n,\mathbb{F}_q)$ - orthogonal matrices with spinor norm = 1.

g) $ND(n,\mathbb{F}_q)$ - orthogonal matrices with (spinor Norm) $\times$ (det) = 1.

h) $SN(n,\mathbb{F}_q) = SO(n,\mathbb{F}_q) \cap N(n,\mathbb{F}_q)$.

i) $Spin(n,\mathbb{F}_q)$ - spinor groups.

j) $Sp(2n,\mathbb{F}_q)$ - symplectic group: matrices preserving the alternating form $\langle x,y\rangle = \Sigma_{i=1}^n (x_{2i}y_{2i-1} - x_{2i-1}y_{2i})$.

k) $U(n,\mathbb{F}_q)$ - unitary group: matrices over $\mathbb{F}_{q^2}$ preserving the Hermitian form $\langle x,y\rangle = \Sigma_{i=1}^n x_i y_i^q$.

ℓ) $SU(n,\mathbb{F}_{q^2})$ - special unitary group: unitary matrices with determinant = 1.

The following fiber sequences define the spaces X (as fibers) corresponding to the groups π.

a') $JU(q) \longrightarrow BU \xrightarrow{\psi^q - 1} BU$

b') $JSU(q) \longrightarrow BSU \xrightarrow{\psi^q - 1} BSU$

c') $JO(q) \longrightarrow BO \xrightarrow{\psi^q - 1} BSO$

d') $J\bar{O}(q) \longrightarrow BO \xrightarrow{\psi^q - 1} BO$

e') $JSO(q) \longrightarrow BSO \xrightarrow{\psi^q - 1} BSO$

f',g') $J(q) \longrightarrow BO \xrightarrow{\psi^q - 1} BSpin$

h') $\tilde{J}(q) \longrightarrow BSO \xrightarrow{\psi^q - 1} BSpin$

i') $JSpin(q) \longrightarrow BSpin \xrightarrow{\psi^q - 1} BSpin$

j') $JSp(q) \longrightarrow BSp \xrightarrow{\psi^q - 1} BSp$

k') $JU(-q) \longrightarrow BU \xrightarrow{\psi^{-q} - 1} BU$

ℓ') $JSU(-q) \longrightarrow BSU \xrightarrow{\psi^{-q} - 1} BSU$

We note that the two components of $X = J\bar{O}(q)$ correspond to $\pi = O_{\pm}(n, \mathbf{F}_q)$. Also $\pi = N(n, \mathbf{F}_q)$, $ND(n, \mathbf{F}_q)$ both correspond to $J(q)$.

If q is even, the orthogonal case is somewhat special (see Theorem III 3.1(ℓ-o)).

Finally we note that in addition to Quillen's basic papers [34, 35] and Friedlander's work [18], various aspects of the cohomology of classical groups over finite fields have been studied by Shapiro [43] and DeConcini [42].

Several applications of our work to topology may be found in May's papers [25,27].

Chapter I

Infinite Loop Spaces Associated with ImJ

§1. Introduction

In this chapter we study the infinite loop spaces (and their variants) defined by the fibrations

$$J\overline{O}(q) \to BO \xrightarrow{\psi^q - 1} BO$$

$$JSp(q) \to BSp \xrightarrow{\psi^q - 1} BSp$$

$$JU(\pm q) \to BU \xrightarrow{\psi^{\pm q} - 1} BU$$

where $q > 0$ and ψ^q is the Adams operation [2]. These spaces are of interest to homotopy theorists because they are related to the real, symplectic and complex J homomorphisms as described in the main introduction . For this reason we call them image of J spaces.

Our main purpose is to compute the integral and mod ℓ homology (and cohomology) algebras of the image of J spaces for ℓ a prime not dividing q. It follows from 1.1 below that the homotopy groups of these spaces are finite without torsion dividing q, i.e. localized at q each of these spaces has the homotopy type of a point. For this reason we shall assume all spaces in this chapter are <u>localized away from</u> q. It follows that Ψ^q is an infinite loop map in this case.

(1.1) For reference we record the action of ψ^q in homotopy and cohomology

$$(\psi^{\pm q})_* = \text{multiplication by } (\pm q)^k \text{ on } \pi_{2k}BU = \mathbb{Z}$$

$$(\psi^q)_* = \text{multiplication by } \begin{cases} q^{2k} \text{ on } \pi_{4k}BO = \mathbb{Z} \\ q \text{ on } \pi_{8k+1}BO \approx \pi_{8k+2}BO = \mathbb{Z}/2 \\ q^{2k} \text{ on } \pi_{4k}BSp = \mathbb{Z} \\ q \text{ on } \pi_{8k+5}BSp \approx \pi_{8k+6}BSp = \mathbb{Z}/2 \end{cases}$$

These results follow from Adams [2], Bott periodicity, and the fact that η (resp. η^2) composed with a generator of $\pi_{8k}BO$ generates $\pi_{8k+1}BO$ (resp. $\pi_{8k+2}BO$).

It is now a simple exercise to show

(1.2) $(\psi^{\pm q})^*(c_k) = (\pm q)^k c_k$ on $H^*(BU;\ \mathbb{Z}) = \mathbb{Z}[c_1, c_2, \ldots]$

$(\psi^q)^*(w_k) = q^k w_k$ on $H^*(BO;\ \mathbb{Z}/2) = \mathbb{Z}/2[w_1, w_2, \ldots]$

$(\psi^q)^*(\bar{p}_k) = q^{2k}\bar{p}_k$ on $H^*(BO;\ \mathbb{Z}/\ell) = \mathbb{Z}/\ell[\bar{p}_1, \bar{p}_2, \ldots]$, ℓ odd prime

$(\psi^q)^*(\bar{g}_k) = q^{2k}\bar{g}_k$ on $H^*(BSp;\ \mathbb{Z}) = \mathbb{Z}[\bar{g}_1, \bar{g}_2, \ldots]$.

Here c_k is the Chern class of degree 2k; w_k is the Stiefel-Whitney class of degree k; $\bar{p}_k$ (resp. $\bar{g}_k$) is the real (resp. symplectic) Pontryagin class of degree 4k.

§2. The real image of J spaces

Consider the infinite loop map

$$\psi^q-1\colon BO \to BO$$

Let $r\colon J\bar{O}(q) \to BO$ denote the homotopy fiber of ψ^q-1. Thus we obtain a fibration sequence of infinite loop spaces

$$(2.1) \qquad \cdots \to 0 \xrightarrow{\tau} J\bar{O}(q) \xrightarrow{r} BO \xrightarrow{\psi^q-1} BO \to \cdots$$

From the long exact homotopy sequence of 2.1 and 1.1 we see that

$$\pi_0(J\bar{O}(q)) = \begin{cases} 0 & q \text{ even} \\ \mathbb{Z}/2 & q \text{ odd} \end{cases}$$

Thus $J\bar{O}(q)$ has one or two components depending on whether q is even or odd. In the latter case we shall be interested in the 0-component of $J\bar{O}(q)$ and its connected covers and so for the remainder of this section we assume <u>q odd</u>. From 1.1 we observe that ψ^q-1 lifts uniquely to an infinite loop map

$$\psi^q-1\colon BO \to BSO$$

Let $r\colon JO(q) \to BO$ be the fiber of ψ^q-1. Then we obtain the following fibration sequence of infinite loop spaces

$$(2.2) \qquad \cdots \to SO \xrightarrow{\tau} JO(q) \xrightarrow{r} BO \xrightarrow{\psi^q-1} BSO \to \cdots$$

There is an infinite loop map $j\colon JO(q) \to J\bar{O}(q)$ such that the following diagram is homotopy commutative as infinite loop maps:

$$\begin{array}{ccccccc}
SO & \xrightarrow{\tau} & JO(q) & \xrightarrow{r} & BO & \xrightarrow{\psi^q-1} & BSO \\
\downarrow & & \downarrow j & & \| & & \downarrow \\
O & \longrightarrow & J\bar{O}(q) & \xrightarrow{r} & BO & \xrightarrow{\psi^q-1} & BO
\end{array}$$

and so up to homotopy $j\colon JO(q) \to J\bar{O}(q)$ is just the inclusion of the 0-component in $J\bar{O}(q)$.

From 1.1 and the long exact homotopy sequence of 2.2 we see that $\pi_1 JO(q)$ has order four. We label the nonzero elements of $\pi_1(JO(q))$ by the letters $\delta_1, \delta_2, \delta_3$ where we take $r_{\#}(\delta_1) = r_{\#}(\delta_2) = \rho$ and $\tau_{\#}(\theta) = \delta_3$. Here ρ, θ denote the nonzero elements of $\pi_1(BO)$, $\pi_1(SO)$ respectively. We also note that the inclusion

$$i\colon \mathbb{R}P^\infty = BO(1) \to BO$$

becomes zero when composed with $\psi^q-1\colon BO \to BSO$. Hence i lifts to a map $k_1\colon \mathbb{R}P^\infty \to JO(q)$. We may assume that $k_{1\#}(\iota) = \delta_1$. (Here ι denotes the non-zero element of $\pi_1(\mathbb{R}P^\infty)$.) Let $\gamma\colon \mathbb{R}P^\infty \to SO$ denote any map which is non trivial on π_1. Then $k_2 = k_1 + \tau \circ \gamma$ is another lift of $i\colon \mathbb{R}P^\infty \to JO(q)$ with $k_{2\#}(\iota) = \delta_2$.

We say that a lift $k\colon \mathbb{R}P^\infty \to JO(q)$ of $i\colon \mathbb{R}P^\infty \to BO$ is of <u>type 1</u> if $k_{\#}(\iota) = \delta_1$ and of <u>type 2</u> if $k_{\#}(\iota) = \delta_2$. The existence of a lift k of $i\colon \mathbb{R}P^\infty \to BO$ also implies that

$$\pi_1 JO(q) \approx \mathbb{Z}/2 \oplus \mathbb{Z}/2$$

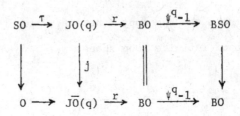

Again using 1.1 we can lift ψ^q-1 further to an infinite loop map

$$\psi^q-1\colon BO \to BSpin$$

Let $\bar{r}\colon J(q) \to BO$ denote the fiber of ψ^q-1. Then the following is also a fibration sequence of infinite loop spaces

$$\cdots \to Spin \xrightarrow{\bar{\tau}} J(q) \xrightarrow{\bar{r}} BO \xrightarrow{\psi^q-1} BSpin \to \cdots$$

We can then find an infinite loop map $f\colon J(q) \to JO(q)$ which makes the following diagram homotopy commutative as infinite loop maps

$$\begin{array}{ccccccc}
Spin & \xrightarrow{\bar{\tau}} & J(q) & \xrightarrow{\bar{r}} & BO & \xrightarrow{\psi^q-1} & BSpin \\
\downarrow & & \downarrow{\scriptstyle f} & & \| & & \downarrow \\
SO & \xrightarrow{\tau} & JO(q) & \xrightarrow{r} & BO & \xrightarrow{\psi^q-1} & BSO
\end{array}$$

If we look at the corresponding diagram of homotopy groups we see that $f_{\#}$ is an isomorphism in all dimensions except

$$\mathrm{im}(f_{\#}\colon \pi_1 J(q) \to \pi_1 JO(q)) = \mathbb{Z}/2$$

generated by δ_t where t is either 1 or 2. Let

$$g_s\colon JO(q) \to \mathbb{R}P^\infty \qquad s = 1,2,3$$

denote the unique map which induces an epimorphism on π_1 and $g_{s_\#}(\delta_s) = 0$. It is easy to see that g_s is an infinite loop map and that $g_t \circ f$ is a trivial infinite loop map. Let $J_s(q) \to JO(q)$ denote the fiber of g_s. Then f factors through an infinite loop map $\overline{f}$

$$\begin{array}{ccccc} & & J(q) & & \\ & \overline{f} \swarrow & \downarrow f & & \\ J_t(q) & \to & JO(q) & \xrightarrow{g_t} & \mathbb{R}P^\infty \end{array}$$

It easily follows that $\overline{f}$ induces an isomorphism on homotopy groups. Hence $\overline{f}$ is an equivalence of infinite loop spaces. In other words $J(q)$ is equivalent as an infinite loop space either to $J_1(q)$ or to $J_2(q)$. However we also have

<u>Proposition</u> 2.3. $J_1(q)$ and $J_2(q)$ are equivalent as infinite loop spaces.

<u>Proof</u>: Let $\Delta\colon BO \to SO$ be the Bott map. Then

$$\varphi = 1 + \tau\Delta r\colon JO(q) \to JO(q)$$

is an equivalence of infinite loop spaces since $2\Delta \simeq 0$ implies

$$\varphi \cdot \varphi \simeq 1$$

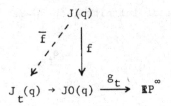

Further since $\Delta_{\#}(\rho) = \theta$ it follows that the diagram

$$\begin{array}{ccc} JO(q) & \xrightarrow{g_1} & \mathbb{R}P^{\infty} \\ \downarrow \varphi & & \downarrow id \\ JO(q) & \xrightarrow{g_2} & \mathbb{R}P^{\infty} \end{array}$$

is homotopy commutative as infinite loop maps. Thus φ induces the desired equivalence of $J_1(q)$ and $J_2(q)$.

Finally we note that

$$h: \mathbb{R}P^{\infty} \times J(q) \to JO(q)$$

given by $h(x,y) = k(x) + f(y)$ where k is a lift of type s, $s \neq t$, induces an isomorphism on all homotopy groups so that

$$JO(q) \approx \mathbb{R}P^{\infty} \times J(q)$$

as spaces although certainly not as infinite loop spaces.

Similarly we can restrict ψ^q-1: BO $\to$ BSO to an infinite loop map

$$\psi^q-1: BSO \to BSO$$

Let r': JSO(q) $\to$ BSO be the fiber of this map. Also we can lift to an infinite loop map

$$\psi^q - 1 \colon BSO \to BSpin$$

We call the fiber of this map $\tilde{r} \colon \tilde{J}(q) \to BSO$. By the same methods as above we can see that there is an infinite loop map $f' \colon JSO(q) \to JO(q)$ which is the fiber of $g_3 \colon JO(q) \to \mathbb{R}P^\infty$. Similarly there is a map $\tilde{f} \colon \tilde{J}(q) \to JO(q)$ which is the fiber of a map $g \colon JO(q) \to K(\mathbb{Z}/2 \oplus \mathbb{Z}/2, 1)$ which induces an isomorphism in π_1.

Similar arguments as before show that as spaces

$$J(q) \approx \mathbb{R}P^\infty \times \tilde{J}(q)$$

$$JSO(q) \approx \mathbb{R}P^\infty \times \tilde{J}(q)$$

$$JO(q) \approx \mathbb{R}P^\infty \times JSO(q)$$

$$JO(q) \approx \mathbb{R}P^\infty \times \mathbb{R}P^\infty \times \tilde{J}(q)$$

Thus as spaces $J(q)$ and $JSO(q)$ are equivalent. However, as we shall see later, $J(q)$ and $JSO(q)$ are very different as infinite loop spaces.

Finally we observe that we can restrict $\psi^q - 1 \colon BSO \to BSpin$ to an infinite loop map

$$\psi^q - 1 \colon BSpin \to BSpin$$

Let $r'' \colon JSpin(q) \to BSpin$ denote the fiber of this map. By the same reasoning as above, we see that there is an infinite loop map $f'' \colon JSpin(q) \to \tilde{J}(q)$ which is the fiber of the infinite loop map $g'' \colon \tilde{J}(q) \to K(\mathbb{Z}_2, 2)$ which induces an isomorphism in π_2. It follows that ff": JSpin(q) → JO(q) is the 2-connected cover of JO(q).

§3. The mod-2 homology of JO(q)

In this section we assume q is odd. From 2.1 we have a map of fibrations

$$\begin{array}{llcccc} (\alpha) & & SO & \xrightarrow{\tau} & JO(q) & \longrightarrow & BO \\ & & \| & & \downarrow & & \downarrow \psi^q - 1 \\ (\beta) & & SO & \longrightarrow & PBSO & \longrightarrow & BSO \end{array}$$

where (β) is the loop-path fibration. The following proposition is due essentially to Clough [14].

Proposition 3.1. The $\mathbb{Z}/2$-homology Serre spectral sequence of the fibration

$$SO \xrightarrow{\tau} JO(q) \xrightarrow{r} BO$$

collapses. Hence, as an algebra

$$H_*(JO(q)) = \mathbb{Z}/2[\hat{v}_1, \hat{v}_2, \ldots] \otimes E[\hat{u}_1, \hat{u}_2, \ldots]$$

where the $\hat{v}_i$'s are elements such that $r_*(\hat{v}_i) = \bar{e}_i$ ($\bar{e}_i$'s being the standard algebra generators of $H_*(BO) = \mathbb{Z}/2[\bar{e}_1, \bar{e}_2, \ldots]$) and $\hat{u}_i = \tau_*(u_i)$ where the u_i's are the standard algebra generators of $H_*(SO) = E[u_1, u_2, \ldots]$.

Proof: The fibration (α) is induced from (β) by $\psi^q - 1$. Since the homology and cohomology Serre spectral sequences of (β), ($E_{*,*}(\beta)$, $E^{*,*}(\beta)$) obviously have trivial local coefficients the same is true of (α).

We have $H^*(SO) = \mathbb{Z}/2[a_1, a_3, a_5, \ldots]$ where a_{2n-1} transgresses to the Stiefel-Whitney class ω_{2n} in $E^{*,*}(\beta)$. By naturality a_{2n-1} is transgressive in $E^{*,*}(\alpha)$ and transgresses to $(\psi^q - 1)^*(\omega_{2n}) = 0$ (by 1.2). Thus $E^{*,*}(\alpha)$ must collapse. By duality $E_{*,*}(\alpha)$ also collapses.

If we pick $\hat{v}_i$'s such that $r_*(\hat{v}_i) = \bar{e}_i$ and set $\hat{u}_i = \tau_*(u_i)$ it follows that $H_*(JO(q))$ is generated as an algebra by $\hat{v}_i, \hat{u}_i$, $i \geq 1$. We have $\hat{u}_i^2 = \tau_*(u_i^2) = 0$. For dimensional reasons there can be no other algebraic relations between the $\hat{v}_i$'s and $\hat{u}_i$'s. Q.E.D.

We can describe the generators $\hat{v}_i$ more explicitly as $\hat{v}_i = k_*(x_i)$ where $x_i \in H_i(\mathbb{R}P^\infty)$ denotes the nonzero element and $k\colon \mathbb{R}P^\infty \to JO(q)$ is a lift of type 1 (see §2) of the inclusion $i\colon \mathbb{R}P^\infty \to BO$. We note that this definition of the $\hat{v}_i$'s is independent of choice of k. For if we take $k'\colon \mathbb{R}P^\infty \to JO(q)$ to be another type 1 lift of i, then $k-k'$ lifts to a map ℓ

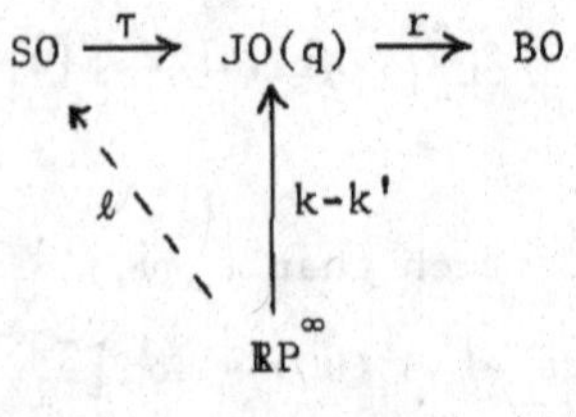

Since $(k-k')_{\#}(\iota) = \delta_1 - \delta_1 = 0$, it follows that $\ell_{\#}(\iota) = 0$. Hence by the lemma below $\ell_*: H_*(\mathbb{R}P^\infty) \to H_*(SO)$ is the zero map, which implies $k_* = k_*'$.

Lemma 3.2. There are only two coalgebra maps $H_*(\mathbb{R}P^\infty) \to H_*(SO)$ which preserve the action of the dual Steenrod squares: the trivial map and the map which sends the elements of $H_*(\mathbb{R}P^\infty)$ to the standard algebra generators of $H_*(SO)$.

Proof: Pass to the dual statement: there are only two algebra homomorphisms $H^*(SO) \to H^*(\mathbb{R}P^\infty)$ which are also module homomorphisms over the Steenrod algebra. Then use the known structure of $H^*(SO)$ and $H^*(\mathbb{R}P^\infty)$ as algebras over the Steenrod algebra.

This lemma also has the following important consequence

Lemma 3.3. If $k_1, k_2: \mathbb{R}P^\infty \to JO(q)$ are lifts of type 1,2 respectively of the inclusion $i: \mathbb{R}P^\infty \to BO$ then

$$(k_2 - k_1)_* \ (x_n) = \hat{u}_n.$$

Proof: Since $r \cdot (k_2 - k_1) = 0$, $k_2 - k_1$ lifts to a map $\ell: \mathbb{R}P^\infty \to SO$ (i.e. $k_2 - k_1 = \tau \circ \ell$). Since

$$\tau_\# \ell_\# (\iota) = (k_2 - k_1)_\# (\iota) = \delta_2 - \delta_1 = \delta_3 = \tau_\# (\theta)$$

it follows that $\ell_\#(\iota) = \theta$. By Lemma 3.2, $\ell_*(x_n) = u_n$. Hence

$$(k_2 - k_1)_*(x_n) = \tau_* \ell_* (x_n) = \tau_*(u_n) = \hat{u}_n. \qquad \text{Q.E.D.}$$

As an immediate consequence of these results we have

Proposition 3.4. The dual Steenrod operations in the mod-2 homology algebra $H_*JO(q) = \mathbb{Z}/2[\hat{v}_1,\hat{v}_2,\ldots] \otimes E[\hat{u}_1,\hat{u}_2,\ldots]$ are given by

$$Sq^i_*(\hat{v}_n) = (i,n-2i)\hat{v}_{n-i} \qquad Sq^i_*(\hat{u}_n) = (i,n-2i)\hat{u}_{n-i}$$

Proof. According to the above remarks $\hat{v}_i = k_{1*}(x_i)$, $\hat{u}_i = (k_2-k_1)_*(x_i)$ where x_i is the generator of $H_i\mathbb{R}P^\infty = H_iB\mathbb{Z}/2$. The results now follow from Prop. VI 2.4(c).

Arguments similar to Proposition 3.1 show

Proposition 3.5. The $\mathbb{Z}/2$-homology Serre spectral sequences of the fibrations

$$\text{Spin} \xrightarrow{\bar{\tau}} J(q) \xrightarrow{\bar{r}} BO$$

$$SO \xrightarrow{\tau'} JSO(q) \xrightarrow{r'} BSO$$

$$\text{Spin} \xrightarrow{\tilde{\tau}} \tilde{J}(q) \xrightarrow{\tilde{r}} BSO \quad \text{collapse.}$$

§4. The symplectic and complex image of J spaces

In this section we define $JSp(q)$, $JU(\pm q)$, and compute their mod 2 homology algebras in case q is odd. If ξ,η denote real and symplectic vector bundles respectively then $\eta \otimes \eta$ is real while $\xi \otimes \eta$ is symplectic. Thus for q odd one can define Adams operations ψ^q in quaterionic K theory $\widetilde{KSp}(\cdot)$. Localized away from q these operations are stable because tensoring with a generator of $\widetilde{KO}(S^8)$ defines a Bott periodicity isomorphism $I: \widetilde{KSp}(X) \xrightarrow{\approx} \widetilde{KSp}(S^8X)$ such that $\psi^q \circ I = q^4 I \circ \psi^q$ (cf. Adams[2, p. 618]. Thus for q odd we have an infinite loop map

$$\psi^q - 1 \colon BSp \to BSp$$

For q even we use the equivalence $BO \simeq BSp$ (recall all spaces are localized away from q) to define $\psi^q - 1$. Let $r\colon JSp(q) \to BSp$ denote the fibre of this map. We thus obtain a fibre sequence of infinite loop spaces (for all q)

$$(4.1) \qquad \cdots \to Sp \xrightarrow{\tau} JSp(q) \xrightarrow{r} BSp \xrightarrow{\psi^q - 1} BSp \to \cdots$$

For q odd we have

<u>Proposition</u> 4.2. The $\mathbb{Z}/2$-homology Serre spectral sequence of the fibration

$$Sp \xrightarrow{\tau} JSp(q) \xrightarrow{r} BSp$$

collapses. Hence, as an algebra

$$H_*(JSp(q)) = \mathbb{Z}/2[\hat{g}_1, \hat{g}_2, \ldots] \otimes E[\hat{h}_1, \hat{h}_2, \ldots]$$

where the $\hat{g}_i$'s are elements of degree 4i such that $r_*(\hat{g}_i) = g_i$ (g_i's being the standard algebra generators of $H_*(BSp) = \mathbb{Z}/2[g_1, g_2, \ldots]$) and $\hat{h}_i = \tau_*(h_i)$ where the h_i's are the standard algebra generators of $H_*(Sp) = E[h_1, h_2, \ldots]$; the degree of h_i is

4i-1. The dual Steenrod operations on the exterior part of $H_*(JSp(q))$ are given by $Sq^{4i}{}_*(\hat{h}_j) = (i,j-2i-1)\hat{h}_{j-i}$, $Sq^m{}_*(\hat{h}_j) = 0$ if $m \not\equiv 0 \pmod 4$.

Proof. The proof is entirely analogous to that of Prop. 3.1. The formulas for the dual Steenrod operations on the $\hat{h}_i$'s follow from the corresponding statement about dual Steenrod operations in $H_*(Sp) = E[h_1, h_2, \ldots,]$ (cf [7]).

We now turn to the complex image of J spaces. For <u>any</u> $q > 0$ define infinite loop spaces $JU(q)$, $JU(-q)$ by the fibrations

(4.3)
$$JU(q) \to BU \xrightarrow{\psi^q - 1} BU$$
$$JU(-q) \to BU \xrightarrow{\psi^q - 1} BU$$

<u>Remark</u>. We have not considered real and symplectic image of J spaces associated with $\psi^{-q}-1$ because $\psi^{-q} = \psi^{-1}\psi^q$ and ψ^{-1} is the identity on BO and BSp.

Because of their similarity we can often treat $JU(q)$ and $JU(-q)$ simultaneously. Thus, for example, we have fibration sequence of infinite loop spaces

(4.4)
$$\cdots \to U \xrightarrow{\tau} JU(\pm q) \xrightarrow{r} BU \xrightarrow{\psi^{\pm q}-1} BU \to \cdots$$

For q odd we have

<u>Proposition</u> 4.5. The $\mathbb{Z}/2$-homology Serre spectral sequences of the fibrations

$$U \xrightarrow{\tau} JU(\pm q) \xrightarrow{r} BU$$

collapse. Hence as algebras,

$$H_*(JU(\pm q)) = \mathbb{Z}/2[\hat{a}_1, \hat{a}_2, \ldots] \otimes E[\hat{b}_1, \hat{b}_2, \ldots]$$

where the $\hat{a}_i$'s are elements of degree 2i such that $r_*(\hat{a}_i) = a_i$ (a_i's being the standard algebra generators of $H_*(BU) = \mathbb{Z}/2[a_1, a_2, \ldots]$) and $\hat{b}_i = \tau_*(b_i)$ where the b_i's are the standard algebra generators of $H_*(U) = E[b_1, b_2, \ldots]$; the degree of b_i is 2i-1.

The dual Steenrod operations on the exterior part of $H_*(JU(\pm q))$ are given by $Sq^{2i}{}_*(\hat{b}_j) = (i, j-2i-1)\hat{b}_{j-i}$; $Sq^m{}_*(\hat{b}_j) = 0$ $m \not\equiv 0 \pmod 2$.

Proof. The proof is entirely analogous to that of Prop. 3.1. The formulas for the dual Steenrod operations follow from the corresponding formulas in $H_*(U)$ (cf [7]).

We shall also need the following result on homology operations (cf Chap. IX §1) for future use.

Lemma 4.6. $Q^2(\hat{b}_1) = \hat{b}_2$

Proof. Since $\hat{b}_i = \tau_*(b_i)$ and τ is an infinite loop map, it suffices to show that $Q^2(b_1) = b_2$. Since $H_3(U) \cong \mathbb{Z}/2$ on generator b_2, it suffices to show that $Q^2(b_1) \neq 0$.

To prove this we use the fact that homology operations commute with homology suspension (cf. Chap. IX 1.8). Since the homology suspension $\sigma_*: H_i(U) \to H_{i+1}(BU)$ sends the generator b_i to the generator a_i, we have by Chap. IX 1.5

$$\sigma_* Q^2(b_1) = Q^2\sigma_*(b_1) = Q^2(a_1) = a_1^2 \neq 0$$

Hence $Q^2(b_1) = b_2$ and $Q^2(\hat{b}_1) = \hat{b}_2$.

Similarly we can define infinite loop spaces $JSU(\pm q)$ by means of the fibration sequences

$$JSU(q) \to BSU \xrightarrow{\psi^q - 1} BSU$$

$$JSU(-q) \to BSU \xrightarrow{\psi^{-q} - 1} BSU$$

It is immediate that $JSU(\pm q)$ are the universal covering spaces of $JU(\pm q)$.

§5. The odd primary homology of JO(q)

Let ℓ be an odd prime not dividing q. We will compute the mod ℓ homology algebras $H_*JO(q)$ for q odd and $H_*J\overline{O}(q)$ for q even. To simplify the notation we will write $JO(q)$ for $J\overline{O}(q)$ when q is even. Let d be the order of q^2 mod ℓ, i.e. $d = \min\{k \mid q^{2k} \equiv 1 \bmod \ell\}$.

First we consider cohomology

Proposition 5.1. As an algebra

$$H^*(JO(q)) = \mathbb{Z}_\ell[\overline{s}_1, \overline{s}_2, \overline{s}_3, \ldots] \otimes E[\overline{t}_k \mid k \geq 1]$$

where degree $\overline{s}_k = 4kd$, degree $\overline{t}_k = 4kd - 1$. The $\overline{s}_k, \overline{t}_k$ can be chosen so that $r^*(\overline{p}_{kd}) = \overline{s}_k$ and $\tau^*(\overline{t}_k) = \overline{f}_{kr}$ where $\overline{p}_k$ denotes the Pontryagin class in $H^{4k}(BO)$ and $\overline{f}_k$ denotes the standard exterior generator in $H^{4k-1}(SO)$. Moreover the $\overline{t}_k$'s may be chosen to be primitive.

Proof. We consider the map of fibrations

$$\begin{array}{llcccc} (\alpha) & & SO & \xrightarrow{T} & JO(q) & \xrightarrow{r} & BO \\ & & \| & & \downarrow & & \downarrow \psi^q - 1 \\ (\beta) & & SO & \longrightarrow & PSO & \longrightarrow & BSO \end{array}$$

and the associated map of Serre spectral sequences in cohomology. We see that

$$E_2^{**}(\alpha) = \mathbb{Z}_\ell[\bar{p}_1, \bar{p}_2, \bar{p}_3, \ldots] \otimes E[\bar{f}_k \mid k \geq 1]$$

From the diagram we find that the $\bar{f}_k$'s are transgressive and transgress to

$$(\psi^q - 1)^*(\bar{p}_k) = \Sigma_{i=0}^{k} q^{2i} \bar{p}_i \chi(\bar{p}_{k-i})$$

We claim that this expression is equal to $(q^{2k} - 1)\bar{p}_k$ in $E_{4k}^{4k,0}(\alpha)$. We prove this statement by induction on k. It is obviously true for $k = 1$. Assuming it for $i < k$, we have

$$q^{2i}\bar{p}_i = \bar{p}_i \in E_{4k}^{4i,0}(\alpha) \qquad i < k$$

Hence

$$(\psi^q - 1)^*(\bar{p}_k) = q^{2k}\bar{p}_k + \Sigma_{i=0}^{k-1} q^{2i} \chi(\bar{p}_{k-1}) = q^{2k}\bar{p}_k + \Sigma_{i=0}^{k-1} \bar{p}_i \chi(\bar{p}_{k-i})$$

$$= q^{2k}\bar{p}_k - \bar{p}_k + \Sigma_{i=0}^{k} \bar{p}_i \chi(\bar{p}_{k-i}) = (q^{2k} - 1)\bar{p}_k$$

Hence $E_*^{**}(\alpha)$ breaks up into a tensor product of simple spectral sequences

$$E_*^{**}(\alpha) = \otimes_{i=1}^{\infty} E_*^{**}(\alpha_i)$$

where

$$E_2^{**}(\alpha_i) = \mathbb{Z}/\ell[\bar{p}_k] \otimes E[\bar{f}_k],$$

and the $\bar{f}_k$ transgress to $(q^{2k} - 1)\bar{p}_k$. If k is not a multiple of d, then $q^{2k} - 1 \not\equiv 0 \pmod{\ell}$ so $E_\infty^{**}(\alpha_k) = 0$. If k is a multiple of d then $q^{2k} - 1 \equiv 0 \pmod{\ell}$ so $E_\infty^{**}(\alpha) = E_2^{**}(\alpha)$. Hence

$$E_\infty^{**}(\alpha) = \otimes_{i=1}^{\infty} E_\infty^{**}(\alpha_i) = \otimes_{i=1}^{\infty} E_2^{**}(\alpha_{id})$$

$$= \mathbb{Z}/\ell[\bar{p}_d, \bar{p}_{2d}, \bar{p}_{3d}, \ldots] \otimes E[\bar{f}_{kd} | k \geq 1]$$

Hence there exist elements $\bar{t}_k \in H^*(JO(q))$ such that $\tau^*(t_k) = \bar{f}_{kd}$. It is also clear that $H^*(JO(q))$ is generated by $\{\bar{t}_k\}_{k=1}^{\infty}$ and $\{\bar{s}_k = r^*(\bar{p}_{kd})\}_{k=1}^{\infty}$. Since $H^*(JO(q))$ has the same dimension in each degree as $E_\infty^{**}(\alpha)$ it also follows that

(**) $$H^*(JO(q)) = \mathbb{Z}_\ell[\bar{s}_1, \bar{s}_2, \bar{s}_3, \ldots] \otimes E\{\bar{t}_k | k \geq 1\}.$$

Since $H^*(JO(q))$ is a commutative and cocommutative Hopf algebra, it follows by the Samelson-Leray theorem (see Milnor and Moore [31]) that there are Hopf subalgebras C and D of $H^*(JO(q))$ such that C is an exterior algebra on primitive generators in odd degrees, D is zero in odd degrees and

$$H^*(JO(q)) = C \otimes D$$

Now let B be the subalgebra of $H^*(SO)$ generated by $\{f_{dk}\}_{k=1}^{\infty}$. Then $\tau^*: H^*(JO(q)) \to B$ is an epimorphism. Hence it induces an epimorphism on indecomposables

$$Q(\tau^*): Q(H^*(JO(q))) = Q(C) \oplus Q(D) \to Q(B).$$

Since $Q(B)$ is zero in even degrees, $Q(\tau^*)$ is zero on $Q(D)$. Hence

$$Q(\tau^*): Q(C) \to Q(B)$$

must be an epimorphism. Consequently

$$\tau^*: C \to B$$

is an epimorphism, and we may choose the elements $\bar{t}_k$ to lie in C. Let I be the ideal in $H^*(JO(q))$ generated by $\{\bar{t}_k\}_{k=1}^{\infty}$. Then by (**) it is clear that

$$H^*(JO(q))/I \cong \mathbb{Z}/2[\bar{s}_1, \bar{s}_2, \bar{s}_3, \ldots]$$

On the other hand it is equally clear that

$$H^*(JO(q))/I = (C \otimes D)/I \cong (C/I') \otimes D$$

Since $\mathbb{Z}/2[\bar{s}_1, \bar{s}_2, \ldots]$ has no exterior elements, it follows that $C/I' = \mathbb{Z}/2$. Hence on indecomposables the map induced by inclusion

$$Q(E[\bar{t}_k \mid k \geq 1]) \to Q(C)$$

is an epimorphism and it follows that C is generated by $\{\bar{t}_k\}_{k=1}^{\infty}$. Consequently

$$\tau^*: C \to B$$

is an isomorphism of Hopf algebras. Therefore the $\bar{t}_k$'s are primitive.

This completes the proof.

We recall that as algebras

$$H_*(SO) = E[f_1, f_2, \ldots]$$

$$H_*(BO) = \mathbb{Z}/\ell[p_1, p_2, \ldots]$$

where $f_k = (\bar{f}_k)^*$ and $p_k = (\bar{p}_1^k)^*$.

<u>Proposition</u> 5.2. As an algebra

$$H_*(JO(q)) = \mathbb{Z}/\ell[s_1, s_2, s_3, \ldots] \otimes E[t_i \mid i \geq 1]$$

where degree $s_k = 4kd$, degree $t_k = 4kd - 1$. The s_k, t_k can be so chosen that $\tau_*(f_{dk}) = t_k$ and $r_*(s_k) = p_{kd}$ + decomposables.

The dual Steenrod operations on the exterior part of $H_*(JO(q))$ are given by

$$P_*^n(t_i) = (n, 2id-\ell n-1)t_{i-ns}$$

where $s = \frac{1}{2d}(\ell-1)$. (Note that 2d divides $\ell-1$ by Prop. VIII 2.6.)

Proof. We have

$$H_*(JO(q)) = (H^*(JO(q)))^* = (\mathbb{Z}_\ell[\bar{s}_1,\bar{s}_2,\bar{s}_3,\ldots])^* \otimes (E\{\bar{t}_i\ i \geq 1\})^*$$

But we know that the coproduct ψ in $H^*(JO(q))$ is given by

$$\psi(\bar{t}_i) = \bar{t}_i \otimes 1 + 1 \otimes \bar{t}_i$$

$$\psi(\bar{s}_k) = \psi\cdot r^*(\bar{p}_{kd}) = (r^* \otimes r^*)\psi(\bar{p}_{kd}) = \Sigma_{i+j=kd} r^*(\bar{p}_i) \otimes r^*(\bar{p}_j)$$

$$= \Sigma_{i+j=k} \bar{s}_i \otimes \bar{s}_j$$

It follows then that

$$(E[\bar{t}_i | i \geq 1])^* = E[t_i | i \geq 1]$$

where the t_i's are primitive and that

$$(\mathbb{Z}_\ell[\bar{s}_1,\bar{s}_2,\bar{s}_3,\ldots])^* = \mathbb{Z}_\ell[s_1,s_2,s_3,\ldots].$$

Since

$$\tau^*: E[\bar{t}_i | i \geq 1] \to H^*(SO)$$

maps epimorphically on indecomposables in degrees $4kd - 1$, it follows that

$$\tau_*: H_*(SO) \to E[t_i | i \geq 1]$$

maps monomorphically on primitives in degrees 4kd - 1. Hence $\tau_*(f_{kd}) = t_k$. Similarly since

$$r^*: H^*(BO) \to \mathbb{Z}_\ell[\bar{t}_1, \bar{t}_2, \ldots]$$

maps monomorphically on primitives in degrees 4kd, it follows that

$$r_*: \mathbb{Z}/\ell[s_1, s_2, \ldots] \to H_*(BO)$$

maps epimorphically on indecomposables in degrees 4kd, i.e.

$$r_*(t_k) = p_{kd} + \text{decomposables}.$$

The formulas for the dual Steenrod operations on the t_i's follow from the fact that $\tau_*(f_{kd}) = t_k$ and the formulas for the dual Steenrod operations in $H_*(SO) = E[f_1, f_2, \ldots]$

$$P_*^n f_i = (n, 2i-\ell n-1) f_{i-nm}$$

where $m = \frac{1}{2}(\ell-1)$ (cf [7]).

§6. The odd primary homology of JSp(q) and JU(±q)

Let q be odd and ℓ be an odd prime not dividing q. We begin by showing that JSp(q) is equivalent to JO(q) at ℓ. Let $c: BO \to BU$, $k: BU \to BSp$ be the natural maps obtained by extension of scalars and let $c': BSp \to BU$, $r': BU \to BO$ be the natural maps obtained by restriction of scalars.

Since $r' \circ c' \circ k \circ c = k \circ c \circ r' \circ c' = 4$, the composite $F = k \circ c$: $BO \to BSP$ is an equivalence at the odd prime ℓ. Further ψ^q commutes with F so there is an induced map f

$$\begin{array}{ccccc}
JO(q) & \to & BO & \xrightarrow{\psi^q - 1} & BO \\
\downarrow{\scriptstyle f} & & \downarrow{\scriptstyle F} & & \downarrow{\scriptstyle F} \\
JSp(q) & \to & BSp & \xrightarrow{\psi^q - 1} & BSp
\end{array}$$

Thus we have proved

Proposition 6.1. $f\colon JO(q) \to JSp(q)$ is an equivalence of infinite loop spaces at ℓ.

We now turn to the mod ℓ homology of $JU(\pm q)$ for any $q > 0$ and any odd prime ℓ not dividing q. Let

$$d_+ = \min\{k \mid q^k \equiv 1 \bmod \ell\}$$

$$d_- = \min\{k \mid (-q)^k \equiv 1 \bmod \ell\}$$

Since $(\psi^{\pm q})_*\colon H_{2m}BU \to H_{2m}BU$ is multiplication by $(\pm q)^m$ by 1.2, the mod ℓ (co)-homology Serre spectral sequences of the fibrations

$$U \xrightarrow{\tau} JU(\pm q) \xrightarrow{r} BU$$

are easily computed in the manner of Propositions 5.1 and 5.2. Let $d = d_{\pm}$ depending on the case.

Proposition 6.2. As an algebra

$$H^*(JU(\pm q)) = \mathbb{Z}/\ell[\bar{c}_1, \bar{c}_2, \ldots] \otimes E[\bar{d}_1, \bar{d}_2, \ldots]$$

where degree $\overline{c}_k = 2kd$, degree $\overline{d}_k = 2kd - 1$. The $\overline{c}_k, \overline{d}_k$ can be chosen so that $r^*(c_{kd}) = \overline{c}_k$ and $\tau^*(\overline{d}_k) = d_{kd}$ where c_k denotes the Chern class in $H^{2k}BU$ and d_k denotes the standard exterior generator of $H^{2k-1}U$. Moreover the $\overline{d}_k$'s can be chosen to be primitive.

Proposition 6.3. As an algebra

$$H_*(JU(\pm q)) = \mathbb{Z}_\ell[\hat{a}_1, \hat{a}_2, \ldots] \otimes E[\hat{b}_1, \hat{b}_2, \ldots]$$

where degree $\hat{a}_k = 2kd$, degree $\hat{b}_k = 2kd - 1$. The $\hat{a}_k, \hat{b}_k$ can be chosen so that $\tau_*(\hat{b}_k) = b_{kd}$ where the b_k's are the standard exterior generators of H_*U and $r_*(\hat{a}_k) = a_{kd}$, the a_k's being the standard generators of H_*BU. The dual Steenrod operations on the exterior part of $H_*(JU(\pm q))$ are given by

$$P_*^n \hat{b}_i = (n, ir-\ell n-1)\hat{b}_{i-ns}$$

where $s = \frac{1}{r}(\ell-1)$. (Note that r divides $\ell-1$ by Prop. VIII 2.6).

Proof. The proof is entirely analogous to that of Prop. 5.2. The formulas for the dual Steenrod operations arise from the corresponding formulas in $H_*(U)$

$$P_*^n b_i = (n, i-n\ell-1)b_{i-n(\ell-1)}$$

(cf [7]).

§7. The integral cohomology of JO(q)

Throughout this section all cohomology groups are taken with integral coefficients unless explicitly stated otherwise.

We will use the Serre spectral sequence of the fibration

(α) $$SO \xrightarrow{T} JO(q) \xrightarrow{r} BO$$

to prove

<u>Proposition 7.1.</u> As an algebra

$$H^*(JO(q)) = \mathbf{Z}[\hat{\zeta}_1, \hat{\zeta}_2, \ldots] / \{(\tfrac{q^{2k}-1}{2})\hat{\zeta}_k\}_{k=1}^{\infty}$$

$$\oplus \text{ (2-torsion of } H^*(BO \times SO)).$$

where $\hat{\zeta}_k = r^*(\bar{\zeta}_k)$ and $\bar{\zeta}_k$ is a generator of the integral indecomposable elements in $H^{4k}(BO)$.

It is clear from the statement of the Proposition that life is made unpleasant for us by the presence of 2-torsion in $H^*(SO)$ and $H^*(BO)$. We begin with the following lemma:

<u>Lemma</u> 7.2. The reduction mod 2 of the Serre spectral sequence of (α) is a monomorphism on the 2-torsion elements in $E_2^{**}(\alpha)$. Moreover if $x \in E_2^{**}(\alpha)$ is a 2-torsion element then there is a torsion-free element $u_x \in E_2^{**}(\alpha)$ such that $x + 2u_x$ is an infinite cycle in $E_*^{**}(\alpha)$ and defines a $\mathbf{Z}_2$-summand of $H^*(JO(q))$.

<u>Proof:</u> We have $E_2^{**}(\alpha) = H^*(BO; H^*(SO))$ since in the proof of Proposition 3.1 we showed that $E_*^{**}(\alpha)$ has trivial local coeffi-

cients. Let $C^*(BO)$, $C^*(SO)$ denote the integral cellular cochain complexes and let $\pi: C^*(SO) \to H^*(SO)$ denote a cochain homotopy equivalence. Then by the Künneth formula

$$H^*(1 \otimes \pi): H^*(BO \times SO) = H^*(C^*(BO) \otimes C^*(SO)) \to H^*(C^*(BO) \otimes H^*(SO))$$
$$= H^*(BO; H^*(SO))$$

is an isomorphism. Moreover we have a commutative diagram

$$\begin{array}{ccc} H^*(BO \times SO) & \xrightarrow{\rho} & H^*(BO \times SO; \mathbb{Z}_2) \\ \downarrow \cong & & \downarrow \cong \\ E_2^{**}(\alpha) = H^*(BO; H^*(SO)) & \xrightarrow{\rho} & H^*(BO; \mathbb{Z}_2) \otimes H^*(SO; \mathbb{Z}_2) = E_2^{**}(\alpha; \mathbb{Z}_2) \end{array}$$

Now a 2-torsion element in $E_2^{**}(\alpha)$ reduces to 0 iff the corresponding element of $H^*(BO \times SO)$ reduces to 0. But we have the Bockstein exact sequence

$$H^*(BO \times SO) \xrightarrow{\times 2} H^*(BO \times SO) \xrightarrow{\rho} H^*(BO \times SO; \mathbb{Z}_2)$$

Now ρ is a monomorphism on 2-torsion elements since all 2-torsion elements in $H^*(BO \times SO)$ have order 2 and hence are not in the image of $\times 2$. This proves the first part of the lemma.

To prove the second part of the lemma observe that Proposition 3.1 and Proposition 3.4 show that

$$H_*(JO(q); \mathbb{Z}_2) \cong H_*(BO; \mathbb{Z}_2) \otimes H_*(SO; \mathbb{Z}_2) \cong H_*(BO \times SO; \mathbb{Z}_2)$$

as Hopf algebras over the Steenrod algebra. Dually

$$H^*(JO(q);\ \mathbb{Z}_2) \cong H^*(BO \times SO;\ \mathbb{Z}_2)$$

as algebras over the Steenrod algebra. Now suppose $x \in E_2^{**}(\alpha)$ is a 2-torsion element. Then x generates a $\mathbb{Z}_2$-summand of $H^*(BO \times SO)$. This implies that $\rho(x) = Sq^1 w$ for some $w \in H^*(BO \times SO;\ \mathbb{Z}_2)$. But by the above remark it follows that

$$\rho(x) = Sq^1 w$$

in $H^*(JO(q);\ \mathbb{Z}_2)$. This implies that there is an element $v \in H^*(JO(q))$ such that $\rho(v) = \rho(x)$ and such that v generates a $\mathbb{Z}_2$-summand of $H^*(JO(q))$. The element v defines an element $\bar{v} \in E_\infty^{**}(\alpha)$. Now $\bar{v}$ is the class of some infinite cycle $\hat{v} \in E_2^{**}(\alpha)$. Since $E_*^{**}(\alpha;\ \mathbb{Z}_2)$ collapses we have $\rho(\hat{v}) = \rho(x)$ in $E_2^{**}(\alpha;\ \mathbb{Z}_2)$. Therefore $\hat{v} = x + 2u_x$ and this proves the lemma.

We now show the 2-torsion in $E_2^{**}(\alpha)$ can be factored out at each stage of the spectral sequence $E_*^{**}(\alpha)$.

<u>Lemma</u> 7.3. Let $\bar{E}_2^{**}(\alpha)$ denote the torsion free part of $E_2^{**}(\alpha) \cong H^*(BO \times SO)$ generated by the standard torsion free generators of $H^*(BO)$ and $H^*(SO)$. Then the elements of $\bar{E}_2^{**}(\alpha)$ generate a sub-(spectral sequence) $\bar{E}_*^{**}(\alpha)$. At each level n, $E_n^{**}(\alpha)$ is generated by the elements of $\bar{E}_n^{**}(\alpha)$ and the elements $x + 2u_x$, x a 2-torsion element of $E_2^{**}(\alpha)$, u_x as in Lemma 7.2. Thus

$$E_\infty^{**}(\alpha) \cong \bar{E}_\infty^{**}(\alpha) \oplus (2 \text{ torsion of } H^*(BO \times SO))$$

Proof: We proceed by induction on n. Obviously $\overline{E}_2^{**}(\alpha) \subseteq E_2^{**}(\alpha)$ and $E_2^{**}(\alpha)$ is generated by $\overline{E}_2^{**}(\alpha)$ and the elements $x + 2u_x$. Assume we have shown that this is true for $\overline{E}_n^{**}(\alpha)$. We want to show the statement is true for $\overline{E}_{n+1}^{**}(\alpha)$.

We must first show that d_n sends $\overline{E}_n^{**}(\alpha)$ to $\overline{E}_n^{**}(\alpha)$. If not there is an element $v \in \overline{E}_n^{**}(\alpha)$ such that

$$d_n v = w + (x + 2u_x).$$

Now consider the reduction

$$\rho: E_n^{**}(\alpha) \to E_n^{**}(\alpha; \mathbb{Z}_2)$$

We have

$$d_n \rho(v) = \rho d_n(v) = \rho(w) + \rho(x)$$

Now the left side of the equation is 0 since $E_*^{**}(\alpha; \mathbb{Z}_2)$ collapses. On the other hand $\rho(w) + \rho(x) \neq 0$ since by Lemma 7.2, $\rho(x) \neq 0$ and $\rho(w)$ lies in a complementary direct summand of $E_n^{**}(\alpha, \mathbb{Z}_2) = E_2^{**}(\alpha; \mathbb{Z}_2)$. This contradiction establishes that $d_n: \overline{E}_n^{**}(\alpha) \to \overline{E}_n^{**}(\alpha)$.

Now suppose that $v \in \overline{E}_n^{**}(\alpha)$ is a d_n-boundary in $E_n^{**}(\alpha)$. This means that $d_n w = v$ where $w \in E_n^{**}(\alpha)$. But by induction hypothesis $w = \overline{w} + m(x + 2u_x)$ where $\overline{w} \in \overline{E}_n^{**}(\alpha)$. Thus

$$v = d_n w = d_n \overline{w}$$

since $x + 2u_x$ is an infinite cycle. Hence v is a d_n-boundary also in $\overline{E}_n^{**}(\alpha)$. This establishes that $\overline{E}_{n+1}^{**}(\alpha) \subseteq E_{n+1}^{**}(\alpha)$.

Now let $w \in E_{n+1}^{**}(\alpha)$. Then w is the class of some d_n-cycle $v \in E_n^{**}(\alpha)$. But $v = \overline{v} + m(x + 2u_x)$ where $\overline{v} \in \overline{E}_n^{**}(\alpha)$. Since $x + 2u_x$ is an infinite cycle

$$0 = d_n v = d_n \overline{v}$$

so $w = \overline{v} + m(x + 2u_x)$. This proves that $E_{n+1}^{**}(\alpha)$ is generated by $E_{n+1}^{**}(\alpha)$ and the elements $x + 2u_x$. This completes the induction and proof.

Proof of Proposition 7.1. Consider the map of fibrations

$$\begin{array}{llcccc}
(\alpha) & SO & \xrightarrow{\tau} & JO(q) & \xrightarrow{r} & BO \\
& \| & & \downarrow & & \downarrow \psi^q - 1 \\
(\beta) & SO & \longrightarrow & PSO & \longrightarrow & BSO
\end{array}$$

Let $\overline{f}_* \in H^{4k-1}(SO)$ denote the standard integral exterior generator. According to Cartan [11], $2\overline{f}_k$ transgresses in the Serre spectral sequence $E_*^{**}(\beta)$ to $\overline{p}_k$, where $\overline{p}_k \in H^{4k}(BSO)$ denotes the Pontryagin class. Hence by naturality $2\overline{f}_k$ transgresses in $E^{**}(\alpha)$ to $(\psi^q - 1)*(\overline{p}_k) = (q^{2k} - 1)\overline{p}_k$ in $E_{4k}^{**}(\alpha)$. (cf. Proposition 5.1).

By the preceding lemma it suffices to compute the sub-(spectral sequence) $\overline{E}_*^{**}(\alpha)$ generated by $\{\overline{f}_k, \overline{p}_k\}_{k-1}^{\infty}$. We certainly have

$$\overline{E}_2^{**}(\alpha) = \mathbb{Z}[\overline{p}_1, \overline{p}_2, \ldots] \otimes E[\overline{f}_k \mid k \geq 1]$$

We will show inductively that $\overline{f}_k$ transgresses in $\overline{E}_*^{**}(\alpha)$ to $(\frac{q^{2k}-1}{2})\overline{\zeta}_k$ where

$$\overline{\zeta}_k = \overline{p}_k + \text{decomposables.}$$

It is obvious that $\overline{f}_1$ transgresses to $(\frac{q^2-1}{2})\overline{p}_1$. Now suppose that $\overline{f}_k$ transgresses to $(\frac{q^{2k}-1}{2})\overline{\zeta}_k$ for $k = 1,2,3,\ldots,n$. Let $\overline{E}_*^{**}(\alpha)$ denote the spectral sequence

$$\overline{E}_2^{**}(\alpha_n) = \mathbb{Z}[\overline{\zeta}_1,\overline{\zeta}_2,\ldots,\overline{\zeta}_n] \otimes E[\overline{f}_k \,|\, k \geq 1]$$

and with differentials induced by inclusion. Then by induction hypothesis

$$(*) \qquad \overline{E}_*^{**}(\alpha_n) = F_*^{**}(\alpha_1) \otimes F_*^{**}(\alpha_2) \otimes \cdots \otimes F_*^{**}(\alpha_n)$$

where

$$F_*^{**}(\alpha_k) = \mathbb{Z}[\overline{\zeta}_k] \otimes E[\overline{f}_k]$$

with $\overline{f}_k$ transgressing to $(\frac{q^{2k}-1}{2})\overline{\zeta}_k$. Obviously

$$\overline{E}_\infty^{**}(\alpha_n) = \mathbb{Z}[\overline{\zeta}_1,\overline{\zeta}_2,\ldots,\overline{\zeta}_n]/\{(\tfrac{q^{2k}-1}{2})\overline{\zeta}_k\}_{k=1}^n$$

Now consider the element $\overline{f}_{n+1}$. If it does not transgress, then some differential $d_m\overline{f}_{n+1} = x \neq 0$ with $m < 4(n+1)$. But then $x \in \overline{E}_m^{p,t}(\alpha)$, $t > 0$ which lies in the image of $\overline{E}_m^{p,t}(\alpha_n)$. Since $\overline{E}_\infty^{p,t}(\alpha_n) = 0$, one of two things must happen to x in $\overline{E}_*^{**}(\alpha_n)$.

(1) $d_s x = y \neq 0$ for some differential d_s in $\overline{E}_*^{**}(\alpha_n)$ with $s > m$. But $y \neq 0$ also in $\overline{E}_s^{**}(\alpha)$ since no differentials from outside the range of $\overline{E}_*^{**}(\alpha_n)$ can reach it at that stage. This contradicts the fact that $x = d_m f_{n+1}$

or else

(2) $x = d_m z$ for some $z \in \overline{E}_m^{0,4(n+1)-1}(\alpha_n)$. But using (*) we can check that this would imply $x = d_m z$ is an integral class in $\overline{E}_m^{p,t}(\alpha_n)$ and hence also in $\overline{E}_m^{p,t}(\alpha)$. Thus $2x \neq 0$ but $2x = d_m(2\overline{f}_{n+1})$ which implies $d_m(2\overline{f}_{n+1}) = 2x \neq 0$ and contradicts the fact that $2\overline{f}_{n+1}$ transgresses.

Hence $\overline{f}_{n+1}$ transgresses. If we could show that it transgresses to $(\frac{q^{2k}-1}{2})\overline{\zeta}_{n+1}$ where

$$\overline{\zeta}_{n+1} = \overline{p}_{n+1} + \text{decomposables}$$

then we would complete the induction. The rest of the proof would follow immediately from a calculation similar to (*) above.

Hence the following lemma completes the proof.

Lemma 7.4. There is an element $\overline{\zeta}_{n+1} \in H^{4(n+1)}(BO)$ of the form

$$\overline{\zeta}_{n+1} = \overline{p}_{n+1} + \text{decomposables}$$

such that $\overline{f}_{n+1}$ transgresses to $(\frac{q^{2(n+1)}-1}{2})\overline{\zeta}_{n+1}$.

Proof: Since $\overline{f}_{n+1}$ transgresses and $2\overline{f}_{n+1}$ transgresses to $(q^{2(n+1)} - 1)\overline{p}_{n+1}$ it follows that $\overline{f}_{n+1}$ transgresses to $(\frac{q^{2(n+1)}-1}{2})\overline{p}_{n+1} + x$ where x is a decomposable element of order 2 in $\overline{E}_{4(n+1)}^{4(n+1),0}(\alpha)$.

Write $\frac{q^{2(n+1)-1}}{2} = 2^{\nu}\ell$ where ℓ is odd. We have to show that x is divisable (in $\overline{E}^{**}_{4(n+1)}(\alpha)$) by $\frac{1}{2}(q^{2(n+1)} - 1)$. Since x has order 2, there is no difficulty in dividing it by ℓ. We must show it is divisable by 2^ν.

Now x can be written in the form

$$x = \Sigma\, a_I \bar{\zeta}^I$$

where we use the notation: $I = (m_1, m_2, m_3, \ldots, m_n)$ is a sequence of natural numbers: $a_I \in \mathbf{Z}$ and $\bar{\zeta}^{(m_1, m_2, \ldots, m_n)} = \bar{\zeta}_1^{m_1} \bar{\zeta}_2^{m_2} \cdots \bar{\zeta}_n^{m_n}$. To show that 2^ν divides x we must show that 2^ν divides a_I for all I. Let $2^m = \text{g.c.d.}(2^\nu, (A_I))$. Suppose $m < \nu$. Then we can write

$$x = 2^m \Sigma_{I \in \mathcal{A}}\, b_I \bar{\zeta}^I + 2^{m+1} \Sigma_{J \in \mathcal{B}}\, c_J \bar{\zeta}^J$$

where $\mathcal{A}$ and $\mathcal{B}$ are disjoint, and the b_I's are odd integers. Since

$$d_{4(n+1)} f_{n+1} = 2^\nu \ell \bar{p}_{n+1} + x = 2^m (\Sigma_{I \in \mathcal{A}}\, b_I \bar{\zeta}^I + 2\Sigma_{J \in \mathcal{B}}\, c_J \bar{\zeta}^J + 2^{\nu - m} \ell \bar{p}_{n+1})$$

it follows that in $H^{4(n+1)}(JO(q))$

$$z = \Sigma_{I \in \mathcal{A}}\, b_I \hat{\zeta}^I + 2\Sigma_{J \in \mathcal{B}}\, c_J \hat{\zeta}^J + 2^{\nu - m} \ell \bar{p}_{n+1}$$

generates a cyclic direct summand of order 2^m. We now use the Bockstein spectral sequence of $H^*(JO(q), \mathbf{Z}/2)$ to derive a contradiction.

By our induction hypothesis it follows that

$$H^s(JO(q)) / \text{ 2-torsion of } H^*(BO \times SO)$$
$$= (\mathbf{Z}[\hat{\zeta}_1, \hat{\zeta}_2, \ldots, \hat{\zeta}_n] / \{(\tfrac{q^{2k}-1}{2}) \hat{\zeta}_k\}_{k=1}^n)_s$$

for $s < 4(n+1)$. From this we see that there are indecomposable elements $w_k \in H^{4k-1}(JO(q), \mathbb{Z}/2)$, $1 \leq k \leq n$ such that $d_{t_k}(w_k) = \zeta_k$ where $t_k = \nu_2(\frac{q^{2k}-1}{2})$, ζ_k denotes $\hat{\zeta}$ reduced mod 2, and d_t denotes the differential in the Bockstein spectral sequence (cf. Remark III 6.3).

From these considerations and the fact that x has order 2 in $\bar{E}_{4(n+1)}^{4(n+1),0}(\alpha)$ it follows that there is a decomposable element y in $H^{4n-3}(JO(q); \mathbb{Z}/2)$ such that

$$d_{m+1}(y) = \Sigma_{I \in \mathfrak{a}} b_I \zeta^I = \rho(z)$$

where ρ denotes reduction mod 2. On the other hand z generates a $\mathbb{Z}/2^m$ summand of $H^{4(n+1)}(JO(q))$. Hence there is an element $w \in H^{4n-3}(JO(q); \mathbb{Z}/2)$ such that

$$d_m(w) = \rho(z) = \Sigma_{I \in \mathfrak{a}} b_I \zeta^I$$

Hence it follows that at the $(m+1)$st stage of the Bockstein spectral sequence

$$d_{m+1}(y) = \rho(z) = 0$$

Now consider the coproduct $\Delta_*(y)$. On the one hand we can easily check that $d_{m+1}\Delta_*(y) \neq 0$. On the other hand

$$d_{m+1}\Delta_*(y) = \Delta_* d_{m+1} y = \Delta_*(0) = 0$$

This contradiction establishes the result.

We conclude by recording for future use some results on Bockstein operations implied by Proposition 7.1.

Lemma 7.4. In the mod-2 homology algebra $H_*(JO(q); \mathbb{Z}/2) = \mathbb{Z}/2[\hat{v}_1, \hat{v}_2, \ldots] \otimes E[\hat{u}_1, \hat{u}_2, \ldots]$ we have

$$d_\nu(\hat{v}_2^2) = \hat{u}_1\hat{u}_2$$

where $\nu = \nu(\frac{1}{2}(q^2-1))$ is the largest integer such that 2^ν divides $\frac{1}{2}(q^2-1)$.

Proof. By Prop. 3.4 we have

$$d_1(\hat{v}_n) = Sq_*^1(\hat{v}_n) = (n-1)\hat{v}_{n-1}$$

$$d_1(\hat{u}_n) = Sq_*^1(\hat{u}_n) = (n-1)\hat{u}_{n-1}$$

Using these formulas we can compute the E^2 term of the Bockstein spectral sequence of $H_*(JO(q); \mathbb{Z}/2)$ and obtain

$$E_n^2 = \begin{cases} 0 & \text{if } n = 1,2 \\ \mathbb{Z}/2 \text{ on generator } \hat{u}_1\hat{u}_2 & \text{if } n = 3 \\ \mathbb{Z}/2 \text{ on generator } \hat{v}_2^2 & \text{if } n = 4 \end{cases}$$

On the other hand by Prop. 7.1 we have that $H^n(JO(q); \mathbb{Z})$ is a $\mathbb{Z}/2$ module in degrees $n = 1,2,3$ while

$$H^4(JO(q); \mathbb{Z}) = \mathbb{Z}/\tfrac{1}{2}(q^2-1) \oplus \mathbb{Z}/2 \text{ module}$$

Consequently by the universal coefficient theorem

$$H_3(JO(q);\, \mathbb{Z}) = \mathbb{Z}/\tfrac{1}{2}(q^2-1) \oplus \mathbb{Z}/2 \text{ module}$$

It follows that

$$d_\nu(\hat{v}_2^2) = \hat{u}_1\hat{u}_2$$

(cf. Remark III 6.3)

Lemma 7.5. In the mod-ℓ homology algebra ($\ell > 2$)

$$H_*(JO(q);\, \mathbb{Z}/\ell) = \mathbb{Z}/\ell[s_1,s_2,\ldots] \otimes E[t_1,t_2,\ldots]$$

we have

$$d_\nu(s_i) = t_i + \text{decomposables},\ i = 1,2,\ldots,\ell$$

where $\nu = \nu_\ell(q^{2d}-1)$ is the largest integer for which ℓ^ν divides $q^{2d}-1$ and where d is as in Prop. 5.2.

Proof. From Prop. 7.1 we have that the ℓ-primary component of $H^n(JO(q);\, \mathbb{Z})$ is a free $\mathbb{Z}/\ell^\nu$ module for $1 \le n \le 4\ell d-1$ while

$$H^{4d\ell}(JO(q);\, \mathbb{Z})_{(\ell)} = \mathbb{Z}/\ell^{\nu+1} \oplus \text{free } \mathbb{Z}/\ell^\nu \text{ module}$$

(cf. VIII 2.4). By the universal coefficient theorem, the ℓ-primary component of $H_n(JO(q);\, \mathbb{Z})$ is a free $\mathbb{Z}/\ell^\nu$ module for $1 \le n < 4\ell d - 1$

and $n = 4\ell d$, while

$$H_{4\ell d-1}(JO(q);\ \mathbb{Z})_{(\ell)} = \mathbb{Z}/\ell^{\nu+1} \oplus \text{free } \mathbb{Z}/\ell^{\nu} \text{ module}$$

Consequently in the Bockstein spectral sequences of $H_*(JO(q);\ \mathbb{Z}/\ell)$ we must have (cf. III 6.3)

$$E_n^{\nu}(H_*(JO(q);\ \mathbb{Z}/\ell)) = H_n(JO(q);\ \mathbb{Z}/\ell) \text{ if } 1 \leq n \leq 4\ell d$$

$$(*) \quad E_n^{\nu+1}(H_*(JO(q);\ \mathbb{Z}/\ell)) = \begin{cases} 0 & \text{if } 1 \leq n < 4\ell d - 1 \\ \mathbb{Z}/\ell & \text{if } n = 4\ell d - 1,\ 4\ell d \end{cases}$$

$$(**) \quad E_n^{\nu+2}(H_*(JO(q);\ \mathbb{Z}/\ell)) = 0 \qquad \text{if } 1 \leq n \leq 4\ell d.$$

It follows that we must have

$$d_\nu(s_i) = k_i t_i + \text{decomposables}, k_i \neq 0, \quad i = 1,2,\ldots,\ell - 1$$

Since the generators s_i of Prop. 5.2 are merely defined to be pre-images of generators in $H_*(BO;\ \mathbb{Z}/\ell)$, we can change s_i by a scalar factor and thus obtain

$$d_\nu(s_i) = t_i + \text{decomposables} \qquad i = 1,2,\ldots,\ell - 1$$

It remains to show that $d_\nu(s_\ell)$ is indecomposable. Assume the contrary that $d_\nu(s_\ell)$ is decomposable. Then

$$d_\nu(s_\ell) = \alpha s_1^{\ell-1} t_1 + y$$

where y is a d_ν boundary. If $\alpha = 0$, then t_ℓ and $s_1^{\ell-1} t_1$ are nonzero classes in $E_{4\ell d-1}^{\nu+1}(H_*(JO(q);\ \mathbb{Z}/\ell))$ contradicting (*). If $\alpha \neq 0$ then according to May [28; Prop. 6.8]

$$d_{\nu+1}(s_1^{\ell}) = s_1^{\ell-1}t_1 = 0 \text{ in } E^{\nu+1}_{4\ell d-1}(H_*(JO(q);\ \mathbf{Z}/\ell))$$

so that s_1^{ℓ} is a nonzero class in $E^{\nu+2}_{4\ell d}(H_*(JO(q);\ \mathbf{Z}/\ell))$ contradicting (**).

§8. The integral cohomology of JSp(q) and JU(±q)

Throughout this section all cohomology groups are taken with integral coefficients.

Proposition 8.1. As algebras

$$H^*(JSp(q)) = \mathbf{Z}[\hat{g}_1, \hat{g}_2, \ldots]/\{(q^{2k}-1)\hat{g}_k\}_{k=1}^{\infty}$$

$$H^*(JU(\pm q)) = \mathbf{Z}[\hat{c}_1, \hat{c}_2, \ldots]/\{((\pm q)^{k}-1)\hat{c}_k\}_{k=1}^{\infty}$$

where $\hat{g}_k = r^*(\bar{g}_k)$ and $\bar{g}_k \in H^{4k}(BSp)$ is the Pontryagin class and where $\hat{c}_k = r^*(\bar{c}_k)$ and $\bar{c}_k \in H^{2k}BU$ is the Chern class.

Proof: The proofs are easy and virtually identical in both cases. Hence we will only briefly treat the case of JSp(q).

We consider the map of fibrations

$$\begin{array}{lccccc}
(\alpha) & Sp & \xrightarrow{\tau} & JSp(q) & \xrightarrow{r} & BSp \\
 & \| & & \downarrow & & \downarrow{\scriptstyle \psi^q-1} \\
(\beta) & Sp & \longrightarrow & PBSp & \longrightarrow & BSp
\end{array}$$

Let $\bar{t}_k \in H^{4k-1}(Sp)$ denote the standard exterior generator. Then according to Cartan [11], $\bar{t}_k$ transgresses in $E^{*,*}(\beta)$ to $\bar{g}_k$. By naturality $\bar{t}_k$ transgress in $E^{**}(\alpha)$ to $(\psi^q-1)(\bar{g}_k)$. By the same reasoning as in Proposition 5.1, $(\psi^q-1)^*(\bar{g}_k) = (q^{2k}-1)\bar{g}_k$ in $E^{**}(\alpha)$. Hence

$$E^{**}(\alpha) = E^{**}(\alpha_1) \otimes E^{**}(\alpha_2) \otimes \cdots \otimes E^{**}(\alpha_k) \otimes \cdots$$

where $E^{**}(\alpha_k) = \mathbb{Z}[\bar{g}_k] \otimes E[\bar{t}_k]$ with $\bar{t}_k$ transgressing to $(q^{2k}-1)\bar{g}_k$. The result is now immediate.

As in §7 we conclude by deriving some results on Bockstein operations for future use.

Lemma 8.2. In the mod-2 homology algebras

$$H_*(JSp(q);\ \mathbb{Z}/2) = \mathbb{Z}/2[\hat{g}_1,\hat{g}_2,\ldots] \otimes E[\hat{h}_1,\hat{h}_2,\ldots]$$

$$H_*(JU(\pm q);\ \mathbb{Z}/2) = \mathbb{Z}/2[\hat{a}_1,\hat{a}_2,\ldots] \otimes E[\hat{b}_1,\hat{b}_2,\ldots]$$

we have

(i) $$d_\nu(\hat{g}_i) = \hat{h}_i + \text{decomposables} \qquad i = 1,2$$

where $\nu = \nu(q^2-1)$ is the largest integer such that 2^ν divides q^2-1

(ii) $$d_\mu(\hat{a}_1) = \hat{b}_1$$

where $\mu = \nu(\pm q-1)$ is the largest integer such that 2^μ divides $\pm q - 1$.

Proof. According to Prop. 8.1, $H^n(JSp(q);\ \mathbb{Z})$ is a free $\mathbb{Z}/(q^2-1)$ module in degrees $1 \leq n \leq 7$, while

$$H^8(JSp(q);\ \mathbb{Z}) = \mathbb{Z}/(q^4-1) \oplus \mathbb{Z}/(q^2-1)$$

By the universal coefficient theorem

$$H_n(JSp(q);\ \mathbb{Z}) = \begin{cases} \mathbb{Z}/(q^2-1)\text{-module} & \text{if } 1 \leq n \leq 6 \text{ or } n = 8 \\ \mathbb{Z}/(q^4-1) \oplus \mathbb{Z}/(q^2-1) & \text{if } n = 7 \end{cases}$$

Consequently in the Bockstein spectral sequence of $H_*(JSp(q);\ \mathbb{Z}/2)$ we must have (cf. III 6.3)

$$E_n^{\nu}(H_*(JSp(q);\ \mathbb{Z}/2)) = H_n(JSp(q);\ \mathbb{Z}/2)$$

for $1 \leq n \leq 7$ while

$$(*) \qquad E_n^{\nu+1}(H_*(JSp(q);\ \mathbb{Z}/2)) = \begin{cases} 0 & \text{if } 1 \leq n \leq 6 \\ \mathbb{Z}/2 & \text{if } n = 7,8 \end{cases}$$

$$(**) \qquad E_n^{\nu+2}(H_*(JSp(q);\ \mathbb{Z}/2)) = 0 \quad \text{if} \quad 1 \leq n \leq 8$$

Hence we must have $d_\nu(\hat{g}_1) = \hat{h}_1$. To prove that

$$d_\nu(\hat{g}_2) = \hat{h}_2 + \text{decomposables}$$

assume the contrary. Then we would have to have $d_\nu(\hat{g}_1) = \alpha\hat{g}_1\hat{h}_1$

$\alpha = 0$ or 1. If $\alpha = 0$, then $\hat{h}_2$ and $\hat{g}_1\hat{h}_1$ would be nonzero classes in $E_7^{\nu+1}(H_*(JSp(q); \mathbb{Z}/2))$ contradicting (*). If $\alpha = 1$, then since according to May [28; Prop. 6.8(iii)] $d_{\nu+1}(\hat{g}_1^2) = g_1 d_\nu(\hat{g}_1) = \hat{g}_1\hat{h}_1$, we would have that $\hat{g}_1^2$ is a nonzero class in $E_8^{\nu+2}(H_*(JSp(q); \mathbb{Z}/2))$ contradicting (**). This proves (i). The proof of (ii) is similar

Lemma 8.3. In the mod-ℓ homology algebra ($\ell > 2$)

$$H_*(JU(\pm q); \mathbb{Z}/\ell) = \mathbb{Z}/\ell[\hat{a}_1, \hat{a}_2, \ldots] \otimes E[\hat{b}_1, \hat{b}_2, \ldots]$$

we have

$$d_\nu(\hat{a}_i) = \hat{b}_i + \text{decomposables} \qquad i = 1,2,\ldots,\ell$$

where $\nu = \nu_\ell((\pm q)^r - 1)$ is the largest integer for which ℓ^ν divides $(\pm q)^r - 1$ and where r is as in Prop. 6.3.

Proof. Entirely analogous to that of Lemma 7.5.

Chapter II

Permutative Categories of Classical Groups over Finite Fields

§1. Introduction

In this chapter we observe that each family $\mathcal{A} = \{G(n), n \geq 0\}$ of classical groups with direct sum homomorphisms $G(m) \times G(n) \to G(m+n)$, commutative up to natural isomorphism, forms a category with particularly nice properties. Such categories are called permutative. The classifying space of such a category (roughly $\coprod BG(n)$) is then converted into an infinite loop space $\Gamma B\mathcal{A}$ by means of a group completion process. This procedure is described in detail in §2. The spaces $\Gamma B\mathcal{A}$ are homologically equivalent to $BG(\infty) \times \mathbb{Z}$ and are used in Chapter III to study the ImJ spaces of Chapter I and to derive the homology of G(n).

The ordinary orthogonal groups over $\mathbb{F}_q$, q odd, are treated in §3. In §4 these results are extended to the extraordinary orthogonal groups by considering quadratic forms in more generality. Spinor groups are studied in §5. Section 6 deals with the general linear, symplectic, and unitary groups. In §7 we consider orthogonal groups over $\mathbb{F}_q$, q even. Finally in §8 we consider certain natural functors between the categories constructed in the previous sections.

§2. Permutative categories and their associated infinite loop spaces

In what follows we will use the following notation: If $\mathcal{A}$ is a category, $\mathcal{O}b\,\mathcal{A}$ will denote the class of objects of $\mathcal{A}$, $|\mathcal{O}b\,\mathcal{A}|$ will denote the class of isomorphism classes of $\mathcal{O}b\,\mathcal{A}$ and $\mathcal{M}or\,\mathcal{A}$ will denote the class of morphisms of $\mathcal{A}$.

Definition 2.1. A permutative category $(\mathcal{A}, \square, 0, c)$ is a small category $\mathcal{A}$ with a bifunctor $\square: \mathcal{A} \times \mathcal{A} \to \mathcal{A}$, a distinguished object 0 and a natural isomorphism c: $\square \to \square\tau$ (where τ: $\mathcal{A} \times \mathcal{A} \to \mathcal{A} \times \mathcal{A}$ is transposition) which satisfy the following properties

(i) $\square$ is strictly associative, i.e.

$$\square(\square \times 1_{\mathcal{A}}) = \square(1_{\mathcal{A}} \times \square): \mathcal{A} \times \mathcal{A} \times \mathcal{A} \to \mathcal{A}$$

(ii) 0 is a strict left and right unit for $\square$, i.e.

$$\square(0 \times 1_{\mathcal{A}}) = 1_{\mathcal{A}} = \square(1_{\mathcal{A}} \times 0)$$

(iii) For all $A \in \mathcal{O}b\,\mathcal{A}$: $c(A,0) = 1_A = c(0,A)$

(iv) For all $A,B \in \mathcal{O}b\,\mathcal{A}$: $c(B,A) = c(A,B)^{-1}$

(v) The following diagrams commute for all A,B,C∈ b

$$\begin{array}{ccc} A \,\square\, B \,\square\, C & \xrightarrow{c(A \,\square\, B,C)} & C \,\square\, A \,\square\, B \\ & {\scriptstyle 1_A \,\square\, c(B,C)} \searrow \quad \swarrow {\scriptstyle c(A,C) \,\square\, 1_B} & \\ & A \,\square\, C \,\square\, B & \end{array}$$

Example 2.2. Perhaps the simplest nontrivial example of a permutative category is the following. Let A be an abelian group. Consider A as a category in the usual way: $\mathcal{O}b\, A = *$ a single object while $\mathcal{M}or\, A = A$. Define the bifunctor $\square$: $A \times A \to A$ by

$$\square(*,*) = * \qquad \text{on the object}$$

$$\square(a,b) = a + b \qquad \text{on morphisms}$$

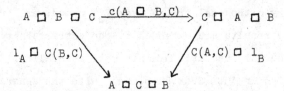

The unit object of A is of course $*$. The natural isomorphism c: $\square \to \square\tau$ is just the element $0 \in A$. It is easily checked that $(A, \square, *, c)$ has all the requisite properties of a permutative category.

Permutative categories are hard to find in nature. Yet they are a special (and extreme) example of a structure which is very common in mathematics: a symmetric monoidal category.

<u>Definition</u> 2.3. A <u>symmetric monoidal</u> category $(\mathcal{A}, \square, 0, a, \ell, r, c)$ is category $\mathcal{A}$ with a bifunctor $\square\colon \mathcal{A} \times \mathcal{A} \to \mathcal{A}$, a distinguished object 0 and four natural isomorphisms:

(i) an associativity isomorphism

$$a\colon \square(1_{\mathcal{A}} \times \square) \cong \square(\square \times 1_{\mathcal{A}})$$

(ii) a left and right unit isomorphism

$$\ell\colon \square(0 \times 1_{\mathcal{A}}) \cong 1_{\mathcal{A}}$$

$$r\colon \square(1_{\mathcal{A}} \times 0) \cong 1_{\mathcal{A}}$$

(iii) a commutativity isomorphism

$$c\colon \square \to \square T$$

which satisfy the following coherence conditions:

(iv) For all $A,B,C,D \in \mathcal{O}b\,\mathcal{A}$ the following diagram commutes

$$\begin{array}{ccccc}
A\square(B\square(C\square D)) & \xrightarrow{a(A,B,C\square D)} & (A\square B)\square(C\square D) & \xrightarrow{a(A\square B,C,D)} & ((A\square B)\square C)\square D \\
\Big\downarrow{\scriptstyle 1_A \square a(B,C,D)} & & & & \Big\downarrow{\scriptstyle a(A,B,C)\square 1_D} \\
A\square((B\square C)\square D) & & \xrightarrow{\quad a(A,B\square C,D) \quad} & & (A\square(B\square C))\square D
\end{array}$$

(v) $\ell_0 = r_0\colon 0 \square 0 \cong 0$

(vi) For all $A,B \in \mathcal{O}b\,\mathcal{A}$ the following diagram commutes

$$\begin{array}{ccc} A \Box (0 \Box B) & \xrightarrow{a(A,0,B)} & (A \Box 0) \Box B \\ & {\searrow}^{1_A \Box \ell_B} \quad {\swarrow}^{r_A \Box 1_B} & \\ & A \Box B & \end{array}$$

(vii) For any $A \in \mathcal{O}b\,\mathcal{A}$

$$r_A = \ell_A c(A,0) : A \Box 0 \cong A$$

(viii) For any $A,B \in \mathcal{O}b\,\mathcal{A}$, $c(B,A) = c(A,B)^{-1}$

(ix) For any $A,B,C \in \mathcal{O}b\,\mathcal{A}$ the following diagram commutes

$$\begin{array}{ccccc} A \Box (B \Box C) & \xrightarrow{a(A,B,C)} & (A \Box B) \Box C & \xrightarrow{c(A \Box B,C)} & C \Box (A \Box B) \\ \downarrow{\scriptstyle 1_A \Box c(B,C)} & & & & \downarrow{\scriptstyle a(C,A,B)} \\ A \Box (C \Box B) & \xrightarrow{a(A,C,B)} & (A \Box C) \Box B & \xrightarrow{c(A,C) \Box 1_B} & (C \Box A) \Box B \end{array}$$

Remark. It can be shown that the coherence conditions 2.3(iv)-2.3(ix) imply that all possible diagrams built from the natural isomorphisms a, ℓ, r and c commute. For details see [30; p. 161-6].

Remark. A permutative category is a special case of a symmetric monoidal category where the associativity and unit isomorphisms are actually identities. Thus coherence conditions 2.3(iv)-2.3(vi) become vacuous, while conditions 2.3(vii)-2.3(ix) reduce to conditions 2.1(iii)-2.1(v) respectively.

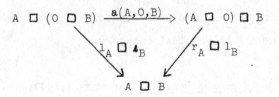

Example 2.4. The basic example of a symmetric monoidal category we encounter in algebraic K-theory is $(\mathcal{P}(R), \oplus, 0, a, \ell, r, c)$ where $\mathcal{P}(R)$ is

the category of finitely generated projective modules over a ring R and all isomorphisms between them, $\oplus$ is direct sum, 0 is the zero module, and a, ℓ, r, c are the obvious isomorphisms

$$a: M \oplus (N \oplus P) \cong (M \oplus N) \oplus P$$

$$\ell: 0 \oplus M \cong M \qquad r: M \oplus 0 \cong M$$

$$c: M \oplus N \cong N \oplus M$$

The reader may be tempted to think that all this fuss about associativity, unit, and commutation isomorphisms is really unnecessary, and could be avoided by picking one object in each isomorphism class, identifying all other objects with these representative objects and taking a, ℓ, r, c to be the identity maps. Unfortunately this approach does not work (cf. [30, p. 160]). However a more delicate argument due to Isbell [21] shows that every symmetric monoidal category is naturally equivalent to a permutative category.

Theorem 2.5. Let ($\mathcal{A}$, $\square$, 0, a, ℓ, r, c) be a symmetric monoidal category such that $|\mathcal{O}b\mathcal{A}|$ is a set. Then there is a naturally equivalent permutative category ($\mathcal{B}$, $\boxtimes$, *, c).

It should be remarked that ($\mathcal{A}$, $\square$, 0, a, ℓ, r, c) and ($\mathcal{B}$, $\boxtimes$, *, c) are equivalent in the sense of symmetric monoidal categories (cf. Def. 2.10 and succeeding remark). The category $\mathcal{B}$ is obtained by a sort of free monoid construction on the category $\mathcal{A}$. Details may be found in May [26].

Of course in many specific instances the Isbell construction is lavishly extravagant. In many cases the conversion of a symmetric monoidal category into an equivalent permutative category can be accomplished in a much more economical fashion.

Example 2.6. Consider the symmetric monoidal category $(\mathcal{P}(\mathbb{F}), \oplus, 0, a, \ell, r, c)$ of Example 2.4 consisting of all finitely generated projective modules over a field $\mathbb{F}$ together with all isomorphisms between them. Since over a field any projective module is free we can easily construct an equivalent permutative category $(\mathcal{GL}(\mathbb{F}), \oplus, 0, c)$ as follows:

Define $\mathcal{O}b\,\mathcal{GL}(\mathbb{F})$ to be the set $\mathbb{N}$ of nonnegative integers. Define the morphisms of $\mathcal{GL}(\mathbb{F})$ by

$$\hom(m,n) = \begin{cases} \emptyset & \text{if } m \neq n \\ GL(n,\mathbb{F}) & \text{if } m = n \end{cases}$$

where $GL(n,\mathbb{F})$ denotes the group of $n \times n$ nonsingular matrices over $\mathbb{F}$. The bifunctor $\oplus$ is given on objects by

$$\oplus(m,n) = m + n$$

and on morphisms by

$$\oplus(A,B) = \begin{pmatrix} A & 0 \\ 0 & B \end{pmatrix}$$

Then $\oplus$ is obviously strictly associative. It is equally obvious that 0 is a two sided unit for $\oplus$. The function

$$c\colon \mathcal{O}b\,\mathcal{GL}(\mathbb{F}) \times \mathcal{O}b\,\mathcal{GL}(\mathbb{F}) \to \mathcal{M}or\,\mathcal{GL}(\mathbb{F})$$

is given by

$$c(m,n) = \begin{pmatrix} 0 & I_n \\ I_m & 0 \end{pmatrix}$$

where I_k denotes the $k \times k$ identity matrix. We readily check that the following diagram commutes for all $A \in hom(m,n)$, $B \in hom(n,n)$

$$\begin{array}{ccc} \oplus(m,n) & \xrightarrow{c(m,n)} & \oplus(n,m) \\ \downarrow{\scriptstyle A \oplus B} & & \downarrow{\scriptstyle B \oplus A} \\ \oplus(m,n) & \xrightarrow{c(m,n)} & \oplus(n,m) \end{array}$$

so that c is a natural transformation from $\oplus$ to $\oplus \cdot T$. It is equally easy to check that diagram 2.1(vi) commutes.

It is not hard to see that $\mathcal{GL}(\mathbb{F})$ is equivalent to $\mathcal{P}(\mathbb{F})$. In fact $\mathcal{GL}(\mathbb{F})$ is isomorphic to the skeleton of $\mathcal{P}(\mathbb{F})$ obtained by identifying isomorphic objects.

Example 2.7. The full subcategory $\mathcal{GL}^{ev}(\mathbb{F})$ of $\mathcal{GL}(\mathbb{F})$ whose objects are all the even nonnegative integers is again a permutative category.

Example 2.8. Suppose $\mathbb{F}$ is a field of characteristic $\neq 2$. Then the subcategory $\mathcal{O}(\mathbb{F})$ of $\mathcal{GL}(\mathbb{F})$ which has the same objects but whose morphisms are orthogonal matrices (i.e. matrices A such that $AA^t = I$) is a permutative category. For it is obvious that the direct sum of two orthogonal matrices is again orthogonal. Also for each m,n it is obvious that $c(m,n)$ is an orthogonal matrix.

Remark. The above definition makes perfect sense even if $\mathbb{F}$ has characteristic 2. However for reasons which will be made clear in §7, we prefer to define $\mathcal{O}(\mathbb{F})$ in another way for fields of characteristic 2.

Definition 2.9. In any category $\mathcal{A}$ there is a natural equivalence relation on $\mathcal{O}b\mathcal{A}$. We say two objects A,B in $\mathcal{A}$ are connected if there is a chain of morphisms joining A to B

$$A \to C_1 \leftarrow C_2 \to C_3 \leftarrow C_4 \to \cdots \to C_k \leftarrow B$$

The equivalence classes of $\mathcal{O}b\,\mathcal{A}$ under this relation are called <u>components</u> of $\mathcal{A}$. The class of components of $\mathcal{A}$ is denoted by $\pi_0\mathcal{A}$.

Most of the categories $\mathcal{A}$ we will consider will have the property that $\mathcal{M}or\,\mathcal{A}$ consists exclusively of isomorphisms. In such a case it is obvious that two objects in $\mathcal{A}$ will be connected iff they are isomorphic. Hence we then have $\pi_0\mathcal{A} = |\mathcal{O}b\,\mathcal{A}|$.

In case $\mathcal{A}$ is a symmetric monoidal category, $\pi_0\mathcal{A}$ is endowed with a binary operation defined by

$$[A] + [B] = [A \,\square\, B]$$

Since the bifunctor $\square$ is associative and commutative and has a unit up to natural isomorphsims, it is clear that $\pi_0\mathcal{A}$ is a commutative monoid.

We can also define an ordering on $\pi_0\mathcal{A}$ by

$$[B] \leq [A] \text{ iff } [A] = [B] + [C] \text{ for some } C \in \mathcal{O}b\,\mathcal{A}.$$

Under this ordering $\pi_0\mathcal{A}$ becomes a directed set.

It is obvious that for the examples $\mathcal{GL}(\mathrm{E})$ and $\mathcal{O}(\mathrm{E})$ $\pi_0\mathcal{A} = |\mathcal{O}b\,\mathcal{A}|$ is the additive monoid of the natural numbers with the usual ordering.

We will also require a notion of morphism of permutative categories and a corresponding notion of natural transformation.

<u>Definition</u> 2.10. A functor Λ: $(\mathcal{A}, \square, 0, c) \to (\mathcal{B}, \square, 0, c)$ between permutative categories is said to be a <u>permutative functor</u> if

i) $\Lambda\cdot\square = \square\cdot(\Lambda \times \Lambda)$

ii) $\Lambda(0) = 0$

iii) $\Lambda[c(A,A')] = c(\Lambda A,\Lambda A')$, $A,A'\in \mathcal{O}b\,\mathcal{A}$

A permutative natural transformation between two permutative functors Λ, Λ': $(\mathcal{A},\square, 0, c) \to (\mathcal{B},\square, 0, c)$ is a natural transformation $\eta: \Lambda \to \Lambda'$ such that

iv) $\eta_{A\square A'} = \eta_A \square \eta_{A'}$ $A,A'\in \mathcal{O}b\,\mathcal{A}$

v) $\eta_0 = 1_0$

Remark 2.11. The reader is invited to formulate for himself the corresponding notions of symmetric monoidal functors and natural transformations (cf. also [30, p. 152].)

It should also be noted that given equivalences of categories $F: \mathcal{A} \xrightarrow{\cong} \mathcal{B}$, $G: \mathcal{B} \xrightarrow{\cong} \mathcal{A}$ where $\mathcal{A}$ is symmetric monoidal, there is a unique symmetric monoidal structure on $\mathcal{B}$ under which the functors F,G and the natural isomorphisms $G\cdot F \cong 1_{\mathcal{A}}$, $F\cdot G \cong 1_{\mathcal{B}}$ are symmetric monoidal.

Example 2.12. Let $A = \begin{pmatrix} a & b \\ c & d \end{pmatrix}$ be a 2×2 nonsingular matrix over a field $\mathbb{F}$. Let $A_n = \oplus_{i=1}^{n} A$. Define a permutative functor

$$\Lambda_A: \mathcal{GL}^{ev}(\mathbb{F}) \to \mathcal{GL}^{ev}(\mathbb{F})$$

as follows: on objects

$$\Lambda_A(2m) = 2m,$$

on morphisms by

$$\Lambda_A(B) = A_m B A_m^{-1} \quad \text{if} \quad B \in GL(2m,\mathbb{F})$$

It is then clear that conditions 2.10i, 2.10ii are satisfied. To see that 2.10iii also holds we observe that

$$c(2m,2n)^{-1}A_{n+m}c(2m,2n) = c(2m,2n)^{-1}(A_n \oplus A_m)c(2m,2n) = A_m \oplus A_n = A_{m+n}$$

so

$$\Lambda_A(c(2m,2n)) = A_{n+m}c(2m,2n)A_{n+m}^{-1} = c(2m,2n)$$

Hence Λ_A is a permutative functor.

There is also a permutative natural transformation $\eta\colon 1_{\mathcal{GL}^{ev}(\mathbb{F})} \to \Lambda_A$ given by

$$\eta_{2m} = A_m$$

For it is clear that following diagram commutes for any $B \in GL(2m,\mathbb{F})$

$$\begin{array}{ccc} 2m & \xrightarrow{\eta_{2m}} & 2m \\ \Big\downarrow B & & \Big\downarrow \Lambda_A(B) = A_m B A_m^{-1} \\ 2m & \xrightarrow{\eta_{2m}} & 2m \end{array}$$

It is also clear that 2.10iv and 2.10v hold.

Example 2.13. Consider the group $\mathbb{F}^*$ of units of a field $\mathbb{F}$ as a permutative category as in 2.2. Then we can define a permutative functor

$$\det\colon \mathscr{GL}^{ev}(\mathbb{F}) \to \mathbb{F}^*$$

as follows: on objects

$$\det(n) = *$$

on morphisms

$$\det(A) = \text{determinant of the matrix } A.$$

It is easily checked that det is a permutative functor.

Observe that the corresponding functor

$$\det\colon \mathscr{GL}(\mathbb{F}) \to \mathbb{F}^*$$

is not permutative since

$$\det(c(2m-1,2n-1)) = -1 \neq 1 = c(*,*) = c(\det(2m-1),\det(2n-1))$$

In certain cases our permutative categories will carry an additional structure - a topology.

Definition 2.14. A topological cateogory $\mathcal{A}$ is a small category in which both $\mathcal{O}b\,\mathcal{A}$ and $\mathcal{M}n\,\mathcal{A}$ are topological spaces and the structural functions source, target, identity and composition are continuous. A functor between topological categories is said to be continuous if it is continuous as a function on objects and morphisms. Similarly a natural transformation between continuous functors is said to be continuous if the induced map from objects of the domain to morphisms of the range is continuous.

A topological permutative category $(\mathcal{A}, \Box, 0, c)$ is a permutative category such that $\mathcal{A}$ is topological and $\square$ and c are continuous.

It is obvious that an (untopologized) category may be regarded as a topological category with the discrete topology.

Example 2.15. Let $\mathbb{F}$ be a topological field. Then there is a topological permutative category $(\mathcal{GL}(\mathbb{F}), \oplus, 0, c)$ defined as in Example 2.6 with

$$\mathcal{O}b\,\mathcal{GL}(\mathbb{F}) = \text{nonnegative integers}$$

being given the discrete topology and

$$\mathcal{M}or\,\mathcal{GL}(\mathbb{F}) = \coprod GL(n,\mathbb{F})$$

being given the topology of a disjoint union where $GL(n,\mathbb{F})$ is regarded as a topological group with the topology induced by $\mathbb{F}$.

Similarly the permutative category $(\mathcal{O}(\mathbb{F}), \oplus, 0, c)$ of Example 2.8 becomes a topological category.

2.16. The bar construction. Given a small category $\mathcal{A}$ we can convert it into a topological space $B\mathcal{A}$ by performing the bar construction on it. Briefly one first constructs a semisimplicial complex $B_*\mathcal{A}$ whose n-simplices are chains of morphisms of the form

$$A_0 \xrightarrow{f_1} A_2 \xrightarrow{f_3} \cdots \xrightarrow{f_{n-1}} A_{n-1} \xrightarrow{f_n} A_n$$

The i-th edge map is defined in the usual way by removing the i-th object in the chain and composing the corresponding morphisms. Similarly the degeneracy maps are defined by inserting the identity morphisms in appropriate places. One then takes $B\mathcal{A}$ to be the geometric realization of $B_*\mathcal{A}$. If $\mathcal{A}$ is a topological category one endows $B_*\mathcal{A}$ with a topology consistent with that of Mor $\mathcal{A}$ and takes this into account when performing geometric realization.

The bar construction has the following important properties

(i) A functor $F: \mathcal{A} \to \mathcal{B}$ induces a map

$$BF: B\mathcal{A} \to B\mathcal{B}$$

(ii) A natural transformation $F \xrightarrow{\eta} G: \mathcal{A} \to \mathcal{B}$ induces a homotopy

$$BF \overset{B\eta}{\simeq} BG: B\mathcal{A} \to B\mathcal{B}$$

(iii) If $\mathcal{A}$ and $\mathcal{B}$ are equivalent categories then $B\mathcal{A}$ and $B\mathcal{B}$ are homotopy equivalent (This is a consequence of (i) and (ii)).

(iv) $B(\mathcal{A} \times \mathcal{B})$ is homeomorphic to $B\mathcal{A} \times B\mathcal{B}$

(v) $\pi_0 B\mathcal{A} = \pi_0\mathcal{A}$ (cf. Def. 2.9)

(vi) If G is a topological group regarded as a topological category with one object with automorphism group G, then BG is the standard classifying space of the group G.

(vii) $B\mathcal{A} \cong \coprod_{\alpha\in\pi_0\mathcal{A}} B\mathcal{A}_\alpha$ where $\mathcal{A}_\alpha$ denotes the full subcategory of whose objects are in the component α (cf. Def. 2.9). If all the morphisms of $\mathcal{A}$ are isomorphisms then $B\mathcal{A} \simeq \coprod_{P\in|\mathcal{O}b\mathcal{A}|} B\mathrm{Aut}(P)$ where $\mathrm{Aut}(P)$ is the group of automorphisms of a representative object in an isomorphism class.

(viii) $H_*(B\mathcal{A}; A) = \oplus_{\alpha\in\pi_0\mathcal{A}} H_*(B\mathcal{A}_\alpha; A)$. In particular

$$H_0(B\mathcal{A}; A) \cong \oplus_{\alpha\in\pi_0\mathcal{A}} A$$

and we will denote the generator of $H_0(B\mathcal{A}_\alpha; A) \cong A$ by $[\alpha]$.

(This is an immediate consequence of (vii)).

Because of 2.16(vi), $B\mathcal{A}$ is usually referred to as the classifying space of the category $\mathcal{A}$. Similarly $BF: B\mathcal{A} \to B\mathcal{B}$ is called the classifying map of the functor $F: \mathcal{A} \to \mathcal{B}$.

For further details on the bar construction on categories the reader is referred to Segal [36] and Quillen [35].

2.17. E_∞ spaces, group completions and infinite loop spaces. If $\mathcal{A}$ is a category with structure, then this structure is reflected in its classifying space $B\mathcal{A}$. For instance if $(\mathcal{A}, \square, 0, a, \ell, r, c)$ is a symmetric monoidal category, then $B\mathcal{A}$ becomes a homotopy commutative H-space with multiplication

$$B\square : B\mathcal{A} \times B\mathcal{A} \cong B(\mathcal{A} \times \mathcal{A}) \to B\mathcal{A}$$

If $\mathcal{A}$ is a permutative category then this multiplication becomes strictly associative (not just homotopy associative) and the unit becomes an honest unit (instead of a homotopy unit). Thus $B\mathcal{A}$ is then a topological monoid with homotopy commutative multiplication.

In fact the classifying space of permutative cateogry carries the far richer structure of an E_∞ space (cf. [29] and [26, p. 84].) In [26] May constructs a functor which assigns to each E_∞ space X an infinite loop space ΓX and a natural map $i: X \to \Gamma X$ of E_∞ spaces. This map has the characteristic property that

$$i_*: H_*(X; k)[\pi_0 X]^{-1} \to H_*(\Gamma X; k)$$

is an isomorphism for all commutative coefficient rings k. In particular if $\pi_0 X$ is a group then i is a homotopy equivalence (ΓX is referred to as the group completion of X).

We can thus apply the composite functor ΓB to a permutative category $\mathcal{A}$ and obtain an infinite loop space $\Gamma B\mathcal{A}$. We summarize here for future reference some important properties of this construction:

(i) A permutative functor $F: \mathcal{A} \to \mathcal{B}$ induces an infinite loop map

$$\Gamma BF: \Gamma B\mathcal{A} \to \Gamma B\mathcal{B}$$

(ii) A permutative natural transformation $\Lambda: F \to G: \mathcal{A} \to \mathcal{B}$ induces a homotopy $\Gamma BF \simeq \Gamma BG$.

(iii) $\pi_0 \Gamma B\mathcal{A}$ is the group completion $K_0\mathcal{A}$ of the monoid $\pi_0\mathcal{A}$ (cf. Def. 2.9), i.e. $K_0\mathcal{A} = \pi_0\Gamma B\mathcal{A}$ can be obtained from the monoid $\pi_0\mathcal{A}$ by a Grothendieck construction: one takes the free abelian group on $\pi_0\mathcal{A}$ and divides out by relations of the form $[\alpha + \beta] - [\alpha] - [\beta]$.

(iv) As a space (but not as an infinite loop space)

$$\Gamma B\mathcal{A} \simeq \Gamma_0 B\mathcal{A} \times K_0\mathcal{A}$$

where $\Gamma_0 B\mathcal{A}$ is the component of the basepoint. (This is in fact immediate since all path components of a loop space are homotopy equivalent).

(v) The natural map $i: B\mathcal{A} \to \Gamma B\mathcal{A}$ induces a map $\tau: \varinjlim_{\alpha\in\pi_0\mathcal{A}} B\mathcal{A}_\alpha \to \Gamma_0 B\mathcal{A}$ (cf. Def. 2.9 and 2.16(vii)) such that $\tau_*: H_*(\varinjlim_{\alpha\in\pi_0\mathcal{A}} B\mathcal{A}_\alpha) \to H_*(\Gamma_0 B\mathcal{A})$ is an isomorphism.

(vi) If M is a submonoid of $\pi_0\mathcal{A}$ and $\mathcal{A}_M$ is the full subcategory $\mathcal{A}_M = \bigcup_{\alpha\in M} \mathcal{A}_\alpha$, then $\mathcal{A}_M$ is a permutative subcategory of $\mathcal{A}$ and the inclusion $\mathcal{A}_M \to \mathcal{A}$ induces an infinite loop map

$$\Gamma B\mathcal{A}_M \to \Gamma B\mathcal{A}$$

Furthermore if M is cofinal in $\pi_0\mathcal{A}$ then this map induces an inclusion on zeroth homotopy groups:

$$K_0 \mathcal{A}_M \subseteq K_0 \mathcal{A},$$

and for each $\alpha \in K_0 \mathcal{A}_M$ the restriction to the corresponding path component

$$\Gamma_\alpha B\mathcal{A}_M \to \Gamma_\alpha B\mathcal{A}$$

is a homotopy equivalence. (This is a consequence of (iv) and (v).)

Thus $\Gamma B\mathcal{A}_M$ may be identified with the subspace

$$\Gamma_{K_0 \mathcal{A}_M} B\mathcal{A} = \bigcup_{\alpha \in K_0 \mathcal{A}_M} \Gamma_\alpha B\mathcal{A}.$$

Further information about the construction $\Gamma B\mathcal{A}$ may be found in May [26].

Example 2.18. Consider the symmetric monoidal category $\mathcal{P}(R)$ of finitely generated projective modules over a ring R (cf. 2.4). By Theorem 2.5 there is an equivalent permutative category, which by abuse of notation we also denote $\mathcal{P}(R)$. Applying the bar construction we get a topological monoid

$$B\mathcal{P}(R) = \coprod BAut(P)$$

where Aut(P) is the automorphism group of R and P ranges over all isomorphism classes of finitely generated projective R modules (cf. 2.16(vii)). Consequently when we take homology with coefficients in a ring k we get a graded ring

$$H_*(B\mathcal{P}(R); k) \cong \oplus H_*(BAut(P); k)$$

with product (which we denote by $*$) given by

$$H_*(\mathrm{BAut}(P);\, k) \times H_*(\mathrm{BAut}(Q);\, k) \xrightarrow{(B\Box)_*} H_*(\mathrm{BAut}(P \oplus Q);\, k)$$

When we apply the group completion functor Γ we get

$$\Gamma B\mathcal{P}(R) \simeq \Gamma_0 B\mathcal{P}(R) \times K_0R \quad \text{(as spaces)}$$

by 2.17(iii and iv), where K_0R is the standard Grothendieck group of R. By 2.17(v) there is a map

$$\tilde{i}: \varinjlim_{P \in \pi_0 \vec{\mathcal{a}} = |\mathcal{O}b\, \mathcal{a}|} \mathrm{BAut}(P) \to \Gamma_0 B\mathcal{P}(R)$$

which induces an isomorphism in homology. But the free modules R^n are cofinal in $\pi_0\, \mathcal{a}$. Since $\mathrm{Aut}(R^n) = GL(n,R)$, it follows that we get a map

$$\tilde{i}: BGL(\infty,R) = \varinjlim BGL(n,R) \to \Gamma_0 B\mathcal{P}(R)$$

which induces an isomorphism in homology. By the universal property of Quillen's plus construction (cf. Wagoner [41]) we get a map

$$\bar{i}: BGL(\infty,R)^+ \to \Gamma_0 B\mathcal{P}(R)$$

which induces an isomorphism in homology. Since $BGL(\infty,R)^+$ and $\Gamma_0 B\mathcal{P}(R)$ are simple spaces, $\bar{i}$ is a homotopy equivalence. Consequently

$$\Gamma B\mathcal{P}(R) \simeq BGL(\infty,R)^+ \times K_0R \quad \text{(as spaces)}$$

Thus the homotopy groups of $\Gamma B\mathcal{P}(R)$ are Quillen's K-groups of R. Also we have

$$H_*(\Gamma B\mathcal{P}(R);\ k) \cong H_*(BGL(\infty,R)^+;\ k) \otimes H_*(K_0R;\ k)$$
$$\cong H_*(BGL(\infty,R);\ k) \otimes k[K_0R]$$

where $k[K_0R]$ is the group ring of K_0R over k. (This isomorphism is in fact an isomorphism of graded algebras). The generator of $k[K_0R]$ corresponding to the element $\alpha \in K_0(R)$ will be denoted by $[\alpha]$.

A similar analysis applies as regards the spaces $\Gamma B\mathcal{GL}(\mathbb{F})$, $\Gamma B\mathcal{O}(\mathbb{F})$ and $\Gamma B\mathcal{A}$ for most permutative categories $\mathcal{A}$ we shall encounter.

Example 2.19. Let A be an abelian group considered as a permutative category (cf. 2.2). Then according to 2.16(vi), BA is the standard classifying space of the group A and hence $BA = K(A,1)$ the Eilenberg-MacLane space. Since $\pi_0 A = *$ is already a group, the group completion map

$$i: BA \to \Gamma BA$$

is an equivalence. Hence $\Gamma B\mathcal{A} \simeq BA \simeq K(A,1)$ as infinite loop spaces.

If $\mathbb{F}$ is a field, the permutative functor det: $\mathcal{GL}^{ev}(\mathbb{F}) \to \mathbb{F}^*$ (cf. 2.13) induces an infinite loop map

$$\Gamma B(\det): \Gamma B\mathcal{GL}^{ev}(\mathbb{F}) \to B\mathbb{F}^*$$

Example 2.20. If $\mathbb{C}$ denotes the field of complex numbers with the usual topology then the same analysis as in Example 2.18 shows that

$$\Gamma B\mathcal{GL}(\mathbb{C}) \simeq BGL(\infty,\mathbb{C})^+ \times \mathbb{Z}$$

Since $BGL(\infty,\mathbb{C}) \simeq BU$ is a simple space already, the plus construction does not change its homotopy type. Hence

$$\Gamma B\mathcal{GL}(\mathbb{C}) \simeq BU \times \mathbb{Z}$$

In fact a more delicate argument (cf. May [25]) shows that $\Gamma B\mathcal{GL}(\mathbb{C})$ is the infinite loop space associated with connective complex K-theory.

A similar argument shows that if $\mathbb{R}$ is the field of real numbers with the usual topology, then

$$\Gamma B\mathcal{O}(\mathbb{R}) \simeq BO \times \mathbb{Z}$$

is the infinite loop space associated with connective real K-theory.

§3. The orthogonal groups and their subgroups

In this section we shall study in detail the permutative categories associated with the orthogonal groups over a field $\mathbb{F}_q$ with $q = p^k$ elements, p an odd prime.

3.1. The category $\mathcal{O}(\mathbb{F}_q)$. Our starting point is the permutative category $\mathcal{O}(\mathbb{F}_q)$ defined in 2.8. Recall that the objects of that category are the nonnegative integers and the morphisms are

$$\hom(m,n) = \begin{cases} \emptyset & \text{if } m \neq n \\ O(n,\mathbb{F}_q) & \text{if } m = n \end{cases}$$

where $O(n,\mathbb{F}_q)$ is the group of $n \times n$ matrices A over $\mathbb{F}_q$ such that $AA^t = I$. The groups $O(n,\mathbb{F}_q)$ are finite with orders

$$|O(2m-1,\mathbb{F}_q)| = 2N_m$$

$$|O(2m,\mathbb{F}_q)| = 2(q^m-\epsilon^m)q^{m-1}N_m$$

where $N_m = \Pi_{i=1}^{m-1}(q^{2i}-1)q^{2i-1}$ and $\epsilon = \pm 1$ as $q \equiv \pm 1 \pmod 4$. See Dickson [16] for proofs.

It follows from the discussion of 2.18 that

$$B\mathcal{O}(\mathbb{F}_q) = \coprod_{n=0}^{\infty} BO(n,\mathbb{F}_q)$$

and that as a space

$$\Gamma B\mathcal{O}(\mathbb{F}_q) \simeq \Gamma_0 B\mathcal{O}(\mathbb{F}_q) \times \mathbb{Z} \simeq (BO(\infty,\mathbb{F}_q))^+ \times \mathbb{Z}$$

where $O(\infty,\mathbb{F}_q) = \lim_{n\to\infty} O(n,\mathbb{F}_q)$ and $(\cdot)^+$ denotes Quillen's plus construction.

The chief purpose of this section will be to define certain important permutative subcategories of $\mathcal{O}(\mathbb{F}_q)$ and to analyze the relationships between them. We first state our main results.

3.2. <u>Important subcategories of $\mathcal{O}(\mathbb{F}_q)$ and their properties.</u>

There exist permutative subcategories of $\mathcal{O}(\mathbb{F}_q)$ of the following type:

(i) $\mathcal{O}^{ev}(\mathbb{F}_q)$ - the full subcategory of $\mathcal{O}(\mathbb{F}_q)$ whose objects are the even nonnegative integers.

(ii) $\mathcal{S}\mathcal{O}^{ev}(\mathbb{F}_q)$ - the subcategory of $\mathcal{O}^{ev}(\mathbb{F}_q)$ with the same objects but whose morphisms are orthogonal matrices with determinant equal to 1.

(iii) $\mathcal{N}^{ev}(\mathbb{F}_q)$ - the subcategory of $\mathcal{O}^{ev}(\mathbb{F}_q)$ with the same objects but whose morphisms are orthogonal matrices with spinor norm equal to 1.

(iv) $\mathcal{N}\mathcal{D}^{ev}(\mathbb{F}_q)$ - the subcategory of $\mathcal{O}^{ev}(\mathbb{F}_q)$ with the same objects but whose morphisms are orthogonal matrices with spinor norm times determinant equal to 1.

(v) $\mathcal{S}\mathcal{N}^{ev}(\mathbb{F}_q)$ - the subcategory of $\mathcal{O}^{ev}(\mathbb{F}_q)$ with the same objects but whose morphisms are orthogonal matrices with both spinor norm and determinant equal to 1.

We shall also see that under certain circumstances the categories $\mathcal{N}^{ev}(\mathbb{F}_q)$ and $\mathcal{N}\mathcal{D}^{ev}(\mathbb{F}_q)$ can be extended to permutative categories $\mathcal{N}(\mathbb{F}_q)$ and $\mathcal{N}\mathcal{D}(\mathbb{F}_q)$ having as objects all the nonnegative integers.

In addition we also have

(vi) There is a permutative functor $\Phi\colon \mathcal{O}^{ev}(\mathbb{F}_q) \to \mathcal{O}^{ev}(\mathbb{F}_q)$ which maps $\mathcal{N}^{ev}(\mathbb{F}_q)$ isomorphically onto $\mathcal{N}\mathcal{D}^{ev}(\mathbb{F}_q)$

(vii) There are fibrations of infinite loop spaces

$$\Gamma_0 B\mathcal{S}\mathcal{O}^{ev}(\mathbb{F}_q) \to \Gamma_0 B\mathcal{O}(\mathbb{F}_q) \xrightarrow{g_1} B\mathbb{Z}/2$$

$$\Gamma_0 B\mathcal{N}^{ev}(\mathbb{F}_q) \to \Gamma_0 B\mathcal{O}(\mathbb{F}_q) \xrightarrow{g_2} B\mathbb{Z}/2$$

$$\Gamma_0 B\mathcal{N}\mathcal{D}^{ev}(\mathbb{F}_q) \to \Gamma_0 B\mathcal{O}(\mathbb{F}_q) \xrightarrow{g_3} B\mathbb{Z}/2$$

$$\Gamma_0 B\mathcal{S}\mathcal{N}^{ev}(\mathbb{F}_q) \to \Gamma_0 B\mathcal{O}(\mathbb{F}_q) \xrightarrow{(g_1,g_2)} B\mathbb{Z}/2 \times B\mathbb{Z}/2$$

(viii) $\pi_1(\Gamma_0 B\mathcal{O}(\mathbb{F}_q)) = \mathbb{Z}/2 \oplus \mathbb{Z}/2.$

We now proceed to details and proofs.

<u>Remark</u> 3.3. It is obvious from definition that $\mathcal{O}^{ev}(\mathbb{F}_q)$ is a permutative subcategory of $\mathcal{O}(\mathbb{F}_q)$. It follows moreover from 2.17(vi) that under the inclusion $\mathcal{O}^{ev}(\mathbb{F}_q) \to \mathcal{O}(\mathbb{F}_q)$ the infinite loop space $\Gamma B\mathcal{O}^{ev}(\mathbb{F}_q)$ maps via a homotopy equivalence (of infinite loop spaces) onto $\Gamma_{2\mathbb{Z}} B\mathcal{O}(\mathbb{F}_q)$.

<u>Remark</u> 3.4. It is equally obvious that $\mathcal{S}\mathcal{O}^{ev}(\mathbb{F}_q)$ is a permutative subcategory of $\mathcal{O}(\mathbb{F}_q)$. We should note however that the corresponding category $\mathcal{S}\mathcal{O}(\mathbb{F}_q)$, having as objects all the nonnegative integers and as morphisms all orthogonal matrices with determinant 1, is not a

permutative category. The reason for this is that if m and n are odd then the commutation matrix

$$c(m,n) = \begin{pmatrix} 0 & I_n \\ I_m & 0 \end{pmatrix} \quad (\text{cf. } 2.6)$$

has determinant -1.

We turn now to the definition of the spinor norm and the construction of permutative categories associated with it. First however we need a few preliminaries.

<u>Definition</u> 3.5. A matrix $A \in O(n, \mathbb{F}_q)$ is said to be a <u>symmetry</u> if there is a vector $y \in \mathbb{F}_q^n$ of nonzero length such that

1) $Ay = -y$

2) $A\big|_{(\mathbb{F}y)^\perp} = 1_{(\mathbb{F}y)^\perp}$

We then write $A = T_y$.

We now quote the following theorem:

<u>Theorem</u> 3.6. Any matrix $A \in O(n, \mathbb{F}_q)$ is a product of symmetries. Moreover if

$$A = T_{u_1} \cdot T_{u_2} \cdot \cdots \cdot T_{u_m} = T_{v_1} \cdot T_{v_2} \cdot \cdots \cdot T_{v_p}$$

then

$$Q(u_1)Q(u_2)\cdots Q(w_m) \equiv Q(v_1)Q(v_2)\cdots Q(v_p) \pmod{\mathbb{F}_q^{*2}}$$

where $Q(x) = x_1^2 + \cdots + x_n^2$ is the standard quadratic form on $\mathbb{F}_q^n$.

For a proof of this theorem see O'Meara [30, pp. 102-103 and pp. 136-137].

Theorem 3.6 enables us to define the following invariant on $\mathcal{O}(\mathbb{F})$.

Definition 3.7. Let θ: $O(n, \mathbb{F}_q) \to \mathbb{F}_q^*/(\mathbb{F}_q^*)^2 = \mathbb{Z}/2$ be the map given as follows: If $A = T_{u_1} \cdot T_{u_2} \cdot \cdots \cdot T_{u_m}$

$$\theta(A) = Q(u_1)\,Q(u_2)\cdots Q(u_m)\,(\mathbb{F}_q^*)^2$$

$\theta(A)$ is called the spinor norm of A.

It is easily checked that θ is a group homomorphism. Moreover if

$$A = T_{u_1} \cdot T_{u_2} \cdot \cdots \cdot T_{u_m}$$

$$B = T_{v_1} \cdot T_{v_2} \cdot \cdots \cdot T_{v_p}$$

then

$$A \oplus B = T_{(u_1,0)} \cdot T_{(u_2,0)} \cdot \cdots \cdot T_{(u_m,0)} \cdot T_{(0,v_1)} \cdot T_{(0,v_2)} \cdot T_{(0,v_p)}$$

so

$$\begin{aligned}\theta(A \oplus B) &= Q(u_1)\,Q(u_2)\cdots Q(u_m)\,Q(v_1)\,Q(v_2)\cdots Q(v_p)\,(\mathbb{F}_q^*)^2 \\ &= \theta(A)\ \theta(B)\end{aligned}$$

For further information see O'Meara [30]. It is clear that we can construct a permutative functor θ: $\mathcal{O}^{ev}(\mathbb{F}_q) \to \mathbb{Z}/2$ as in 2.13.

It is now possible to define the following subgroups of $O(n, \mathbb{F}_q)$.

1) $N(n,\mathbb{F}_q) = \{A \in O(n,\mathbb{F}_q) : \theta(A) = 1\}$

2) $ND(n,\mathbb{F}_q) = \{A \in O(n,\mathbb{F}_q) : \theta(A)\det(A) = 1\}$

3) $SN(n,\mathbb{F}_q) = \{A \in O(n,\mathbb{F}_q) : \theta(A) = 1 \text{ and } \det(A) = 1\}$

We can define corresponding subcategories of $\mathcal{O}(\mathbb{F}_q)$ as follows:

1) $\mathcal{N}(\mathbb{F}_q)$ subcategory of $\mathcal{O}(\mathbb{F}_q)$ with objects the nonnegative integers and morphisms $\coprod_{n=0}^{\infty} N(n, \mathbb{F}_q)$
2) $\mathcal{N}\mathcal{D}(\mathbb{F}_q)$ subcategory of $\mathcal{O}(\mathbb{F}_q)$ with objects the nonnegative integers and morphisms $\coprod_{n=0}^{\infty} ND(n, \mathbb{F}_q)$
3) $\mathcal{S}\mathcal{N}(\mathbb{F}_q)$ subcategory of $\mathcal{O}(\mathbb{F}_q)$ with objects the nonnegative integers and morphisms $\coprod_{n=0}^{\infty} SN(n, \mathbb{F}_q)$

We can also define $\mathcal{N}^{ev}(\mathbb{F}_q)$, $\mathcal{N}\mathcal{D}^{ev}(\mathbb{F}_q)$ and $\mathcal{S}\mathcal{N}^{ev}(\mathbb{F}_q)$ to be the full subcategories of $\mathcal{N}(\mathbb{F}_q)$, $\mathcal{N}\mathcal{D}(\mathbb{F}_q)$ and $\mathcal{S}\mathcal{N}(\mathbb{F}_q)$ respectively whose objects are the even nonnegative integers.

Theorem 3.8. The categories $\mathcal{N}^{ev}(\mathbb{F}_q)$, $\mathcal{N}\mathcal{D}^{ev}(\mathbb{F}_q)$, and $\mathcal{S}\mathcal{N}^{ev}(\mathbb{F}_q)$ are permutative categories.

Proof. Since

$$\theta(A \oplus B) = \theta(A)\theta(B) \text{ and } \det(A \oplus B) = \det(A)\det(B)$$

it is obvious that $\mathcal{N}^{ev}(\mathbb{F}_q)$, $\mathcal{N}\mathcal{D}^{ev}(\mathbb{F}_q)$, and $\mathcal{S}\mathcal{N}^{ev}(\mathbb{F}_q)$ are closed under $\oplus$. It remains only to check that the natural transformation c lies in this category.

We first note that any even permutation matrix has spinor norm equal to 1. For any even permutation σ of $\{1,2,\ldots,n\}$ can be written as a product of an even number of 2-cycles, i.e.

$$\sigma = (i_1\ i_2)(i_3\ i_4)\cdots(i_{2k-1}\ i_{2k})$$

with k even. Then for the corresponding matrix

$$L_\sigma = L_{(i_1 i_2)} L_{(i_3 i_4)} \cdots L_{(i_{2k-1} i_{2k})}$$

But

$$L_{(j_1 j_2)} = T_{e_{j_1} - e_{j_2}}$$

so

$$\begin{aligned}\theta(L_\sigma) &= \Pi_{r=1}^{k} Q(e_{i_{2r-1}} - e_{i_{2r}})(\mathbb{F}_q^*)^2 = 2^k(\mathbb{F}_q^*)^2 \\ &= 1(\mathbb{F}_q^*)^2\end{aligned}$$

Now if m,n are both even then c(m,n) is an even permutation matrix so $\theta(c(m,n)) = 1$ and $\det(c(m,n)) = 1$. Thus the natural transformation c lies in $\mathcal{N}^{ev}(\mathbb{F}_q)$, $\mathcal{ND}^{ev}(\mathbb{F}_q)$ and $\mathcal{SN}^{ev}(\mathbb{F}_q)$. This completes the proof.

Theorem 3.9. If $q \equiv \pm 3 \pmod 8$ then $\mathcal{ND}(\mathbb{F}_q)$ is a permutative category.

Proof: Again it suffices to check that the natural transformation c lies in $\mathcal{ND}(\mathbb{F}_q)$. If we can show that any permutation matrix lies in $\mathcal{ND}(\mathbb{F}_q)$, this will automatically hold. Since any permutation is a product of transpositions, it suffices to show that any transposition matrix, $L_{(j_1 j_2)}$ lies in $\mathcal{ND}(\mathbb{F}_q)$. But

$$L_{(j_1j_2)} = T_{e_{j_1}-e_{j_2}}$$

and

$$\theta(L_{(j_1j_2)})\det(L_{(j_1j_2)}) = [Q(e_{j_1}-e_{j_2})(\mathbb{F}_q^*)^2](-1)$$
$$= [2(\mathbb{F}_q^*)^2](-1)$$

Now if $q \equiv \pm 3 \pmod 8$ then $q = p^a$ where a is odd, and p is a prime such that $p \equiv \pm 3 \pmod 8$. It is readily seen that in such a field $\mathbb{F}_q$, 2 is not a square. Hence

$$\theta(L_{(j_1j_2)})\det(L_{(j_1j_2)}) = (-1)(-1) = 1$$

This completes the proof.

Theorem 3.10. If $q \equiv \pm 1 \pmod 8$ then $\mathcal{N}(\mathbb{F}_q)$ is a permutative category.

Proof: As in the preceding theorem, it suffices to show that any transposition matrix $L_{(j_1j_2)}$ lies in $\mathcal{N}(\mathbb{F}_q)$. But

$$\theta(L_{(j_1j_2)}) = \theta(T_{e_{j_1}-e_{j_2}}) = Q(e_{j_1}-e_{j_2})(\mathbb{F}_q^*)^2$$
$$= 2(\mathbb{F}_q^*)^2$$

But if $q \equiv \pm 1 \pmod 8$ then $q = p^a$ where p is a prime such that $p \equiv \pm 1 \pmod 8$ or $q = p^a$ where p is a prime such that $p \equiv \pm 3 \pmod 8$ and a is even. In either case it follows that $2 \in (\mathbb{F}_q^*)^2$. Hence

$$\theta(L_{(j_1 j_2)}) = 1$$

which completes the proof.

We next turn to showing that the categories $\mathcal{N}^{ev}(\mathbb{F}_q)$ and $\mathcal{N}\mathcal{D}^{ev}(\mathbb{F}_q)$ are isomorphic. To construct such an isomorphism we first prove the following result

<u>Lemma</u> 3.11. If $\mathbb{F}_q$ is a finite field such that the sum of squares is a square, then $\mathbb{F}_q$ has characteristic 2.

<u>Proof:</u> Suppose $q = p^a$ with p an odd prime. Let S be the set of elements in $\mathbb{F}_q$ which are squares. Then the order of S is

$$|S| = \frac{q-1}{2} + 1 = \frac{q+1}{2}$$

By hypothesis S is closed under addition. Since S is obviously closed under multiplication, S is a semiring. But in a finite field, every subsemiring is a subfield. This is impossible since

$$p^{a-1} < \frac{q+1}{2} < p^a.$$

<u>Theorem</u> 3.12. There is a permutative functor $\Phi\colon \mathcal{O}^{ev}(\mathbb{F}_q) \to \mathcal{O}^{ev}(\mathbb{F}_q)$ such that $\Phi^2 = 1$ and $\Phi(\mathcal{N}^{ev}(\mathbb{F}_q)) = \mathcal{N}\mathcal{D}^{ev}(\mathbb{F}_q)$.

<u>Proof:</u> By Lemma 3.11 there are $a,b \in \mathbb{F}_q$ such that $a^2 + b^2$ is a nonsquare. Let

$$C = \begin{pmatrix} -a & b \\ b & a \end{pmatrix}$$

Let $C_n = \oplus_{i=1}^{n} C$. Define $\Phi\colon \mathcal{O}^{ev}(\mathbb{F}_q) \to \mathcal{O}^{ev}(\mathbb{F}_q)$ to be the identity on objects and to act on morphisms as

$$\Phi(A) = C_n A C_n^{-1} \quad \text{if} \quad A \in O(2n, \mathbb{F}_q)$$

To see that this is well defined, note that if $u^2 = \frac{1}{a^2+b^2}$, $u \in \mathbb{F}_{q^2}$, then $uC \in O(2, \mathbb{F}_{q^2})$ and hence $uC_n \in O(2n, \mathbb{F}_{q^2})$. Therefore

$$\Phi(A) = (uC_n)A(uC_n)^{-1} \in O(2n, \mathbb{F}_{q^2}) \cap GL(2n, \mathbb{F}_q) = O(2n, \mathbb{F}_q)$$

It is evident that $\Phi(A \oplus B) = \Phi(A) \oplus \Phi(B)$. Also since

$$c(2m,2n)^{-1} C_{n+m} c(2m,2n) = c(2m,2n)^{-1}(C_n \oplus C_m) c(2m,2n) = C_m \oplus C_n = C_{m+n}$$

it follows that

$$\Phi(c(2m,2n)) = C_{n+m} c(2m,2n) C_{n+m}^{-1} = c(2m,2n)$$

so Φ is a permutative functor.

Since $C^2 = (a^2 + b^2)\left(\begin{smallmatrix}1 & 0\\ 0 & 1\end{smallmatrix}\right)$, likewise $C_n^2 = (a^2 + b^2) I_{2n}$. Hence $\Phi^2 = 1$.

Now suppose $A \in N(2n, \mathbb{F}_q)$. Write A as a product of symmetries, say

$$A = T_{u_1} \cdot T_{u_2} \cdot \cdots \cdot T_{u_m}$$

Then

$$\Phi(A) = C_n A C_n^{-1} = T_{C_n u_1} \cdot T_{C_n u_2} \cdot \cdots \cdot T_{C_n u_m}$$

so

$$\theta(\Phi(A)) = \Pi_{i=1}^{m}\theta(T_{C_n u_i})$$

Now note that $C_n^t C_n = C_n^2 = (a^2+b^2)I_{2n}$. Hence

$$Q\left[C_n\begin{pmatrix}\alpha_1\\ \vdots\\ \alpha_{2n}\end{pmatrix}\right] = \left[C_n\begin{pmatrix}\alpha_1\\ \vdots\\ \alpha_{2n}\end{pmatrix}\right]^t\left[C_n\begin{pmatrix}\alpha_1\\ \vdots\\ \alpha_{2n}\end{pmatrix}\right]$$

$$= (\alpha_1,\dots,\alpha_{2n})C_n^t C_n\begin{pmatrix}\alpha_1\\ \vdots\\ \alpha_{2n}\end{pmatrix} = (a^2+b^2)(\alpha_1,\dots,\alpha_{2n})\begin{pmatrix}\alpha_1\\ \vdots\\ \alpha_{2n}\end{pmatrix}$$

$$= (a^2+b^2)Q\left[\begin{pmatrix}\alpha_1\\ \vdots\\ \alpha_{2n}\end{pmatrix}\right]$$

Since a^2+b^2 is a nonsquare, $\theta(T_{C_n x}) = (a^2+b^2)Q(x)(\mathbb{F}_q^*)^2 = -\theta(T_x)$. Thus

$$\theta(\Phi(A)) = \Pi_{i=1}^{m}\theta(T_{C_n u_i}) = \Pi_{i=1}^{m}(-\theta(T_{u_i})) = (-1)^m\theta(A) = (-1)^m = \det A$$

Hence

$$\theta(\Phi(A))\det(\Phi(A)) = (\det A)^2 = 1$$

It follows that $\Phi(\mathcal{N}^{ev}(\mathbb{F}_q)) = \mathcal{N}\mathcal{D}^{ev}(\mathbb{F}_q)$.

<u>Corollary</u> 3.13. $\mathcal{N}^{ev}(\mathbb{F}_q)$ and $\mathcal{N}\mathcal{D}^{ev}(\mathbb{F}_q)$ are equivalent permutative categories.

<u>Remark</u> 3.14. If $q \equiv \pm 3 \pmod 8$ then as mentioned in Theorem 3.9, $2 = 1^2 + 1^2$ is not a square in $\mathbb{F}_q$. Hence we may take $a = b = 1$ and

$$C = \begin{pmatrix} -1 & 1 \\ 1 & 1 \end{pmatrix}$$

Remark 3.15. The choice of elements a,b such that $a^2 + b^2$ is a nonsquare does not affect the definition of $\Phi\colon \Gamma B\mathcal{O}^{ev}(\mathbb{F}_q) \to \Gamma B\mathcal{O}^{ev}(\mathbb{F}_q)$. For if we choose another pair of elements $a', b' \in \mathbb{F}_q$ such that $a'^2 + b'^2$ is a nonsquare and then used the matrix

$$C' = \begin{pmatrix} -a' & b' \\ b' & a' \end{pmatrix}$$

to define a permutative functor $\Phi'\colon \mathcal{O}^{ev}(\mathbb{F}_q) \to \mathcal{O}^{ev}(\mathbb{F}_q)$, we would have $\Phi' \simeq \Phi\colon \Gamma B\mathcal{O}^{ev}(\mathbb{F}_q) \to \Gamma B\mathcal{O}^{ev}(\mathbb{F}_q)$.

For since $(a'^2 + b'^2)(a^2 + b^2)^{-1}$ is a square in $\mathbb{F}_q$ say $r^2 = (a'^2 + b'^2)(a^2 + b^2)^{-1}$, it follows that

$$r^{-1}C'C^{-1} = [a'^2 + b'^2]^{-1/2}C'([a^2 + b^2]^{-1/2}C)^{-1} \in O(2,\mathbb{F}_q)$$

Hence there is permutative natural transformation of functors

$$\lambda\colon \Phi \to \Phi'$$

given by

$$\lambda_{2n} = \oplus_{i=1}^{n} r^{-1}C'C^{-1}$$

(compare with Example 2.12). By 2.17(ii) it follows that λ induces a homotopy between the infinite loop maps $\Phi, \Phi'\colon \Gamma B\mathcal{O}^{ev}(\mathbb{F}_q) \to \Gamma B\mathcal{O}^{ev}(\mathbb{F}_q)$.

Finally we turn to the relationship between the infinite loop spaces associated with these various subcategories of $\mathcal{O}(\mathbb{F}_q)$.

In what follows we use the notation

$$g_1, g_2, g_3 : \Gamma_0 B\mathcal{O}(\mathbb{F}_q) \longrightarrow B\mathbb{Z}/2$$

$$g_4 : \Gamma_0 B\mathcal{N}^{ev}(\mathbb{F}_q) \longrightarrow B\mathbb{Z}/2$$

$$g_5 : \Gamma_0 B\mathcal{N}\mathcal{D}^{ev}(\mathbb{F}_q) \longrightarrow B\mathbb{Z}/2$$

$$g_6 : \Gamma_0 B\mathcal{S}\mathcal{O}^{ev}(\mathbb{F}_q) \longrightarrow B\mathbb{Z}/2$$

$$g_7 : \Gamma_0 B\mathcal{O}(\mathbb{F}_q) \longrightarrow B\mathbb{Z}/2 \times B\mathbb{Z}/2$$

for the maps $g_1 = \Gamma_0 B\theta$, $g_2 = \Gamma_0 B(\theta \cdot \det)$, $g_3 = \Gamma_0 B(\det)$, $g_4 = \Gamma_0 B(\det)$, $g_5 = \Gamma_0 B(\det)$, $g_6 = \Gamma_0 B\theta$, $g_7 = \Gamma_0 B(\theta, \det)$. (cf. 2.19).

Theorem 3.16. The following are fibrations of infinite loop spaces

(1) $\Gamma_0 B\mathcal{N}^{ev}(\mathbb{F}_q) \xrightarrow{i_1} \Gamma_0 B\mathcal{O}(\mathbb{F}_q) \xrightarrow{g_1} B\mathbb{Z}/2$

(2) $\Gamma_0 B\mathcal{N}\mathcal{D}^{ev}(\mathbb{F}_q) \xrightarrow{i_2} \Gamma_0 B\mathcal{O}(\mathbb{F}_q) \xrightarrow{g_2} B\mathbb{Z}/2$

(3) $\Gamma_0 B\mathcal{S}\mathcal{O}^{ev}(\mathbb{F}_q) \xrightarrow{i_3} \Gamma_0 B\mathcal{O}(\mathbb{F}_q) \xrightarrow{g_3} B\mathbb{Z}/2$

(4) $\Gamma_0 B\mathcal{S}\mathcal{N}^{ev}(\mathbb{F}_q) \xrightarrow{i_4} \Gamma_0 B\mathcal{N}^{ev}(\mathbb{F}_q) \xrightarrow{g_4} B\mathbb{Z}/2$

(5) $\Gamma_0 B\mathcal{S}\mathcal{N}^{ev}(\mathbb{F}_q) \xrightarrow{i_5} \Gamma_0 B\mathcal{N}\mathcal{D}^{ev}(\mathbb{F}_q) \xrightarrow{g_5} B\mathbb{Z}/2$

(6) $\Gamma_0 B\mathcal{S}\mathcal{N}^{ev}(\mathbb{F}_q) \xrightarrow{i_6} \Gamma_0 B\mathcal{S}\mathcal{O}^{ev}(\mathbb{F}_q) \xrightarrow{g_6} B\mathbb{Z}/2$

(7) $\Gamma_0 B\mathcal{S}\mathcal{N}^{ev}(\mathbb{F}_q) \xrightarrow{i_7} \Gamma_0 B\mathcal{O}(\mathbb{F}_q) \xrightarrow{g_7} B\mathbb{Z}/2 \times B\mathbb{Z}/2$

where the i_k's denote the maps induced by inclusion of permutative categories.

Proof: We shall confine ourselves to proving (3) is a fibration. The proofs of the other statements are similar.

We know that

(3') $$BSO(\infty,\mathbb{F}_q) \xrightarrow{\tilde{i}_3} BO(\infty,\mathbb{F}_q) \xrightarrow{B\,\det} B\mathbb{Z}/2$$

is indeed a fibration. Let $\tilde{j}_3\colon F_3 \to \Gamma_0 B\mathcal{O}(\mathbb{F}_q)$ be the fiber of g_3. Then there is a map of fibrations

$$\begin{array}{ccccc}
BSO(\infty,\mathbb{F}_q) & \xrightarrow{\tilde{i}_3} & BO(\infty,\mathbb{F}_q) & \xrightarrow{B\,\det} & B\mathbb{Z}/2 \\
\downarrow \tilde{J} & & & & \\
\Gamma B\,\mathcal{SO}^{ev}(\mathbb{F}_q) & & \downarrow & & \| \\
\downarrow f & & & & \\
F_3 & \xrightarrow{\tilde{j}_3} & \Gamma_0 B\mathcal{O}(\mathbb{F}_q) & \xrightarrow{g_3} & B\mathbb{Z}/2
\end{array}$$

(cf. 2.17(v)). Since g_3 restricts to the trivial infinite loop map on $\Gamma_0 B\mathcal{SO}^{ev}(\mathbb{F}_q)$, f can be taken as an infinite loop map. The fibration at the bottom of the diagram is one of infinite loop spaces and hence has trivial local coefficients in homology.

Next we show that the fibration sequence (3') also has trivial local coefficients in integral homology.

Now the local coefficients in (3') are given as follows: Given $\alpha_1 \in \pi_1(B\mathbb{Z}_2) = \mathbb{Z}/2$ we pick $\tilde{\alpha} \in \det^{-1}(\alpha)$. Then the action of α on $H_*(BSO(\mathbb{F}_q);\, \mathbb{Z})$ is $[B(\tilde{\alpha}^{-1}(\ )\tilde{\alpha})]_*$. But

$$H_*(BSO(\infty,\mathbb{F}_q),\mathbb{Z}) = \varinjlim H_*(BSO(n,\mathbb{F}_q);\, \mathbb{Z})$$

and given ι the generator of $\pi_1(B\mathbb{Z}/2)$, we may pick

$$\begin{pmatrix} I_n & 0 \\ 0 & -1 \end{pmatrix} \in \det^{-1}(\iota)$$

so ι and hence $\pi_1(B\mathbb{Z}/2)$ act trivially on $H_*(BSO(\infty,\mathbb{F}_q);\, \mathbb{Z})$. Thus (3') has trivial local coefficients.

Now apply the comparison theorem to the Eilenberg Moore spectral sequence in homology with $\mathbb{Z}$-coefficients of the diagram (*). We obtain that

$f_* \cdot \bar{J}_*$ is an isomorphism. Since $\bar{J}_*$ is already an isomorphism, it follows that f_* is an isomorphism. Since f is an infinite loop map it follows by the Whitehead theorem that f is a homotopy equivalence. This completes the proof.

We conclude with a calculation of $\pi_1 \Gamma_0 B\mathcal{O}(\mathbb{F}_q)$. Let c_1 be the image of the generator $\iota \in \pi_1(BZ/2)$ under the composite map

$$B\mathbb{Z}/2 = BO(1,\mathbb{F}_q) \to BO(\infty,\mathbb{F}_q) \xrightarrow{\bar{I}} \Gamma_0 B\mathcal{O}(\mathbb{F}_q)$$

Let $c_2 = \Phi_*(c_1)$ and let $c_3 = c_2 - c_1$. We then have the following result

<u>Theorem</u> 3.17. $\pi_1(\Gamma_0 B\mathcal{O}(\mathbb{F}_q) = \mathbb{Z}/2 \oplus \mathbb{Z}/2$ with nonzero elements c_1, c_2, c_3.

<u>Proof</u>: We have

$$\pi_1(\Gamma_0 B\mathcal{O}(\mathbb{F}_q)) \cong H_1(\Gamma_0 B\mathcal{O}(\mathbb{F}_q)) \cong H_1(BO(\mathbb{F}_q)) \cong O(\mathbb{F}_q)/[O(\mathbb{F}_q),\ O(\mathbb{F}_q)]$$

But by O'Meara [30, p. 280]

$$[O(\mathbb{F}_q), O(\mathbb{F}_q)] = SN(\mathbb{F}_q)$$

Thus the map

$$O(\mathbb{F}_q) \xrightarrow{(\det,\theta)} \mathbb{Z}/2 \oplus \mathbb{Z}/2$$

has kernel $[O(\mathbb{F}_q), O(\mathbb{F}_q)]$. Hence

$$\pi_1(\Gamma_0 B\mathcal{O}(\mathbb{F}_q)) \cong \mathbb{Z}/2 \oplus \mathbb{Z}/2$$

When we apply the map $g_2 = \Gamma_0 B(\theta\cdot\det)$: $\Gamma_0 B\mathcal{O}(\mathbb{F}_q) \to B\mathbb{Z}/2$ we get

$$g_{2*}(c_1) = \iota, \quad g_{2*}(c_2) = 0, \quad g_{2*}(c_3) = \iota$$

When we apply the map $g_1 = \Gamma_0 B\theta$: $\Gamma_0 B\mathcal{O}(\mathbb{F}_q) \to B\mathbb{Z}/2$ we get

$$g_{1*}(c_1) = 0, \quad g_{1*}(c_2) = \iota, \; g_{1*}(c_3) = \iota.$$

Hence c_1, c_2, c_3 are the distinct nonzero elements of $\pi_1\Gamma_0 B\mathcal{O}(\mathbb{F}_q)$.

§4. Quadratic forms and extraordinary orthogonal groups

In this section we extend our discussion of §3 to a more general framework. Throughout this section as in §3 we will take $\mathbb{F}_q$ to be a field of $q = p^k$ elements, p an odd prime.

Definition 4.1. A quadratic form on a vector space V over $\mathbb{F}_q$ is a function $Q: V \to \mathbb{F}_q$ such that

(i) $Q(\alpha x) = \alpha^2 Q(x) \quad \alpha\in\mathbb{F}_q,\ x\in V$

(ii) $Q(x + y) - Q(x) - Q(y)$ is bilinear in x and y.

A quadratic space (V,Q) is a finite dimensional vector space over $\mathbb{F}_q$ together with a quadratic form $Q: V \to \mathbb{F}_q$.

The radical of a quadratic space (V,Q) is defined by

$$\mathrm{Rad}(V,Q) = \{x\in V \mid Q(v + x) = Q(v) \text{ for all } v\in V\}$$

It is readily seen that $\mathrm{Rad}(V,Q)$ is a subspace of V. A quadratic form is said to be nondegenerate if $\mathrm{Rad}(V,Q) = 0$.

A map $f: (V,Q) \to (W,Q')$ between quadratic spaces is said to be an isometric isomorphism if

(a) f: $V \to W$ is a linear isomorphsim

(b) $Q'(f(v)) = Q(v)$ for every $v \in V$.

We denote the automorphism group of a quadratic space (V, Q) by $O(V, Q)$.

Remark 4.2. An equivalent structure on a vector space V over $\mathbb{F}_q$ is a symmetric bilinear form b: $V \times V \to \mathbb{F}_q$. For given a quadratic form $Q: V \to \mathbb{F}_q$ we can associate to it the symmetric blinear form

$$b_Q(x,y) = \frac{1}{2}[Q(x+y) - Q(x) - Q(y)]$$

Conversely given a symmetric bilinear form b: $V \times V \to \mathbb{F}_q$ we can associate to it the quadratic form

$$Q_b(x) = b(x,x)$$

It is easily seen that these constructions give a 1-1 correspondence between quadratic forms and symmetric blinear forms on V.

The radical of a quadratic space (V, Q) can be equally well described in terms of the associated symmetric bilinear form:

$$\mathrm{Rad}(V,Q) = \{x \in V \mid b_Q(x,v) = 0 \text{ for all } v \in V\}$$

Similarly a map f: $(V, Q) \to (W, Q')$ is an isometric isomorphism iff

(a) f: $V \to W$ is a linear isomorphism

(b) $b_{Q'}(f(x), f(y)) = b_Q(x,y)$ for all $x, y \in V$.

We can now define a symmetric monoidal category corresponding to quadratic spaces and isometric isomorphisms.

Definition 4.3. The symmetric monoidal category $(\tilde{\mathcal{O}}(\mathbb{F}_q), \oplus, 0, a, \ell, r, c)$ is the category with objects all nondegenerate quadratic spaces (V,Q) over $\mathbb{F}_q$ and with morphisms all isometric isomorphisms between such spaces. The direct sum of two quadratic spaces is defined by

$$(V_1,Q_1) \oplus (V_2,Q_2) = (V_1 \oplus V_2, Q_1 \oplus Q_2)$$

where $(Q_1 \oplus Q_2)(x,y) = Q_1(x) + Q_2(y)$. On morphisms direct sum is defined in the usual way, i.e.

$$(f \oplus g)(x,y) = (f(x),g(y)).$$

The 0 module is equipped with the trivial form. The natural transformations a, ℓ, r, c are the obvious (isometric) isomorphisms

$$a\colon (V_1,Q_1) \oplus [(V_2,Q_2) \oplus (V_3,Q_3)] \cong [(V_1,Q_1) \oplus (V_2,Q_2)] \oplus (V_3,Q_3)$$

$$\ell\colon 0 \oplus (V,Q) \cong (V,Q) \qquad r\colon (V,Q) \oplus 0 \cong (V,Q)$$

$$c\colon (V_1,Q_1) \oplus (V_2,Q_2) \cong (V_2,Q_2) \oplus (V_1,Q_1)$$

Definition 4.4. Let μ be a nonsquare element of $\mathbb{F}_q^*$. Define the symmetric monoidal functor $\Phi\colon \tilde{\mathcal{O}}(\mathbb{F}_q) \to \tilde{\mathcal{O}}(\mathbb{F}_q)$ by

$$\Phi(V,Q) = (V,\mu Q)$$

on objects and by

$$\Phi(f\colon (V_1,Q_1) \to (V_2,Q_2)) = f\colon (V_1,\mu Q_1) \to (V_2,\mu Q_2)$$

on morphisms.

4.5. The structure of the category $\tilde{\mathcal{O}}(\mathbb{F}_q)$

It can be shown that in each positive dimension there are exactly two isomorphism classes of nondegenerate quadratic spaces, namely quadratic spaces isomorphic to

$$(\mathbb{F}_q^n, Q_+), \quad Q_+(x_1,\ldots,x_n) = \Sigma_{i=1}^n x_i^2$$

and quadratic spaces isomorphic to

$$(\mathbb{F}_q^n, Q_-), \quad Q_-(x_1,\ldots,x_n) = \mu x_1^2 + \Sigma_{i=2}^n x_i^2$$

where μ is a nonsquare element of $\mathbb{F}_q$. (Different choices of μ give isomorphic quadratic spaces as long as μ is nonsquare). The automorphisms of $(\mathbb{F}_q^n, Q_+)$ are commonly denoted by $O_+(n,\mathbb{F}_q)$ or just $O(n,\mathbb{F}_q)$. They are precisely the orthogonal groups studied in §3. The automorphism groups of $(\mathbb{F}_q^n, Q_-)$ are frequently called the <u>extraordinary orthogonal groups</u> and are denoted by $O_-(n,\mathbb{F}_q)$. The orders of these groups are given by

$$|O_+(2m-1,\mathbb{F}_q)| = |O_-(2m-1,\mathbb{F}_q)| = 2N_m$$

$$|O_+(2m,\mathbb{F}_q)| = 2(q^m-\epsilon^m)q^{m-1}N_m$$

$$|O_-(2m,\mathbb{F}_q)| = 2(q^m+\epsilon^m)q^{m-1}N_m$$

where $N_m = \Pi_{i=1}^{m-1}(q^{2i}-1)q^{2i-1}$ and $\epsilon = \pm 1$ as $q \equiv \pm 1 \pmod 4$. See Dickson [16] or Dieudonné [15] for proofs.

Under direct sum these quadratic forms behave as follows:

$$(\mathbb{F}_q^m, Q_+) \oplus (\mathbb{F}_q^n, Q_\pm) \cong (\mathbb{F}_q^{m+n}, Q_\pm)$$

$$(\mathbb{F}_q^m, Q_-) \oplus (\mathbb{F}_q^n, Q_\pm) \cong (\mathbb{F}_q^{m+n}, Q_\mp)$$

When we apply the functor Φ: $(V,Q) \to (V,\mu Q)$ of 4.4 we obtain

$$\Phi(\mathbb{F}_q^n, Q_{\pm}) \cong (\mathbb{F}_q^n, Q_{\pm}) \quad \text{if } n \text{ is even}$$

$$\Phi(\mathbb{F}_q^n, Q_{\pm}) \cong (\mathbb{F}_q^n, Q_{\mp}) \quad \text{if } n \text{ is odd}$$

Thus for n odd, Φ induces an isomorphism $O_+(n,\mathbb{F}_q) \cong O_-(n,\mathbb{F}_q)$. If n is even, Φ induces automorphisms of $O_{\pm}(n,\mathbb{F}_q)$. We shall later see that this automorphism on the groups $O_+(2m,\mathbb{F}_q)$ is essentially the same as that of Theorem 3.12.

There is an invariant called the discriminant which distinguishes between $(\mathbb{F}_q^n, Q_+)$ and $(\mathbb{F}_q^n, Q_-)$. It is defined as follows: If (V,Q) is a quadratic space, choose any basis $\{e_i\}_{i=1}^n$ for V. The determinant $\det(b_Q(e_i,e_j))$ is nonzero iff Q is nondegenerate. In that case the residue class

$$\Delta(Q) = \det(b_Q(e_i,e_j))(\mathbb{F}_q^*)^2$$

in $\mathbb{F}_q^*/(\mathbb{F}_q^*)^2 = \mathbb{Z}/2$ depends only on Q and not on the choice of basis $\{e_i\}_{i=1}^n$. This invariant is called the <u>discriminant</u> of Q. If $\Delta(Q)$ is a square the (V,Q) is isomorphic to $(\mathbb{F}_q^n, Q_+)$. If $\Delta(Q)$ is a nonsquare then (V,Q) is isomorphic to $(\mathbb{F}_q^n, Q_-)$.

<u>Example</u> 4.6. Let $\mathbb{F}_{q^r}$ be an extension field of $\mathbb{F}_q$. Then $\mathbb{F}_{q^r}$ can be considered as an r-dimensional vector space over $\mathbb{F}_q$. Moreover $\mathbb{F}_{q^r}$ comes equipped with a natural quadratic form

$$Q(x) = \mathrm{tr}(x^2)$$

where tr: $\mathbb{F}_{q^r} \to \mathbb{F}_q$ is the <u>trace</u> map $\mathrm{tr}(x) = x + x^q + \cdots + x^{q^{r-1}}$. The associated symmetric bilinear form is

$$b_Q(x,y) = tr(xy)$$

It is nondegenerate, since there is a $u \in \mathbb{F}_{q^r}$ for which $tr(u) \neq 0$ ($tr(u) = 0$ has only q^{r-1} roots). Consequently

$$b_Q(x,x^{-1}u) = tr(u) \neq 0 \text{ if } x \neq 0$$

Let us now compute the discriminant of Q. Let $\{e_i\}_{i=1}^r$ be a basis for $\mathbb{F}_{q^r}$ over $\mathbb{F}_q$. Then

$$\Delta(Q) = det(tr(e_ie_j)) = det(e_ie_j + e_i^qe_j^q + \cdots + e_i^{q^{r-1}}e_j^{q^{r-1}}) = D^2$$

where

$$D = det\begin{bmatrix} e_1 & e_1^q & \cdots & e_1^{q^{r-1}} \\ e_2 & e_2^q & \cdots & e_2^{q^{r-1}} \\ \cdots & \cdots & \cdots & \cdots \\ e_r & e_r^q & \cdots & e_r^{q^{r-1}} \end{bmatrix}$$

Hence $\Delta(Q)$ is a square in $\mathbb{F}_q^*$ iff $D \in \mathbb{F}_q^*$. But

$$D^q = det\begin{bmatrix} e_1^q & e_1^{q^2} & \cdots & e_1^{q^{r-1}} & e_1 \\ e_2^q & e_2^{q^2} & \cdots & e_2^{q^{r-1}} & e_2 \\ \cdots & \cdots & \cdots & \cdots & \cdots \\ e_r^q & e_r^{q^2} & \cdots & e_r^{q^{r-1}} & e_r \end{bmatrix} = (-1)^{r-1}D$$

Hence $D \in \mathbb{F}_q^*$ iff r is odd. Thus $\Delta(Q)$ is a square in $\mathbb{F}_q^*$ iff r is odd and

$$(\mathbb{F}_{q^r}, Q) \cong \begin{cases} (\mathbb{F}_q^r, Q_+) & \text{if } r \text{ is odd} \\ (\mathbb{F}_q^r, Q_-) & \text{if } r \text{ is even} \end{cases}$$

Alternatively we can endow $\mathbb{F}_{q^r}$ with the quadratic form

$$Q'(x) = \operatorname{tr}(\mu x^2)$$

where μ is a nonsquare in $\mathbb{F}_{q^r}$. There is no loss in generality in assuming that μ is a generator of $\mathbb{F}_{q^r}^*$. Then calculating the discriminant as above we see that

$$\Delta(Q') = N(\mu) D^2 \in \mathbb{F}_q^*/(\mathbb{F}_q^*)^2$$

where D is as above and $N(\mu)$ is the norm of μ over $\mathbb{F}_q$, i.e.

$$N(\mu) = \mu\mu^q\mu^{q^2} \cdots \mu^{q^{r-1}} = \mu^{(\frac{q^r-1}{q-1})}$$

Moreover $N(\mu)$ is a nonsquare in $\mathbb{F}_q^*$, since if $N(\mu) = v^2$, $v \in \mathbb{F}_q^*$ then

$$1 = v^{q-1} = (v^2)^{\frac{1}{2}(q-1)} = (N(\mu))^{\frac{1}{2}(q-1)} = \mu^{\frac{1}{2}(q^r-1)}$$

contradicting the fact that μ is a generator of $\mathbb{F}_{q^r}^*$.

Consequently

$$(\mathbb{F}_{q^r}, Q') \cong \begin{cases} (\mathbb{F}_q^r, Q_+) & \text{if } r \text{ is even} \\ (\mathbb{F}_q^r, Q_-) & \text{if } r \text{ is odd} \end{cases}$$

4.7. Notation. We shall find it convenient to establish separate notation for the set $|\mathcal{O}b(\tilde{\mathcal{O}}(\mathbb{F}_q))|$ of isomorphism classes of objects in $\tilde{\mathcal{O}}(\mathbb{F}_q)$. We will denote

$$\tilde{\mathbb{N}} = |\mathcal{O}b(\tilde{\mathcal{O}}(\mathbb{F}_q))|$$

The isomorphism class of $(\mathbb{F}_q^n, Q_+)$ will be denoted by n; the isomorphism class of $(\mathbb{F}_q^n, Q_-)$ by $\bar{n}$.

The set $\tilde{\mathbb{N}}$ is a monoid under direct sum. The behavior of quadratic forms under $\oplus$ analyzed in 4.5 gives the following addition table for $\tilde{\mathbb{N}}$

$$m + n = m + n,\; m + \bar{n} = \overline{m + n},\; \bar{m} + n = \overline{m + n},\; \bar{m} + \bar{n} = m + n$$

where the operation $+$ on the right hand side of the equations is ordinary addition of nonnegative integers. Also multiplication of elements of $\tilde{\mathbb{N}}$ by nonnegative integers makes sense in terms of iterated addition

$$na = \underbrace{a + a + \cdots + a}_{n \text{ times}} \qquad n \in \mathbb{N},\; a \in \tilde{\mathbb{N}}.$$

The ordering of $\tilde{\mathbb{N}}$ defined in 2.9 can be described as follows:

$$\left.\begin{matrix} m \le n \\ \bar{m} \le \bar{n} \end{matrix}\right\} \text{ iff } m \le n \qquad\qquad \left.\begin{matrix} \bar{m} \le n \\ m \le \bar{n} \end{matrix}\right\} \text{ iff } m < n$$

Clearly the set $\mathbb{N}$ of nonnegative integers is a subset of $\tilde{\mathbb{N}}$ and the induced addition and ordering is the usual one. Moreover $\mathbb{N}$ is a cofinal subset of the ordered set $\tilde{\mathbb{N}}$. We shall denote by $\tilde{\mathbb{N}}^{ev}$ the submonoid $\{2m, \overline{2m} \mid m \in \mathbb{N}\}$.

We shall find it convenient to denote the automorphism group of the quadratic space (V, Q) by $O(a,\mathbb{F}_q)$ where $a \in \tilde{\mathbb{N}}$ denotes the isomorphism class of (V, Q). Thus in the notation of 4.5

$$O(n,\mathbb{F}_q) = O_+(n,\mathbb{F}_q), \quad O(\overline{n},\mathbb{F}_q) = O_-(n,\mathbb{F}_q).$$

The advantage of this notation is that the direct sum $O(a,\mathbb{F}_q) \oplus O(b,\mathbb{F}_q)$ lies in $O(a+b,\mathbb{F}_q)$ for all $a,b \in \tilde{\mathbb{N}}$. Similarly the n-fold direct sum $O(a,\mathbb{F}_q)^n$ lies in $O(na,\mathbb{F}_q)$.

We next turn to the problem of constructing a permutative category equivalent to $\tilde{\mathcal{O}}(\mathbb{F}_q)$. Unfortunately, unlike in preceding examples, the simple expedient of passing to the skeleton of the category does not work here. Perhaps the simplest construction is the following:

4.8. The permutative category $\bar{\mathcal{O}}(\mathbb{F}_q)$. The objects of $\bar{\mathcal{O}}(\mathbb{F}_q)$ are quadratic spaces $\{(\mathbb{F}_q^n, Q)\}_{n \geq 0}$, where Q denotes a nondegenerate diagonal quadratic form on $\mathbb{F}_q^n$, i.e.

$$Q(x_1,\ldots,x_n) = \Sigma_{i=1}^n \alpha_i x_i^2, \quad \alpha_i \in \mathbb{F}_q^*.$$

The morphisms of $\bar{\mathcal{O}}(\mathbb{F}_q)$ are all isometric isomorphisms between these spaces. The bifunctor $\oplus$: $\bar{\mathcal{O}}(\mathbb{F}_q) \times \bar{\mathcal{O}}(\mathbb{F}_q) \to \bar{\mathcal{O}}(\mathbb{F}_q)$ is defined exactly as in 4.3 except that $\mathbb{F}_q^m \oplus \mathbb{F}_q^n$ is identified with $\mathbb{F}_q^{m+n}$ in the usual way. The commutativity transformation c is the map

$$c((\mathbb{F}_q^m, Q_1),(\mathbb{F}_q^n, Q_2)): (\mathbb{F}_q^{m+n}, Q_1 \oplus Q_2) \to (\mathbb{F}_q^{m+n}, Q_2 \oplus Q_1)$$

whose matrix with respect to the standard basis of $\mathbb{F}_q^{m+n}$ is

$$\begin{pmatrix} 0 & I_n \\ I_m & 0 \end{pmatrix}$$

It is clear that $\bar{\bar{\mathcal{O}}}(\mathbb{F}_q)$ is a permutative category equivalent to $\tilde{\bar{\mathcal{O}}}(\mathbb{F}_q)$. It is also clear that the functor $\Phi: \tilde{\bar{\mathcal{O}}}(\mathbb{F}_q) \to \tilde{\bar{\mathcal{O}}}(\mathbb{F}_q)$ of 4.4 can be considered a permutative functor $\Phi: \bar{\bar{\mathcal{O}}}(\mathbb{F}_q) \to \bar{\bar{\mathcal{O}}}(\mathbb{F}_q)$.

4.9. Important permutative subcategories of $\bar{\bar{\mathcal{O}}}(\mathbb{F}_q)$

(a) $\mathcal{O}(\mathbb{F}_q)$ - The category $\mathcal{O}(\mathbb{F}_q)$ treated in §3 may be considered as the full subcategory of $\bar{\bar{\mathcal{O}}}(\mathbb{F}_q)$ whose objects are $\{(\mathbb{F}_q^n, Q_+)\}_{n\geq 0}$. Thus all the subcategories of $\mathcal{O}(\mathbb{F}_q)$ analyzed in §3 may also be regarded as subcategories of $\bar{\bar{\mathcal{O}}}(\mathbb{F}_q)$

(b) $\bar{\bar{\mathcal{O}}}_+(\mathbb{F}_q)$ - The full subcategory of $\bar{\bar{\mathcal{O}}}(\mathbb{F}_q)$ whose objects are quadratic spaces isomorphic to some $(\mathbb{F}_q^n, Q_+)$

(c) $\bar{\bar{\mathcal{O}}}^{ev}(\mathbb{F}_q)$ - The full subcategory of $\bar{\bar{\mathcal{O}}}(\mathbb{F}_q)$ whose objects are even dimensional quadratic spaces

(d) $\bar{\bar{\mathcal{O}}}_+^{ev}(\mathbb{F}_q) = \bar{\bar{\mathcal{O}}}_+(\mathbb{F}_q) \cap \bar{\bar{\mathcal{O}}}^{ev}(\mathbb{F}_q)$

Remark 4.10. It is clear from the discussion of 2.16 and 4.5 that

$$B\bar{\bar{\mathcal{O}}}(\mathbb{F}_q) = B\tilde{\bar{\mathcal{O}}}(\mathbb{F}_q) \simeq \coprod_{a\in\tilde{\mathbb{N}}} BO(a,\mathbb{F}_q)$$

We next turn to the question of identifying the infinite loop space $\Gamma B\bar{\bar{\mathcal{O}}}(\mathbb{F}_q)$.

Proposition 4.11. The infinite loop space $\Gamma B\bar{\bar{\mathcal{O}}}(\mathbb{F}_q)$ is homotopy equivalent as a space to

$$\Gamma_0 B\mathcal{O}(\mathbb{F}_q) \times \mathbb{Z}/2 \times \mathbb{Z} \simeq \Gamma B\mathcal{O}(\mathbb{F}_q) \times \mathbb{Z}/2$$

The inclusion $\mathcal{O}(\mathbb{F}_q) \to \bar{\bar{\mathcal{O}}}(\mathbb{F}_q)$ induces an infinite loop map $\Gamma_0 B\mathcal{O}(\mathbb{F}_q) \to \Gamma_0 B\bar{\bar{\mathcal{O}}}(\mathbb{F}_q)$ which is an equivalence.

Proof. From the discussion of 2.18 it follows that as a space

$$\Gamma B\bar{\mathcal{O}}(\mathbb{F}_q) \simeq \Gamma_0 B\bar{\mathcal{O}}(\mathbb{F}_q) \times K_0(\bar{\mathcal{O}}(\mathbb{F}_q))$$

where $K_0(\bar{\mathcal{O}}(\mathbb{F}_q))$ is the Grothendieck group obtained from the monoid $\tilde{\mathbb{N}} = |\mathcal{O}\ell(\bar{\mathcal{O}}(\mathbb{F}_q))|$. But clearly

$$K_0(\bar{\mathcal{O}}(\mathbb{F}_q)) = \mathbb{Z} \times \mathbb{Z}/2$$

with the $\mathbb{Z}$-summand generated by 1 and the $\mathbb{Z}/2$ summand generated by $\bar{1} - 1$.

By 2.17(vi) and the fact that $\mathbb{N} = |\mathcal{O}\ell(\mathcal{O}(\mathbb{F}_q))|$ is a cofinal subset of $\tilde{\mathbb{N}} = |\mathcal{O}\ell(\bar{\mathcal{O}}(\mathbb{F}_q)))$ it follows that $\Gamma_0 B\mathcal{O}(\mathbb{F}_q) \to \Gamma_0 B\bar{\mathcal{O}}(\mathbb{F}_q)$ is an equivalence of infinite loop spaces. The result is now clear.

4.12. Notation. We shall find it convenient to label the torsion element of $K_0(\bar{\mathcal{O}}(\mathbb{F}_q)) = \mathbb{Z}/2 \times \mathbb{Z}$ as γ. Thus

$$\gamma = \bar{1} - 1 = 1 - \bar{1}$$

We shall denote

$$\Gamma_0^{\pm} B\bar{\mathcal{O}}(\mathbb{F}_q) = \Gamma_0 B\bar{\mathcal{O}}(\mathbb{F}_q) \cup \Gamma_\gamma B\bar{\mathcal{O}}(\mathbb{F}_q)$$

It is clear that $\Gamma_0^{\pm} B\bar{\mathcal{O}}(\mathbb{F}_q)$ is an infinite loop space.

Lastly we consider the permutative functor $\Phi\colon \bar{\mathcal{O}}(\mathbb{F}_q) \to \bar{\mathcal{O}}(\mathbb{F}_q)$ (cf. 4.4) and the associated infinite loop map $\Phi\colon \Gamma B\bar{\mathcal{O}}(\mathbb{F}_q) \to \Gamma B\bar{\mathcal{O}}(\mathbb{F}_q)$. We summarize its properties:

4.13. Important properties of Φ

(a) The definition of Φ is independent of the choice of nonsquare element μ

(b) $\Phi^2 = 1$

(c) $2\Phi = 2$

(d) When restricted to the subcategory $\mathcal{O}_+^{ev}(\mathbb{F}_q)$, Φ is equivalent to the functor Φ: $\mathcal{O}^{ev}(\mathbb{F}_q) \to \mathcal{O}^{ev}(\mathbb{F}_q)$ constructed in 3.12.

It is easy to check the first three assertions:

(a) If μ_1, μ_2 are two nonsquares in $\mathbb{F}_q$ then $\mu_1 = \alpha^2\mu_2$ for some $\alpha \in \mathbb{F}_q$. Then the map $x \to \alpha x$

$$(\mathbb{F}_q^n, \mu_1 Q) \to (\mathbb{F}_q^n, \mu_2 Q)$$

defines a natural isomorphism between the functors Φ_1, Φ_2: $\bar{\mathcal{O}}(\mathbb{F}_q) \to \bar{\mathcal{O}}(\mathbb{F}_q)$ corresponding to μ_1, μ_2. Therefore the associated infinite loop maps Φ_1, Φ_2: $\Gamma B\bar{\mathcal{O}}(\mathbb{F}_q) \to \Gamma B\bar{\mathcal{O}}(\mathbb{F}_q)$ are homotopic.

(b) The map $x \to \mu x$

$$(\mathbb{F}_q^n, \mu^2 Q) \to (\mathbb{F}_q^n, Q)$$

defines a natural equivalence between the functors Φ^2 and the identity of $\bar{\mathcal{O}}(\mathbb{F}_q)$. Hence $\Phi^2 \simeq 1$: $\Gamma B\bar{\mathcal{O}}(\mathbb{F}_q) \to \Gamma B\bar{\mathcal{O}}(\mathbb{F}_q)$

(c) Pick elements $a, b \in \mathbb{F}_q$ such that $a^2 + b^2 = \mu^{-1}$. Then the map $(x,y) \to (-ax + by, bx + ay)$

$$(\mathbb{F}_q^n \oplus \mathbb{F}_q^n, \mu Q \oplus \mu Q) \to (\mathbb{F}_q^n \oplus \mathbb{F}_q^n, Q \oplus Q)$$

defines a natural isomorphism between the functors 2Φ and 2 on $\bar{\mathcal{O}}(\mathbb{F}_q)$. Hence $2\Phi \simeq 2$: $\Gamma B\bar{\mathcal{O}}(\mathbb{F}_q) \to \Gamma B\bar{\mathcal{O}}(\mathbb{F}_q)$.

The proof of 4.13(d) is more delicate. We first restate the result more precisely.

Proposition 4.14. There is a commutative diagram of permutative categories and functors

$$\begin{array}{ccc} \mathcal{O}^{ev}(\mathbb{F}_q) & \xrightarrow{\tilde{K}} & \bar{\mathcal{O}}_+^{ev}(\mathbb{F}_q) \\ \downarrow \Phi & & \downarrow \Phi \\ \mathcal{O}^{ev}(\mathbb{F}_q) & \xleftarrow{\tilde{L}} & \bar{\mathcal{O}}_+^{ev}(\mathbb{F}_q) \end{array}$$

where Φ: $\mathcal{O}^{ev}(\mathbb{F}_q) \to \mathcal{O}^{ev}(\mathbb{F}_q)$ is the functor defined in 3.12. The functors $\tilde{K}$ and $\tilde{L}$ are inverse equivalences: $\tilde{K}$ is the inclusion, $\tilde{L}$ is a functor such that $\tilde{L}\tilde{K} = 1_{\mathcal{O}(\mathbb{F}_q)}$ and there is a natural isomorphism Λ: $\tilde{K}\tilde{L} \xrightarrow{\cong} 1_{\bar{\mathcal{O}}_+^{ev}(\mathbb{F}_q)}$.

We will actually construct an equivalence L: $\bar{\mathcal{O}}_+(\mathbb{F}_q) \to \mathcal{O}(\mathbb{F}_q)$ inverse to the inclusion K: $\mathcal{O}(\mathbb{F}_q) \to \bar{\mathcal{O}}_+(\mathbb{F}_q)$ such that $LK = 1_{\mathcal{O}(\mathbb{F}_q)}$ and $KL \xrightarrow{\cong} 1_{\bar{\mathcal{O}}_+(\mathbb{F}_q)}$ via a natural isomorphism Λ. Then the restriction of L to the subcategory $\bar{\mathcal{O}}_+^{ev}(\mathbb{F}_q)$ will be the desired functor $\tilde{L}$.

The construction of the functor L is analogous to the Gram Schmidt process. However to insure that L will be a permutative functor, we have to be extremely careful to do this process consistently for each quadratic space.

Proof of 4.14. First of all for each element $x \in (\mathbb{F}_q^*)^2$ pick one element $y \in \mathbb{F}_q^*$ such that $y^2 = x$. Define $\sqrt{x} = y$.

Now consider an object $V = (\mathbb{F}_q^n, Q)$ of $\bar{\mathcal{O}}_+(\mathbb{F}_q)$. Let $\{e_i\}_{i=1}^n$ be the standard basis for the vector space $\mathbb{F}_q^n$. Define inductively a sequence of pairs of ordered subsets of V, $(S_i^V, T_i^V)_{i=1}^{m_V}$ such that $\emptyset = S_1^V \subseteq S_2^V \subseteq \cdots \subseteq S_{m_V}^V$, $\{e_i\}_{i=1}^n = T_1^V \supseteq T_2^V \supseteq \cdots \supseteq T_{m_V}^V = \emptyset$, for each i, $S_i^V \coprod T_i^V$ is an orthogonal basis for V, and each element of S_i^V has unit quadratic norm. We regard $T_1^V = \{e_i\}_{i=1}^n$ as being given the ordering $e_1 < e_2 < \cdots < e_n$ and regard the rest of the T_i^V's as ordered subsets of T_1^V.

We define $S_1^V = \emptyset$, $T_1^V = \{e_i\}_{i=1}^n$ to start the induction. Having defined S_{i-1}^V, T_{i-1}^V we let e_j be the first element of T_{i-1}^V.

If $Q(e_j)$ is a square in $\mathbb{F}_q$ we define $f = \frac{1}{\sqrt{Q(e_j)}} e_j$ and let $S_i^V = S_{i-1}^V \cup \{f\}$ with ordering given by taking f to be the maximal element of S_i^V. We define $T_i^V = T_{i-1}^V - \{e_j\}$.

If $Q(e_j)$ is a nonsquare in $\mathbb{F}_q$, we let e_k be the next element of T_{i-1}^V such that $Q(e_k)$ is nonsquare. (Such an element e_k must exist since V is isomorphic to $(\mathbb{F}_q^n, Q_+)$). Then $\mu/Q(e_j)$, $\mu/Q(e_k)$ are squares in $\mathbb{F}_q^*$. Let $x = (\sqrt{\mu/Q(e_j)})e_j$, $y = (\sqrt{\mu/Q(e_k)})e_k$. Let $a, b \in \mathbb{F}_q^*$ be the elements chosen in §2 such that $a^2 + b^2 = \mu$. Define $f = \frac{1}{\mu}(-ax + by)$, $f' = \frac{1}{\mu}(bx + ay)$. Let $S_i^V = S_{i-1}^V \cup \{f, f'\}$ with ordering determined by the requirements that $f < f'$ and that f be larger than any element in S_{i-1}^V. Define $T_i^V = T_{i-1}^V - \{e_j, e_k\}$.

It is clear from construction that $S_{m_V}^V = \{f_i\}_{i=1}^n$ with $f_1 < f_2 < \cdots < f_n$ is an ordered orthonormal basis for V. Define an isomorphism of quadratic spaces $\Lambda(V)\colon V \to (\mathbb{F}_q^n, Q_+)$ by $\Lambda(V)(f_i) = e_i$. Define the functor $L\colon \bar{\mathcal{O}}_+(\mathbb{F}_q) \to \bar{\mathcal{O}}(\mathbb{F}_q)$ on objects as follows:

$$L(\mathbb{F}_q^n, Q) = (\mathbb{F}_q^n, Q_+)$$

On morphisms define L as follows: if $A\colon V \to W$ is a morphism in $\bar{\mathcal{O}}_+(\mathbb{F}_q)$ let

$$L(A) = \Lambda(W) \cdot A \cdot \Lambda(V)^{-1}$$

It is clear that $\Lambda(\mathbb{F}_q^n, Q_+) = 1_{(\mathbb{F}_q^n, Q_+)}$. Hence it follows that $L \cdot K = 1_{(\mathbb{F}_q)}$. It is also clear that Λ is a natural transformation between $1_{\bar{\mathcal{O}}(\mathbb{F}_q)}$ and $K \cdot L$.

If we look at the above construction of the S_i^V's and T_i^V's we see that $m_{V \oplus W} = m_V + m_W$ and that

$$S_i^{V \oplus W} = \begin{cases} S_i^V & \text{if } i \leq m_V \\ S_{m_V}^V \cup T_{i-m_V}^W & \text{if } i > m_V \end{cases}$$

$$T_i^{V \oplus W} = \begin{cases} T_i^V \cup T_1^W & \text{if } i \leq m_V \\ T_{i-m_V}^W & \text{if } i > m_V \end{cases}$$

where we regard V and W to be imbedded in $V \oplus W$ in the usual way and $S_{V \oplus W}^{V \oplus W}(T_1^{V \oplus W})$ are ordered so that any element of $S_{m_V}^V(T_1^V)$ is smaller than any element of $S_{m_W}^W(T_1^W)$. Consequently $\Lambda(V \oplus W) = \Lambda(V) \oplus \Lambda(W)$.

It follows that

$$L\oplus = \oplus(L \times L)$$

Also since we have the commutative diagram

$$\begin{array}{ccc} V \oplus W & \xrightarrow{C(V,W)} & W \oplus V \\ \Big\downarrow {\scriptstyle \Lambda(V) \oplus \Lambda(W)} & & \Big\downarrow {\scriptstyle \Lambda(W) \oplus \Lambda(V)} \\ LV \oplus LW & \xrightarrow{c(LV,LW)} & LW \oplus LV \end{array}$$

it follows that

$$L(c(V,W)) = \Lambda(W \oplus V) \cdot c(V,W) \cdot \Lambda(V \oplus W)^{-1}$$

$$= [\Lambda(W) \oplus \Lambda(V)] \cdot c(V,W) \cdot [\Lambda(V) \oplus \Lambda(W)]^{-1}$$

$$= c(LV,LW)$$

Hence L is a permutative functor and $\Lambda\colon 1_{\bar{\mathcal{O}}_+(\mathbb{F}_q)} \to KL$ is an isomorphism of permutative functors. Thus K and L are equivalences of permutative categories.

It is obvious that L and Λ restrict to functors $\tilde{L}$ and $\tilde{\Lambda}$ on $\bar{\mathcal{O}}_+^{ev}(\mathbb{F}_q)$ and that $\tilde{K}\colon \mathcal{O}^{ev}(\mathbb{F}_q) \to \bar{\mathcal{O}}_+^{ev}(\mathbb{F}_q)$, $\tilde{L}\colon \bar{\mathcal{O}}_+^{ev}(\mathbb{F}_q) \to \mathcal{O}^{ev}(\mathbb{F}_q)$ are equivalences of permutative categories.

Finally to verify that the diagram of Prop. 4.14 commutes, it suffices to check that $\tilde{L} \cdot \Phi \cdot \tilde{K}\colon O(2,\mathbb{F}_q) \to O(2,\mathbb{F}_q)$ is the same as the automorphism $\Phi\colon O(2,\mathbb{F}_q) \to O(2,\mathbb{F}_q)$ defined in §2. To see this we observe that $\Lambda(\Phi \cdot \tilde{K}(\mathbb{F}_q^2, Q_+))\colon (\mathbb{F}_q^2,\mu Q_+) \to (\mathbb{F}_q^2, Q_+)$ has matrix

$$\begin{pmatrix} -a & b \\ b & a \end{pmatrix}$$

with respect to the standard basis $\{e_1,e_2\}$ of $\mathbb{F}_q^2$. Hence

$$\tilde{L} \cdot \Phi \cdot \tilde{K}(A) = \Lambda(\mathbb{F}_q^2,\mu Q_+) \cdot A \cdot \Lambda(\mathbb{F}_q^2,\mu Q_+)^{-1}$$

$$= \begin{pmatrix} -a & b \\ b & a \end{pmatrix} A \begin{pmatrix} -a & b \\ b & a \end{pmatrix}^{-1}$$

which is precisely the definition of Φ given in 3.12.

§5. Spinor Groups Over Finite Fields

Let V be a vector space of finite dimension over a field $\mathbb{F}$ of characteristic $\neq 2$. Let $Q\colon V \to \mathbb{F}$ be a nondegenerate quadratic form on

V.

Define the Clifford algebra $C = C(V,Q)$ in the usual way as the universal algebra over $\mathbb{F}$ generated by V subject to the relations $v^2 = Q(v)$. It is $\mathbb{Z}/2$-graded: $C = C_0 \oplus C_1$. It admits an automorphism $\alpha: C \to C$ such that $\alpha(v) = -v$ for $v \in V$, and an antiautomorphism $x \to \overline{x}$ such that $\overline{v} = v$ for $v \in V$. We define $\Gamma = \Gamma(V,Q) \subseteq C$ to be the set of elements x such that

(i) x is invertible

(ii) $v \in V$ implies $(\alpha x) v x^{-1} \in V$

Then Γ is a group under multiplication, and we can define a homomorphism $\pi: \Gamma \to GL(V)$ by

$$(\pi x) v = (\alpha x) v x^{-1}$$

The basic properties of Clifford algebras are summarized in Proposition 5.1 below. Proofs may be found in Dieudonné [15] or in Atiyah, Bott and Shapiro [5]. The latter prove it only for the case $\mathbb{F} = \mathbb{R}$ but the proof for the general case is virtually identical.

Proposition 5.1. (i) If $v, w \in V$ and $v \perp w$ then $vw = -wv$

(ii) If $v \in V$ and $Q(v) \neq 0$ then $v \in \Gamma$. Moreover $\pi(v) = T_v$, the symmetry with respect to v.

(iii) The image of $\pi: \Gamma \to GL(V)$ is $O(V,Q)$ the orthogonal group of (V,Q). The kernel of π is $\mathbb{F}^*$.

(iv) Γ is generated by $\mathbb{F}^*$ and $\{v \mid Q(v) \neq 0\}$

(v) $\Gamma = \Gamma_0 \cup \Gamma_1$ where $\Gamma_0 = \Gamma \cap C_0$ and $\Gamma_1 = \Gamma \cap C_1$. The image of Γ_0 under $\pi: \Gamma \to GL(V)$ is $SO(V,Q)$, the special orthogonal group of (V,Q).

(vi) If $x \in \Gamma$, then $x\overline{x} = \overline{x}x \in \mathbb{F}^*$. Hence there is a homomorphism $H: \Gamma \to \mathbb{F}^*$ such that $H(x) = x\overline{x} = \overline{x}x$.

Lemma 5.2. Let θ: $O(V,Q) \to \mathbb{F}^*/(\mathbb{F}^*)^2$ denote the spinor norm (cf. 3.7). Then the following diagram commutes

$$\begin{array}{ccc} \Gamma & \xrightarrow{H} & \mathbb{F}^* \\ \pi \downarrow & & \downarrow p \\ O(V,Q) & \xrightarrow{\theta} & \mathbb{F}^*/(\mathbb{F}^*)^2 \end{array}$$

Proof: It suffices to check on generators. If $\gamma \in \mathbb{F}^*$ then

$$pH(\gamma) = p(\gamma\bar{\gamma}) = p(\gamma^2) = 1 = \theta(1_V) = \theta\,\pi(\gamma)$$

If $v \in V$ and $Q(v) \neq 0$ then

$$pH(v) = p(v\bar{v}) = p(v^2) = p(Q(v)) = \theta(T_v) = \theta\,\pi(v)$$

Definition 5.3. We define $Pin(V,Q)$ to be the kernel of $\Gamma \xrightarrow{H} \mathbb{F}^*$ and $Spin(V,Q)$ to be $Pin(V,Q) \cap \Gamma_0$. We also define

$$N(V,Q) = \{A \in O(V,Q) \mid \theta(A) = 1\}$$

$$SN(V,Q) = \{A \in O(V,Q) \mid \theta(A) = 1 \text{ and } \det A = 1\}$$

(cf. 3.2).

Lemma 5.4. There are exact sequences

$$1 \to \mathbb{Z}/2 \to Pin(V,Q) \xrightarrow{\pi} N(V,Q) \to 1$$

$$1 \to \mathbb{Z}/2 \to Spin(V,Q) \xrightarrow{\pi} SN(V,Q) \to 1$$

Proof: In view of Proposition 5.1 and Lemma 5.2, we need only show that π is epimorphic. Suppose $A \in N(V,Q)$. There is a $y \in \Gamma$ such that $\pi(y) = A$. We have $1 = \theta(A) = \theta\pi(y) = pH(y)$. Hence $H(y) = \gamma^2$. Let $z = \gamma^{-1}y$. Then $\pi(z) = \pi(\gamma^{-1}y) = \pi(y) = A$. Moreover

$$H(z) = H(\gamma^{-1}y) = H(\gamma^{-1})H(y) = \gamma^{-2}\gamma^2 = 1$$

Hence $z \in Pin(V,Q)$.

The other statement is similarly proved.

From now on let $F = F_q$ be the field with q elements. Let $V = F_q^n$ and let Q be the standard form on V. Define $\Gamma(n,F_q) = \Gamma(V,Q)$, $Pin(n,F_q) = Pin(V,Q)$, $Spin(n,F_q) = Spin(V,Q)$, etc.

Proposition 5.5. $[SN(n,F_q), SN(n,F_q)] = SN(n,F_q)$, $n \geq 5$.

Proof: It is shown in O'Meara [30] that

$$SN(n,F_q) = [O(n,F_q), O(n,F_q)], \; n \geq 2$$

Also $[O(n,F_q), O(n,F_q)]$ is generated by commutators of the form

$$[T_u,T_v] = T_uT_vT_u^{-1}T_v^{-1} \qquad Q(u) \neq 0, \; Q(v) \neq 0$$

Now since $n \geq 5$, we can find vectors x,y such that $x,y \in \{u,v\}^{\perp}$, $x \perp y$ and

$$Q(x) = Q(u) \qquad \text{mod}(F_q)^2$$

$$Q(y) = Q(v) \qquad \text{mod}(F_q)^2$$

Then T_x,T_y commute with both T_w and T_v and $T_xT_y = T_yT_x$. Hence

$$[T_u, T_v] = [T_uT_x, T_vT_y] \in [SN(n,\mathbb{F}_q),\ SN(n,\mathbb{F}_q)]$$

Hence

$$SN(n,\mathbb{F}_q) = [O(n,\mathbb{F}_q),\ O(n,\mathbb{F}_q)] \subseteq [SN(n,\mathbb{F}_q),\ SN(n,\mathbb{F}_q)]$$

This completes the proof of the theorem.

<u>Proposition</u> 5.6. $[\mathrm{Spin}(n,\mathbb{F}_q),\ \mathrm{Spin}(n,\mathbb{F}_q)] = \mathrm{Spin}(n,\mathbb{F}_q)$ if $n \geq 5$.

<u>Proof:</u> Let $\{e_i\}$ be the standard basis for $\mathbb{F}_q^n$. Then $e_1e_2, e_3e_2 \in \mathrm{Spin}(n,\mathbb{F}_q)$. Then

$$[e_1e_2, e_3e_2] = (e_1e_2)(e_3e_2)(e_2e_1)(e_2e_3) = e_1e_2e_3e_1e_2e_3 = e_1e_2e_1e_2e_3^2$$
$$= e_1e_2e_1e_2 = -e_1^2e_2^2 = -1$$

Hence $-1 \in [\mathrm{Spin}(n,\mathbb{F}_q),\ \mathrm{Spin}(n,\mathbb{F}_q)]$.

By Lemma 5.4, there is a short exact sequence

$$1 \to \mathbb{Z}/2 \to \mathrm{Spin}(n,\mathbb{F}_q) \xrightarrow{\pi} SN(n,\mathbb{F}_q) \to 1$$

Since $-1 \in [\mathrm{Spin}(n,\mathbb{F}_q),\ \mathrm{Spin}(n,\mathbb{F}_q)]$, the projection p

$$\begin{array}{ccc}
\mathrm{Spin}(n,\mathbb{F}_q) & \xrightarrow{\quad\pi\quad} & SN(n,\mathbb{F}_q) \\
\downarrow p & \swarrow \ell & \downarrow \bar{p} \\
\mathrm{Spin}(n,\mathbb{F}_q)/[\mathrm{Spin}(n,\mathbb{F}_q),\mathrm{Spin}(n,\mathbb{F}_q)] & \xleftarrow{k} & SN(n,\mathbb{F}_q)/[SN(n,\mathbb{F}_q),SN(n,\mathbb{F}_q)]
\end{array}$$

factors as $p = \ell\pi = k\bar{p}\cdot\pi = 0$. Hence

$$\mathrm{Spin}(n,\mathbb{F}_q)/[\mathrm{Spin}(n,\mathbb{F}_q),\mathrm{Spin}(n,\mathbb{F}_q)] = 0$$

We now define a category $\mathcal{S}\mathrm{pin}^{\mathrm{oct}}(\mathbb{F}_q)$ as follows: The objects of $\mathcal{S}\mathrm{pin}^{\mathrm{oct}}(\mathbb{F}_q)$ are $8\mathbb{N}$, nonnegative integers that are divisible by 8. The morphisms of $\mathcal{S}\mathrm{pin}^{\mathrm{oct}}(\mathbb{F}_q)$ are the following

$$\hom(8m,8n) = \begin{cases} \emptyset & \text{if } m \neq n \\ \mathrm{Spin}(8n,\mathbb{F}_q) & \text{if } m = n \end{cases}$$

Let $\mathcal{SN}^{\mathrm{oct}}(\mathbb{F}_q)$ denote the full subcategory of $\mathcal{SN}(\mathbb{F}_q)$ whose objects are $8\mathbb{N}$. Define a functor π: $\mathcal{S}\mathrm{pin}^{\mathrm{oct}}(\mathbb{F}_q) \to \mathcal{SN}^{\mathrm{oct}}(\mathbb{F}_q)$ to be the identity on objects and the homomorphism

$$\pi\colon \mathrm{Spin}(8n,\mathbb{F}_q) \to SN(8n,\mathbb{F}_q)$$

on morphisms.

Our efforts will now be directed towards proving the following theorem

Theorem 5.7. (a) $\mathcal{S}\mathrm{pin}^{\mathrm{oct}}(\mathbb{F}_q)$ is a permutative category
(b) π: $\mathcal{S}\mathrm{pin}^{\mathrm{oct}}(\mathbb{F}_q) \to \mathcal{SN}^{\mathrm{oct}}(\mathbb{F}_q)$ is a permutative functor.

We define a bifunctor $\oplus$: $\mathcal{S}\mathrm{pin}^{\mathrm{oct}}(\mathbb{F}_q) \times \mathcal{S}\mathrm{pin}^{\mathrm{oct}}(\mathbb{F}_q) \to \mathcal{S}\mathrm{pin}^{\mathrm{oct}}(\mathbb{F}_q)$ as follows: Let $i_1\colon \mathbb{F}_q^{8m} \to \mathbb{F}_q^{8m} \oplus \mathbb{F}_q^{8n} = \mathbb{F}_q^{8(m+n)}$, $i_2\colon \mathbb{F}_q^{8n} \to \mathbb{F}_q^{8m} \oplus \mathbb{F}_q^{8n} = \mathbb{F}_q^{8(m+n)}$ be the standard inclusions

$$i_1(e_i) = (e_i,0) = e_i$$

$$i_2(e_j) = (0,e_j) = e_{j+8m}$$

Let $\bar{i}_1$: $\Gamma(8m,\mathbb{F}_q) \to \Gamma(8(m+n),\mathbb{F}_q)$, $\bar{i}_2$: $\Gamma(8n,\mathbb{F}_q) \to \Gamma(8(m+n),\mathbb{F}_q)$ be the homomorphisms defined on generators as follows:

$$\bar{i}_1(\gamma) = \bar{i}_2(\gamma) = \gamma \qquad \gamma\in\mathbb{F}_q^*$$

$$\bar{i}_1(v) = i_1(v) \qquad v\in\mathbb{F}_q^{8m}$$

$$\bar{i}_2(v) = i_2(v) \qquad v\in\mathbb{F}_q^{8n}$$

Obviously these homomorphisms restrict to the corresponding spinor groups. Moreover it is easy to see that for any $x\in\mathrm{Spin}(8m,\mathbb{F}_q)$, $y\in\mathrm{Spin}(8n,\mathbb{F}_q)$

$$\bar{i}_1(x)\bar{i}_2(y) = \bar{i}_2(y)\bar{i}_1(x)$$

Hence there is a homomorphism $\oplus$: $\mathrm{Spin}(8m,\mathbb{F}_q) \times \mathrm{Spin}(8n,\mathbb{F}_q) \to \mathrm{Spin}(8(m+n),\mathbb{F}_q)$ such that

$$\oplus(x,y) = \bar{i}_1(x)\bar{i}_2(y) = \bar{i}_2(y)\bar{i}_1(x)$$

It is readily seen that the following diagram commutes

$$\begin{array}{ccc}
\mathrm{Spin}(8m,\mathbb{F}_q) \times \mathrm{Spin}(8n,\mathbb{F}_q) & \xrightarrow{\oplus} & \mathrm{Spin}(8(m+n),\mathbb{F}_q) \\
\downarrow{\scriptstyle \pi\times\pi} & & \downarrow{\scriptstyle \pi} \\
SN(8m,\mathbb{F}_q) \times SN(8n,\mathbb{F}_q) & \xrightarrow{\oplus} & SN(8(m+n),\mathbb{F}_q)
\end{array}$$

Now we can define the bifunctor $\oplus$ by

$$\oplus(8m,8n) = 8(m+n) \qquad \text{on objects}$$

$$\oplus(x,y) = \overline{i}_1(x)\overline{i}_2(y) = \overline{i}_2(y)\overline{i}_1(x) \qquad \text{on morphisms}$$

It is readily apparent that $\oplus$ is a strictly associative bifunctor. By definition 0 is the identity for $\oplus$. Moreover the above diagram shows that π is a functor of monoidal categories.

Next we introduce some notation. Suppose $1 \leq i \leq n-1$. Then we define an element $A_i \in \mathrm{Spin}(8n, \mathbb{F}_q)$ by

$$A_i = 2^{-4}\Pi_{j=0}^{7}(e_{8i-j} - e_{8i-j+8})$$

It is readily seen that $A_i^2 = 1$.

Now for each pair of nonnegative integers m,n define an element $\overline{c}(8m,8n) \in \mathrm{Spin}(8(m+n), \mathbb{F}_q)$ by

$$\overline{c}(8m,8n) = \Pi_{i=0}^{m-1}\Pi_{j=0}^{n-1}A_{n+i-j}$$

<u>Lemma</u> 5.8. $\overline{c}(8m,8n)^{-1} = \overline{c}(8n,8m)$

<u>Proof</u>: We proceed by induction on n. For $n = 0$, this is obvious. Assume this for $n - 1$. Then since A_i commutes with A_j for $j < i - 1$, we have

$$\overline{c}(8m,8n) = \Pi_{i=0}^{m-1}\Pi_{j=0}^{n-1}A_{n+i-j} = \Pi_{i=0}^{m-1}A_{n+i}\Pi_{j=1}^{n-1}A_{n+i-j}$$

$$= (\Pi_{k=0}^{m-1}A_{n+k})(\Pi_{i=0}^{m-1}\Pi_{j=1}^{n-1}A_{n+i-j})$$

$$= (\Pi_{k=0}^{m-1}A_{n+k})(\Pi_{i=0}^{m-1}\Pi_{j=0}^{n-2}A_{n-1+i-j})$$

$$= (\Pi_{k=0}^{m-1} A_{n+k})\bar{c}(8m, 8(n-1))$$

Hence using the induction hypothesis, we obtain

$$\begin{aligned}
\bar{c}(8m,8n)^{-1} &= \bar{c}(8m,8(n-1))^{-1}\Pi_{k=0}^{m-1}A_{n+m-1-k}^{-1} \\
&= \bar{c}(8(n-1),8m)\Pi_{j=0}^{m-1}A_{m+n-1-j} \\
&= (\Pi_{i=0}^{n-2}\Pi_{j=0}^{m-1}A_{m+i-j})(\Pi_{j=0}^{m-1}A_{m+n-1-j}) \\
&= \Pi_{i=0}^{n-1}\Pi_{j=0}^{m-1}A_{m+i-j} \\
&= \bar{c}(8n,8m)
\end{aligned}$$

This completes the induction and the proof.

Lemma 5.9. For all $m,n,p \in \mathbb{N}$ the following diagram commutes in $\mathcal{S}pin^{oct}(\mathbb{F}_q)$

$$\begin{array}{ccc}
8m + 8n + 8p & \xrightarrow{\bar{c}(8(m+n),8p)} & 8p + 8m + 8n \\
& {\scriptstyle \oplus(1,\bar{c}(8n,8p))} \searrow \qquad \nearrow {\scriptstyle \oplus(\bar{c}(8m,8p),1)} & \\
& 8m + 8p + 8n &
\end{array}$$

Proof: We have by definition

$$\begin{aligned}
\oplus(1,\bar{c}(8n,8p)) &= \bar{i}_1(1)\bar{i}_2(\bar{c}(8n,8p)) = 1\ \Pi_{i=0}^{n-1}\Pi_{j=0}^{p-1}\bar{i}_2(A_{p+i-j}) \\
&= \Pi_{i=0}^{n-1}\Pi_{j=0}^{p-1}A_{p+m+i-j}
\end{aligned}$$

$$= \Pi_{i=m}^{m+n-1} \Pi_{j=0}^{p-1} A_{p+i-j}$$

$$\oplus(\bar{c}(8m,8p),1) = \bar{i}_1(\bar{c}(8m,8p))\bar{i}_2(1) = \Pi_{i=0}^{m-1}\Pi_{j=0}^{p-1}\bar{i}_1(A_{p+i-j})$$

$$= \Pi_{i=0}^{m-1}\Pi_{j=0}^{p-1}A_{p+i-j}$$

Hence

$$\oplus(\bar{c}(8m,8p),1) \cdot \oplus(1,\bar{c}(8n,8p)) = \Pi_{i=0}^{m+n-1}\Pi_{j=0}^{p-1}A_{p+i-j}$$

$$= \bar{c}(8(m+n),8p)$$

<u>Lemma</u> 5.10. $\pi(\bar{c}(8m,8n)) = c(8m,8n) = \begin{pmatrix} 0 & I_{8n} \\ I_{8m} & 0 \end{pmatrix}$

<u>Proof</u>: Since both $\bar{c}(8m,8n)$ and $c(8m,8n)$ are built up from $\bar{c}(8,8)$ and $c(8,8)$ respectively through the coherence conditions of Lemma 5.8 and 5.9, it suffices to check that $\pi(\bar{c}(8,8)) = c(8,8)$. But

$$\bar{c}(8,8) = A_1 = 2^{-4}\Pi_{j=0}^{7}(e_{8-j} - e_{16-j})$$

so

$$\pi(\bar{c}(8,8)) = \Pi_{j=0}^{7}T_{e_{8-j} - e_{16-j}} = \begin{pmatrix} 0 & I_8 \\ I_8 & 0 \end{pmatrix} = c(8,8).$$

Hence $\pi(\bar{c}(8m,8n)) = c(8m,8n)$ for all $m,n\in\mathbb{N}$.

<u>Lemma</u> 5.11. $\bar{c}\colon \oplus \to \oplus \cdot \tau$ is a natural transformation.

<u>Proof</u>: We have to show that for any $m,n\in\mathbb{N}$ and say

$A \in \mathrm{Spin}(8m,\mathbb{F}_q), B \in \mathrm{Spin}(8n,\mathbb{F}_q)$ the following diagram commutes

$$\begin{array}{ccc}
\oplus(8m,8n) & \xrightarrow{\bar{c}(8m,8n)} & \oplus(8n,8m) \\
\downarrow{\scriptstyle \oplus(A,B)} & & \downarrow{\scriptstyle \oplus(B,A)} \\
\oplus(8m,8n) & \xrightarrow{\bar{c}(8m,8n)} & \oplus(8n,8m)
\end{array}$$

This is equivalent to saying that

$$\oplus(A,B)^{-1}\,\bar{c}(8m,8n)^{-1}(\oplus(B,A))\bar{c}(8m,8n) = 1$$

Let ρ: $\mathrm{Spin}(8(m+n),\mathbb{F}_q) \to \mathrm{Spin}(8(m+n),\mathbb{F}_q)$ denote conjugation by $\bar{c}(8m,8n)^{-1}$. Let f: $\mathrm{Spin}(8m,\mathbb{F}_q) \times \mathrm{Spin}(8n,\mathbb{F}_q) \to \mathrm{Spin}(8(m+n),\mathbb{F}_q)$ denote the homomorphism $f = \rho \circ \oplus \circ \tau$. Let g: $\mathrm{Spin}(8m,\mathbb{F}_q) \times \mathrm{Spin}(8n,\mathbb{F}_q) \to \mathrm{Spin}(8(m+n),\mathbb{F}_q)$ denote the function

$$g(A,B) = \oplus(A,B)^{-1}f(A,B)$$

Then

$$\begin{aligned}
\pi \circ g(A,B) &= \pi\,(\oplus(A,B)^{-1}\,\bar{c}(8m,8n)^{-1}\oplus(B,A)\bar{c}(8m,8n)\,) \\
&= \oplus(\pi A,\pi B)^{-1}\,c(8m,8n)^{-1}\oplus(\pi B,\pi A)c(8m,8n) \\
&= 1
\end{aligned}$$

Hence $\mathrm{im}\, g \subset \ker \pi = \mathbb{Z}/2$.

This means that g is a homomorphism. For we have

$$g(AC,BD) = \oplus(AC,BD)^{-1}f(AC,BD) = \oplus(C,D)^{-1}\oplus(A,B)^{-1}f(A,B)f(C,D)$$

$$= \oplus(C,D)^{-1}g(A,B)f(C,D) = g(A,B)\oplus(C,D)^{-1}f(C,D) = g(A,B)g(C,D)$$

But $[\mathrm{Spin}(k,\mathbb{F}_q), \mathrm{Spin}(k,\mathbb{F}_q)] = \mathrm{Spin}(k,\mathbb{F}_q)$ for $k \geq 5$. Hence im $g = 1$ so

$$g(A,B) = 1$$

for all A,B. But this is exactly what we wanted to prove.

This concludes the proof of Theorem 5.7.

<u>Proposition</u> 5.12. There is a fibration of infinite loop spaces

$$B\mathbb{Z}/2 \to \Gamma_0 B\mathcal{S}\mathrm{pin}^{\mathrm{oct}}(\mathbb{F}_q) \xrightarrow{\pi} \Gamma_0 B\mathcal{S}\mathcal{N}(\mathbb{F}_q)$$

<u>Proof</u>: Let $X \to \Gamma_0 B\mathcal{S}\mathrm{pin}(\mathbb{F}_q)$ denote the fiber of π. Then there is an induced map of fibrations

$$\begin{array}{ccccc} B\mathbb{Z}/2 & \to & B\mathrm{Spin}(\infty,\mathbb{F}_q) & \longrightarrow & BSN(\infty,\mathbb{F}_q) \\ \downarrow{\scriptstyle \ell_3} & & \downarrow{\scriptstyle \ell_1} & & \downarrow{\scriptstyle \ell_2} \\ X & \to & \Gamma_0 B\mathcal{S}\mathrm{pin}(\mathbb{F}_q) & \longrightarrow & \Gamma_0 B\mathcal{S}\mathcal{N}(\mathbb{F}_q) \end{array}$$

Passing to the homology spectral sequence we see that $\pi_1(BSN(\infty,\mathbb{F}_q))$ acts trivially on the fiber since $\mathbb{Z}/2$ is in the center of $\mathrm{Spin}(\infty,\mathbb{F}_q)$. Now ℓ_1,ℓ_2 are homology equivalences. By the comparison theorem so is ℓ_3. Since ℓ_3 is a Hopf map it follows that ℓ_3 is an equivalence.

<u>Theorem</u> 5.13. $\Gamma_0 B\mathcal{S}pin^{oct}(\mathbb{F}_q)$ is the 2-connected cover of $\Gamma_0 B\mathcal{S}\mathcal{N}(\mathbb{F}_q)$ (and hence also of $\Gamma_0 B\mathcal{O}(\mathbb{F}_q)$).

<u>Proof</u>: We look at the homotopy sequence of the fibration of Proposition 5.12. If $n > 2$ then we have the exact sequence

$$0 = \pi_n(B\mathbb{Z}/2) \to \pi_n(\Gamma_0 B\mathcal{S}pin^{oct}(\mathbb{F}_q)) \xrightarrow{\pi} \pi_n(\Gamma_0 B\mathcal{S}\mathcal{N}(\mathbb{F}_q)) \to \pi_{n-1}(B\mathbb{Z}/2) = 0$$

so π is an isomorphism.

If $n = 1$ then

$$\pi_1(\Gamma_0 B\mathcal{S}pin(\mathbb{F}_q)) = Spin(\infty,\mathbb{F}_q)/[Spin(\infty,\mathbb{F}_q),Spin(\infty,\mathbb{F}_q)] = 0$$

For $n = 2$ we have the exact sequence

$$0 = \pi_2(B\mathbb{Z}/2) \to \pi_2(\Gamma_0 B\mathcal{S}pin^{oct}(\mathbb{F}_q)) \xrightarrow{\pi} \pi_2(\Gamma_0 B\mathcal{S}\mathcal{N}(\mathbb{F}_q)) \xrightarrow{\delta} \pi_1(B\mathbb{Z}/2) \to 0$$

Now since $\pi_2(\Gamma_0 B\mathcal{S}\mathcal{N}(\mathbb{F}_q)) = \mathbb{Z}/2$, (cf. II 3.16 and III 3.5), $\pi_1(B\mathbb{Z}/2) = \mathbb{Z}/2$ and δ is epimorphic, it follows that δ is an isomorphism. Hence

$$\pi_2(\Gamma_0 B\mathcal{S}pin^{oct}(\mathbb{F}_q)) = 0$$

This completes the proof.

§6. The general linear, symplectic and unitary groups

In this section we discuss the general linear, symplectic and unitary groups over finite fields and their associated permutative categories and infinite loop spaces. We make no restrictions about the characteristic of the ground field. In particular characteristic 2 is allowed.

6.1. The category $\mathcal{GL}(\mathbb{F}_q)$ and the subcategory $\mathcal{SL}^{ev}(\mathbb{F}_q)$

(a) The category $\mathcal{GL}(\mathbb{F}_q)$ was defined in 2.6. It has as objects the nonnegative integers $\mathbb{N}$ and as morphisms

$$\hom(m,n) = \begin{cases} \emptyset & m \neq n \\ GL(n,\mathbb{F}_q) & m = n \end{cases}$$

where $GL(n,\mathbb{F}_q)$ is the general linear group. The order of $GL(n,\mathbb{F}_q)$ is

$$GL(n,\mathbb{F}_q) = \Pi_{i=1}^{n}(q^i-1)q^{i-1} \qquad \text{(cf. Dickson [16])}$$

(b) As shown in 2.16(vii) and 2.18

$$B\mathcal{GL}(\mathbb{F}_q) \simeq \coprod_{n=0}^{\infty} BGL(n,\mathbb{F}_q)$$

and that as a space

$$\Gamma B\mathcal{GL}(\mathbb{F}_q) \simeq \Gamma_0 B\mathcal{GL}(\mathbb{F}_q) \times \mathbb{Z} \simeq (BGL(\infty,\mathbb{F}_q))^+ \times \mathbb{Z}$$

where $(\cdot)^+$ denotes Quillen's plus construction.

(c) There is a permutative subcategory $\mathcal{SL}^{ev}(\mathbb{F}_q)$ of $\mathcal{GL}(\mathbb{F}_q)$ whose objects are the even nonnegative integers and whose morphisms are matrices with determinant 1. (compare 3.4).

(d) There is a fibration of infinite loop spaces

$$\Gamma_0 B\mathcal{SL}^{ev}(\mathbb{F}_q) \to \Gamma_0 B\mathcal{GL}(\mathbb{F}_q) \xrightarrow{\Gamma B(\det)} B\mathbb{F}_q^* = B\mathbb{Z}/(q-1)$$

This is proved in a similar way as Theorem 3.16.

(e) $\Gamma_0 B\mathcal{SL}^{ev}(\mathbb{F}_q)$ is the universal cover of $\Gamma_0 B\mathcal{GL}(\mathbb{F}_q)$. This follows from (d) and the fact that

$$SL(\infty,\mathbb{F}_q) = [GL(\infty,\mathbb{F}_q),GL(\infty,\mathbb{F}_q)]$$

and

$$SL(\infty,\mathbb{F}_q) = [SL(\infty,\mathbb{F}_q),SL(\infty,\mathbb{F}_q)]$$

We next turn to consideration of symplectic forms, their associated permutative categories and infinite loop spaces

Definition 6.2. A symplectic form on a vector space V over $\mathbb{F}_q$ is a bilinear form $B: V \times V \to \mathbb{F}_q$ such that

$$B(x,x) = 0 \quad \forall\ x \in V$$

If char $\mathbb{F}_q \neq 2$ this is equivalent to saying that A is a antisymmetric, i.e.

$$B(x,y) = -B(y,x) \quad \forall\ x,y \in V.$$

A <u>symplectic space</u> (V,B) is a finite dimensional vector space V over $\mathbb{F}_q$ together with a symplectic form B: $V \times V \to \mathbb{F}_q$.

The <u>radical</u> of a symplectic space (V,B) is defined by

$$\mathrm{Rad}(V,B) = \{x \in V \mid B(v,x) = 0 \text{ for all } v \in V\}$$

Rad(V,B) is a subspace of V. A symplectic form is said to be <u>nondegenerate</u> if Rad(V,B) = 0.

A map f: (V,B) → (W,B') between symplectic spaces is said to be an <u>isometric isomorphism</u> if

(a) f: V → W is a linear isomorphism

(b) $B'(f(u),f(v)) = B(u,v) \ \forall\, u,v \in V$

We denote the automorphism group of a symplectic space (V,B) by Sp(V,B).

6.3. <u>The structure of symplectic spaces</u>. It can be shown that there are no odd dimensional nondegenerate symplectic spaces. Any 2n dimensional nondegenerate symplectic space (V,B) is isomorphic to $(\mathbb{F}_q^{2n}, A)$ where

$$A((x_1,\ldots,x_{2n}),(y_1,\ldots,y_{2n})) = \Sigma_{i=1}^{n}(x_{2i}y_{2i-1} - x_{2i-1}y_{2i})$$

The group of automorphisms of $(\mathbb{F}_q^{2n}, A)$ is denoted by $Sp(2n,\mathbb{F}_q)$ and is called the <u>symplectic group</u>. The order of $Sp(2n,\mathbb{F}_q)$ is given by

$$Sp(2n,\mathbb{F}_q) = \Pi_{i=1}^{n}(q^{2i} - 1)q^{2i-1}$$

cf. Dickson [16, p. 94].

Any element of $Sp(2n,\mathbb{F}_q)$ can be represented by a $2n \times 2n$ matrix over $\mathbb{F}_q$. It can be shown that any such matrix has determinant 1. Hence there is a natural inclusion

$$Sp(2n,\mathbb{F}_q) \subseteq SL(2n,\mathbb{F}_q)$$

Comparing orders we see that in particular $Sp(2,\mathbb{F}_q) = SL(2,\mathbb{F}_q)$.

Definition 6.4. The permutative category $\mathcal{S}p(\mathbb{F}_q)$ is the category whose objects are the even nonnegative integers and whose morphisms are

$$\hom(2m,2n) = \begin{cases} 0 & 2m \neq 2n \\ Sp(2n,\mathbb{F}_q) & 2m = 2n \end{cases}$$

Proposition 6.5. The infinite loop space $\Gamma_0 B\mathcal{S}p(\mathbb{F}_q)$ is simply connected.

Proof: We have

$$\pi_1(\Gamma_0 B\mathcal{S}p(\mathbb{F}_q)) \cong H_1(\Gamma_0 B\mathcal{S}p(\mathbb{F}_q)) \cong H_1(BSp(\infty,\mathbb{F}_q))$$
$$\cong \frac{Sp(\infty,\mathbb{F}_q)}{[Sp(\infty,\mathbb{F}_q),Sp(\infty,\mathbb{F}_q)]} = 0$$

(cf. Dieudonné [15])

Finally we examine hermitian forms, unitary groups, and their associated permutative categories and infinite loop spaces.

Consider a finite field $\mathbb{F}_{q^2}$. Since $\mathbb{F}_{q^2}$ is a quadratic extension of $\mathbb{F}_q$, the Galois group of $\mathbb{F}_{q^2}$ over $\mathbb{F}_q$ is cyclic of order 2. The unique nontrivial automorphism of $\mathbb{F}_{q^2}$ over $\mathbb{F}_q$ is given by $x \to x^q$. We shall find it convenient to write $x^q = \bar{x}$.

<u>Definition</u> 6.6. A <u>hermitian</u> form on a vector space V over $\mathbb{F}_{q^2}$ is a function $H: V \times V \to \mathbb{F}_{q^2}$ such that

i) H is $\mathbb{F}_{q^2}$-linear in the first variable

$$H(\alpha x + \beta y, z) = \alpha H(x,z) + \beta H(y,z) \;\; \forall\; x,y,z \in V, \; \forall\; \alpha,\beta \in \mathbb{F}_{q^2}$$

ii) H is conjugate symmetric

$$H(y,x) = \overline{H(x,y)}$$

A hermitian space (V,H) is a finite dimensional vector space V together with a hermitian form $H: V \times V \to \mathbb{F}_{q^2}$.

The <u>radical</u> of a hermitian space (V,H) is defined by

$$\mathrm{Rad}(V,H) = \{x \in V \mid H(x,v) = 0 \text{ for all } v \in V\}$$

$\mathrm{Rad}(V,H)$ is a subspace of V. A hermitian form is said to be <u>non-degenerate</u> of $\mathrm{Rad}(V,H) = 0$.

A map $f: (V,H) \to (W,H')$ between hermitian spaces is said to be a <u>unitary</u> map if

(a) $f: V \to W$ is a ($\mathbb{F}_{q^2}$) linear isomorphism

(b) $H'(f(u),f(v)) = H(u,v) \ \forall\ u,v \in V$

We denote the automorphism group of a symplectic space (V,H) by U(V,H).

6.7. <u>The structure of hermitian spaces.</u> It can be shown that any n-dimensional hermitian space is isomorphic to $(\mathbb{F}^n_{q^2}, H)$ where

$$H((x_1,\ldots,x_n),(y_1,\ldots,y_n)) = \Sigma^n_{i=1} x_i \bar{y}_i$$

We denote by $U(n,\mathbb{F}_{q^2})$ the group of unitary automorphisms of $(\mathbb{F}_{q^2},H)$. It is easily seen that $U(n,\mathbb{F}_{q^2})$ is isomorphic to the group of $n \times n$ matrices A over $\mathbb{F}_{q^2}$ such that

$$A\bar{A}^t = I$$

Here if $A = (a_{ij})^n_{i,j=1}$, then $\bar{A}^t = (\bar{a}_{ji})^n_{i,j=1}$. In particular $U(1,\mathbb{F}_{q^2})$ is the subgroup of $\mathbb{F}^*_{q^2}$ consisting of those elements x such that $x\bar{x} = 1$. Among other things, this implies $U(1,\mathbb{F}_{q^2})$ is cyclic. Also if A is a unitary matrix then

$$1 = \det I = \det(A\bar{A}^t) = (\det A)(\det \bar{A}^t) = (\det A)(\overline{\det A})$$

This implies that

$$U(n,\mathbb{F}_q) \xrightarrow{\det} U(1,\mathbb{F}_{q^2}) \cong \mathbb{Z}/(q+1)$$

It is easily seen that this is an epimorphism. We denote by $SU(n,\mathbb{F}_{q^2})$ the group of $n \times n$ unitary matrices which have determinant 1.

The order of $U(n,\mathbb{F}_{q^2})$ is given by

$$U(n,\mathbb{F}_{q^2}) = \Pi_{i=1}^{n}(q^i - (-1)^i)q^{i-1}$$

For a proof of this and further information about the unitary groups over $\mathbb{F}_{q^2}$ the reader is advised to consult [16, p. 134] or [15].

<u>Definition</u> 6.8. The permutative category $\mathcal{U}(\mathbb{F}_{q^2})$ is the category whose objects are the nonnegative integers and whose morphisms are

$$\hom(m,n) = \begin{cases} \emptyset & m \neq n \\ U(n,\mathbb{F}_{q^2}) & m = n \end{cases}$$

The bifunctor $\oplus$: $\mathcal{U}(\mathbb{F}_{q^2}) \times \mathcal{U}(\mathbb{F}_{q^2}) \to \mathcal{U}(\mathbb{F}_{q^2})$ and the commutativity transformation c are defined as in 2.6. We also define a permutative subcategory $\mathcal{SU}^{ev}(\mathbb{F}_{q^2})$ of $\mathcal{U}(\mathbb{F}_{q^2})$ by

$$\mathcal{SU}^{ev}(\mathbb{F}_{q^2}) = \mathcal{U}(\mathbb{F}_{q^2}) \cap \mathcal{SL}(\mathbb{F}_{q^2})$$

<u>Proposition</u> 6.9 (i) There is a fibration of infinite loop spaces

$$\Gamma_0 B\mathcal{SU}(\mathbb{F}_{q^2}) \to \Gamma_0 B\mathcal{U}(\mathbb{F}_{q^2}) \to B\mathbb{Z}/(q+1)$$

(ii) $\Gamma_0 B\mathcal{SU}(\mathbb{F}_{q^2})$ is the universal cover of $\Gamma_0 B\mathcal{U}(\mathbb{F}_{q^2})$.

Proof: The proof of (i) is similar to that of Theorem 3.16. Part (ii) follows from (i) and the fact that

$$SU(\infty,\mathbb{F}_{q^2}) = [U(\infty,\mathbb{F}_{q^2}),\ U(\infty,\mathbb{F}_{q^2})]$$

$$SU(\infty,\mathbb{F}_{q^2}) = [SU(\infty,\mathbb{F}_{q^2}),SU(\infty,\mathbb{F}_{q^2})]$$

(cf. Dieudonné [15]).

§7. Orthogonal groups over finite fields of characteristic 2.

When it comes to defining orthogonal groups over finite field of characteristic 2, the situation becomes much more complicated. For one thing, there is no longer the nice correspondence between symmetric bilinear forms and quadratic forms (cf. 4.2). Hence there exists a whole profusion of different orthogonal groups together with their associated permutative categories and infinite loop spaces. In this section we will analyze exhaustively these various possibilities.

In what follows $\mathbb{F}_q$ will be a finite field with $q = 2^k$ elements.

We first analyze the orthogonal groups associated with symmetric bilinear forms.

Definition 7.1. An orthobilinear space (V,B) is a finite-dimensional vector space V over $\mathbb{F}_q$ together with a symmetric bilinear form $B: V \times V \to \mathbb{F}_q$.

The radical of an orthobilinear space (V,B) is defined by

$$\mathrm{Rad}(V,B) = \{x \in V \mid B(x,v) = 0 \ \forall\ v \in V\}.$$

Rad(V,B) is a subspace of V. We say that an orthobilinear space is <u>nondegenerate</u> if Rad(V,B) = 0.

A map $f: (V,B) \to (W,B'')$ between orthobilinear spaces is said to be an <u>isometric isomorphism</u> if

(a) $f: V \to W$ is a linear isomorphism

(b) $B'(f(u),f(v)) = B(u,v)\ \forall\ u,v \in V$.

We define the direct sum of two orthobilinear spaces in the usual way:

$$(V,B_1) \oplus (W,B_2) = (V \oplus W,\ B_1 \oplus B_2)$$

where

$$(B_1 \oplus B_2)((v,w),(v',w')) = B_1(v,v') + B_2(w,w')$$

It is clear that the direct sum of two isometric isomorphisms is again an isometric isomorphism.

7.2. <u>The structure of orthobilinear spaces</u>. It can be shown that any nondegenerate orthobilinear space (V,B) is isomorphic to one of the following spaces

(i) $(\mathbf{F}_q^n, E)$ where E is the standard Euclidean form

$$E((x_1,\ldots,x_n),(y_1,\ldots,y_n)) = \Sigma_{i=1}^n x_i y_i$$

or else

(ii) $(\mathbb{F}_q^{2n}, A)$ where A is the standard symplectic form (cf. 6.3)

$$A((x_1,\ldots,x_{2n}),(y_1,\ldots,y_{2n})) = \Sigma_{i=1}^{n}(x_{2i}y_{2i-1} - x_{2i-1}y_{2i})$$

Under direct sum these forms behave as follows:

$$(\mathbb{F}_q^m,E) \oplus (\mathbb{F}_q^n,E) \cong (\mathbb{F}_q^{m+n},E)$$

$$(\mathbb{F}_q^{2n},A) \oplus (\mathbb{F}_q^{2n},A) \cong (\mathbb{F}_q^{2m+2n},A)$$

$$(\mathbb{F}_q^{2m},A) \oplus (\mathbb{F}_q^{n},E) \cong (\mathbb{F}_q^{2m+n},E)$$

For proofs the reader is referred to Kaplansky [23].

The automorphism group of $(\mathbb{F}_q^{2n},A)$ is of course the symplectic group $Sp(2n,\mathbb{F}_q)$ treated in §6. The automorphism group of $(\mathbb{F}_q^n,E)$ is the Euclidean orthogonal group and will be denoted by $EO(n,\mathbb{F}_q)$. It is isomorphic to the group of $n \times n$ matrices M over $\mathbb{F}_q$ such that $MM^t = I$. The orders of the groups $EO(n,\mathbb{F}_q)$ are given by

$$|EO(2n,\mathbb{F}_q)| = q^{2n-1}\Pi_{i=1}^{n-1}(q^{2i}-1)q^{2i-1}$$

$$|EO(2n+1,\mathbb{F}_q)| = \Pi_{i=1}^{n}(q^{2i}-1)q^{2i-1}$$

Remark 7.3. The isomorphism

$$(\mathbb{F}_q^{2n},A) \oplus (\mathbb{F}_q,E) \cong (\mathbb{F}_q^{2n+1},E)$$

induces an inclusion

$$Sp(2n,\mathbb{F}_q) \hookrightarrow EO(2n+1,\mathbb{F}_q)$$

Since

$$|Sp(2n,\mathbb{F}_q)| = \Pi_{i=1}^{n}(q^{2i}-1)q^{2i-1} = |EO(2n+1,\mathbb{F}_q)|$$

it follows that

$$Sp(2n,\mathbb{F}_q) \cong EO(2n+1,\mathbb{F}_q).$$

Thus there are basically three different permutative categories we can associate with symmetric bilinear forms over the field $\mathbb{F}_q$: the permutative category $\mathcal{Sp}(\mathbb{F}_q)$ defined in 6.4, a permutative category built up from the groups $EO(n,\mathbb{F}_q)$, and a permutative category encompassing both the groups $Sp(2n,\mathbb{F}_q)$ and $EO(n,\mathbb{F}_q)$.

Definition 7.4. The permutative category $\mathcal{EO}(n,\mathbb{F}_q)$ is the category whose objects are the nonnegative integers and whose morphisms are

$$\hom(m,n) = \begin{cases} \emptyset & \text{if } m \neq n \\ EO(n,\mathbb{F}_q) & \text{if } m = n \end{cases}$$

The bifunctor $\oplus$: $\mathcal{EO}(\mathbb{F}_q) \times \mathcal{EO}(\mathbb{F}_q) \to \mathcal{EO}(\mathbb{F}_q)$ and the commutativity transformation c are defined as in 2.6. (Compare with Example 2.8 and subsequent remark).

Definition 7.5. The symmetric monoidal category $(\mathcal{OBL}(\mathbb{F}_q),\oplus,0,a,\ell,r,c)$ is the category with objects all nondegenerate orthobilinear spaces over $\mathbb{F}_q$ and with morphisms all isometric isomorphisms between such spaces. The direct sum of two orthobilinear spaces is defined as in 7.1. The direct sum of two morphisms is defined as usual. The 0-module is equipped with the trivial form. The natural transformations a,ℓ,r,c are defined in the usual way (cf. 2.4 or 4.3).

We construct an equivalent permutative category, eg. by the Isbell construction (cf. 2.5), and by abuse of notation we also denote it $\mathcal{O}\mathcal{B}\mathcal{L}(\mathbb{F}_q)$. (Of course there are more economical ways of constructing such a permutative category. The interested reader may try his hand at it).

It is easily seen that $\pi_0\mathcal{O}\mathcal{B}\mathcal{L}(\mathbb{F}_q) = |\mathcal{O}\mathcal{L}\,\mathcal{O}\mathcal{B}\mathcal{L}(\mathbb{F}_q)|$ is the commutative monoid on two generators (1) (corresponding to $(\mathbb{F}_q, E)$) and $(\overline{2})$ (corresponding to $(\mathbb{F}_q^2, A)$). These generators are subject to a single relation

$$(1) + (\overline{2}) = 3(1). \quad (\text{cf. } 7.2(\text{ii}))$$

It follows immediately that $K_0\mathcal{O}\mathcal{B}\mathcal{L}(\mathbb{F}_q) \cong \mathbb{Z}$ and that under the natural map

$$\pi_0\mathcal{O}\mathcal{B}\mathcal{L}(\mathbb{F}_q) \to K_0\mathcal{O}\mathcal{B}\mathcal{L}(\mathbb{F}_q)$$

the generator $(\overline{2})$ is identified with $2\cdot(1)$.

Although we have these three apparantly different permutative categories associated with symmetric bilinear forms over $\mathbb{F}_q$, after we apply the functor ΓB, they all become equivalent as infinite loop spaces.

Theorem 7.6 (i) The inclusion of categories

$$\mathcal{E}\mathcal{O}(\mathbf{F}_q) \to \mathcal{O}\mathcal{B}\mathcal{L}(\mathbf{F}_q) \quad m \mapsto (\mathbf{F}_q^m, E)$$

induces an equivalence of infinite loop spaces

$$\Gamma B\,\mathcal{E}\mathcal{O}(\mathbf{F}_q) \xrightarrow{\simeq} \Gamma B\,\mathcal{O}\mathcal{B}\mathcal{L}(\mathbf{F}_q)$$

(ii) The inclusion of categories

$$\mathcal{S}p(\mathbf{F}_q) \to \mathcal{O}\mathcal{B}\mathcal{L}(\mathbf{F}_q) \quad 2m \mapsto (\mathbf{F}_q^{2m}, A)$$

induces an equivalence of infinite loop spaces

$$\Gamma B\,\mathcal{S}p(\mathbf{F}_q) \xrightarrow{\simeq} \Gamma_{2\mathbf{Z}} B\,\mathcal{O}\mathcal{B}\mathcal{L}\,(\mathbf{F}_q) \qquad \text{(cf. 2.17(vi))}$$

<u>Proof</u>: (i) This follows immediately from 2.17(vi) because the objects $(\mathbf{F}_q^m, E)$ are cofinal in $\pi_0\,\mathcal{O}\mathcal{B}\mathcal{L}(\mathbf{F}_q)$ and because under the inclusion $K_0\,\mathcal{E}\mathcal{O}(\mathbf{F}_q) \cong K_0\,\mathcal{O}\mathcal{B}\mathcal{L}\,(\mathbf{F}_q) \cong \mathbf{Z}$.

(ii) While 2.17(vi) does not apply directly in this case since the objects $(\mathbf{F}_q^{2m}, A)$ are not cofinal in $\pi_0\,\mathcal{O}\mathcal{B}\mathcal{L}(\mathbf{F}_q)$, a slight modification using 2.17(v) works: On the 0-component we have

$$\begin{aligned} H_*(\Gamma_0 B\,\mathcal{O}\mathcal{B}\mathcal{L}\,(\mathbf{F}_q)) &\cong H_*(\varinjlim_{\alpha\in\pi_0\,\mathcal{O}\mathcal{B}\mathcal{L}\,(\mathbf{F}_q)} B\,\mathcal{O}\mathcal{B}\mathcal{L}\,(\mathbf{F}_q)_\alpha) \\ &\cong H_*(\varinjlim_{m} BEO(2m+1, \mathbf{F}_q)) \\ &\cong H_*(\varinjlim_{m} BSp(2m, \mathbf{F}_q)) \qquad \text{(cf. 7.3)} \end{aligned}$$

Hence the infinite loop map $\Gamma_0 B\mathcal{Sp}(\mathbb{F}_q) \to \Gamma_0 B\,\mathcal{OBL}(\mathbb{F}_q)$ is a homology equivalence and therefore a homotopy equivalence. The rest of (ii) follows from the obvious fact that the map $K_0\,\mathcal{Sp}(\mathbb{F}_q) \to K_0\,\mathcal{OBL}(\mathbb{F}_q)$ is the inclusion $2\mathbb{Z} \subseteq \mathbb{Z}$.

Having disposed of the problems connected with symmetric bilinear forms over $\mathbb{F}_q$, we turn to quadratic forms over $\mathbb{F}_q$.

Quadratic forms over a field of characteristic 2 are defined in exactly the same way as for fields of odd characteristic (cf. 4.1).

<u>Definition</u> 7.7. A <u>quadratic form</u> on a vector space V over $\mathbb{F}_q$ is a function $Q\colon V \to \mathbb{F}_q$ such that

(i) $Q(\alpha x) = \alpha^2 Q(x) \quad x\in\mathbb{F}_q, v\in V$

(ii) $B_Q(x,y) = Q(x+y)-Q(x)-Q(y)$ is bilinear on x and y

A <u>quadratic space</u> (V,Q) is a finite dimensional vector space V over $\mathbb{F}_q$ together with a quadratic form $Q\colon V \to \mathbb{F}_q$.

The <u>radical</u> of a quadratic space (V,Q) is defined by

$$\mathrm{Rad}(V,Q) = \{x\in V \mid Q(v+x) = Q(v) \text{ for all } x\in V\}$$

It is readily seen that $\mathrm{Rad}(V,Q)$ is a subspace of V. A quadratic form is said to be <u>nondegenerate</u> if $\mathrm{Rad}(V,Q) = 0$.

A map $f\colon (V,Q) \to (W,Q')$ between quadratic spaces is said to be an <u>isometric isomorphism</u> (<u>epimorphism</u>) if

(a) $f\colon V \to W$ is a linear isomorphism (epimorphism)

(b) $Q'(f(v)) = Q(v)$ for every $v\in V$

It is readily seen that an isometric epimorphism $f: (V,Q) \to (W,Q')$ maps $\mathrm{Rad}(V,Q)$ onto $\mathrm{Rad}(W,Q')$. We denote the automorphism group of (V,Q) by $O(V,Q)$.

We define the direct sum of two quadratic spaces in the usual way

$$(V,Q_1) \oplus (W,Q_2) = (V \oplus W, Q_1 \oplus Q_2)$$

where

$$(Q_1 \oplus Q_2)(v,w) = Q_1(v) + Q_2(w)$$

It is clear that the direct sum of two isometric isomorphisms (epimorphisms) is again an isometric isomorphism (epimorphism).

7.8. <u>The structure of quadratic spaces</u>. It can be shown that any nondegenerate quadratic space (V,Q) is isomorphic to one of the following spaces

(i) $(\mathbf{F}_q^{2n+1}, Q_+)$ where

$$Q_+(x_1,x_2,\ldots,x_{2n+1}) = \Sigma_{i=1}^n x_{2i-1}x_{2i} + x_{2n+1}^2$$

(ii) $(\mathbf{F}_q^{2n}, Q_+)$ where

$$Q_+(x_1,x_2,\ldots,x_{2n}) = \Sigma_{i=1}^n x_{2i-1}x_{2i}$$

or

(iii) $(\mathbf{F}_q^{2n}, Q_-)$ where

$$Q_-(x_1,x_2,\ldots,x_{2n}) = \Sigma_{i=1}^n x_{2i-1}x_{2i} + \lambda x_{2n-1}^2 + \lambda x_{2n}^2$$

where $\lambda \in \mathbb{F}_q$ is such that the polynomial $\lambda x^2 + xy + \lambda y^2$ is irreducible (Different choices of λ give isomorphic quadratic forms as long as $\lambda x^2 + xy + \lambda y^2$ is irreducible.)

Under direct sum, these forms behave as follows:

$$(\mathbb{F}_q^{2m}, Q_+) \oplus (\mathbb{F}_q^{2n}, Q_\pm) \cong (\mathbb{F}_q^{2m+2n}, Q_\pm)$$

$$(\mathbb{F}_q^{2m}, Q_-) \oplus (\mathbb{F}_q^{2n}, Q_-) \cong (\mathbb{F}_q^{2m+2n}, Q_+)$$

$$(\mathbb{F}_q^{2m+1}, Q_+) \oplus (\mathbb{F}_q^{2n}, Q_\pm) \cong (\mathbb{F}_q^{2m+2n+1}, Q_+)$$

$(\mathbb{F}_q^{2m+1}, Q_+) \oplus (\mathbb{F}_q^{2n+1}, Q_+)$ is a degenerate quadratic space.

For proofs the reader is referred to Dickson [16].

We denote the automorphism group of $(\mathbb{F}_q^n, Q_+)$ by $O_+(n, \mathbb{F}_q)$ and the automorphism group of $(\mathbb{F}_q^{2n}, Q_-)$ by $O_-(2n, \mathbb{F}_q)$. The orders of these groups are given by the following formulas:

$$|O_+(2n+1, \mathbb{F}_q)| = \Pi_{i=1}^{n} (q^{2i}-1) q^{2i-1}$$

$$|O_+(2n, \mathbb{F}_q)| = 2(q^n-1) \Pi_{i=1}^{n-1} (q^{2i}-1) q^{2i}$$

$$|O_-(2n, \mathbb{F}_q)| = 2(q^n+1) \Pi_{i=1}^{n-1} (q^{2i}-1) q^{2i}$$

(cf. [16, p. 206]. Compare 4.5)

There is an invariant called the <u>Arf invariant</u> which distinguishes between $(\mathbb{F}_q^{2n}, Q_+)$ and $(\mathbb{F}_q^{2n}, Q_-)$ playing a role analogous to that of the discriminant in the case q odd (cf. 4.5). Let

$M = M_q$ be the subgroup of $\mathbb{F}_q$ consisting of

$$M = \{x^2 + x \mid x \in \mathbb{F}_q\}$$

Then $\mathbb{F}_q/M \cong \mathbb{Z}/2$.

Now let (V,Q) be a 2n dimensional nondegenerate quadratic space over $\mathbb{F}_q$. Then

$$B_Q(x,y) = Q(x + y) - Q(x) - Q(y)$$

is a nondegenerate symplectic form on V (cf. 7.15). Let $\{e_i\}_{i=1}^{2n}$ be a symplectic basis for (V,B_Q) so that

$$B_Q(e_i,e_j) = \begin{cases} 1 & \text{if } i = 2m-1, j = 2m \text{ or vice versa}, 1 \leq m \leq n \\ 0 & \text{otherwise.} \end{cases}$$

Then the residue class

$$A(Q) = \Sigma_{i=1}^{n} Q(e_{2i-1})Q(e_{2i}) \in \mathbb{F}_q/M \cong \mathbb{Z}/2$$

depends only on Q, not on the choice of symplectic basis $\{e_i\}_{i=1}^{2n}$. This invariant is called the Arf invariant. If $A(Q) \in M$, then $(V,Q) \cong (\mathbb{F}_q^{2n}, Q_+)$. If $A(Q) \notin M$ then $(V,Q) \cong (\mathbb{F}_q^{2n}, Q_-)$. For proofs and further details the reader is referred to Dieudonné [15] or Kaplansky [23].

Example 7.9. We present here some examples of quadratic spaces analogous to those in 4.6. Let $\mathbb{F}_{q^r}$ be an extension field of $\mathbb{F}_q$. We can consider $\mathbb{F}^2_{q^r}$ as a 2r-dimensional vector space over $\mathbb{F}_q$. Moreover we can equip $\mathbb{F}^2_{q^r}$ with a quadratic form over $\mathbb{F}_q$

$$Q(x,y) = \mathrm{tr}(\alpha x^2 + xy + \beta y^2)$$

where tr: $\mathbb{F}_{q^r} \to \mathbb{F}_q$ is the trace map $\mathrm{tr}(x) = x + x^q + \cdots + x^{q^{r-1}}$ and where α, β are any fixed elements of $\mathbb{F}_{q^r}$.

The associated symplectic form on $\mathbb{F}^2_{q^r}$ is given by

$$B_Q((x,y),(w,z)) = \mathrm{tr}(xz + wy)$$

By the same argument as in 4.6, B_Q is nondegenerate. Hence Q is also nondegenerate.

Now B: $\mathbb{F}_{q^r} \times \mathbb{F}_{q^r} \to \mathbb{F}_q$ given by

$$B(x,y) = \mathrm{tr}(xy)$$

is a nondegenerate symmetric form on $\mathbb{F}_{q^r}$ considered as an r-dimensional vector space over $\mathbb{F}_q$. Since B is clearly not a symplectic form, by 7.2 it follows that $(\mathbb{F}_{q^r}, B) \cong (\mathbb{F}^r_q, E)$. Hence $(\mathbb{F}_{q^r}, B)$ has an orthonormal basis $\{g_i\}^r_{i=1}$ such that

$$B(g_i,g_j) = \mathrm{tr}(g_i g_j) = \begin{cases} 0 & \text{if } i \neq j \\ 1 & \text{if } i = j \end{cases}$$

Let us define a basis $\{e_i\}_{i=1}^{2r}$ for $\mathbb{F}^2_{q^r}$ by

$$e_{2i-1} = (g_i,0) \qquad e_{2i} = (0,g_i)$$

It is easily seen that $\{e_i\}_{i=1}^{2r}$ is a symplectic basis for $(\mathbb{F}^2_{q^r},B_Q)$. We now make a case by case analysis of the Arf invariant

$$\begin{aligned} A(Q) &= \Sigma_{i=1}^{r} Q(e_{2i-1})Q(e_{2i}) \\ &= \Sigma_{i=1}^{r} \mathrm{tr}(\alpha g_i^2)\mathrm{tr}(\beta g_i^2) \end{aligned}$$

We note that by 7.8 we can replace α,β by $\bar{\alpha},\bar{\beta}$ without changing the isomorphism class $(\mathbb{F}^2_{q^r},Q)$ as long as

$$\alpha\beta = \bar{\alpha}\bar{\beta} \quad \bmod M_{q^r}$$

(a) If $\alpha\beta \in M_{q^r}$ then we can assume that $\alpha = \beta = 0$ so that

$$A(Q) = \Sigma_{i=1}^{r} \mathrm{tr}(\alpha g_i^2)\mathrm{tr}(\beta g_i^2) = 0 \in M_q$$

Hence by 7.8

$$(\mathbb{F}^2_{q^r},Q) \cong (\mathbb{F}_q^{2r},Q_+)$$

(b) If r is odd, $\alpha\beta \notin M_{q^r}$, then since $\mathbb{F}_q \cap (\mathbb{F}_{q^r} - M_{q^r}) \neq \emptyset$, we can take $\alpha,\beta \in \mathbb{F}_q$. Hence

$$A(Q) = \Sigma_{i=1}^{r} tr(\alpha g_i^2) tr(\beta g_i^2) = \Sigma_{i=1}^{r} \alpha\beta tr(g_i^2) tr(g_i^2)$$

$$= r\alpha\beta \notin M_q$$

since r is odd. Consequently by 7.8

$$(\mathbb{F}_{q^r}^2, Q) \cong (\mathbb{F}_q^{2r}, Q_-)$$

(c) Finally suppose r is even and $\alpha\beta \notin M_{q^r}$. We may suppose that $\beta = 1$. Since $tr(x^2 + x) = (tr(x))^2 + tr(x)$, it follows that $tr\colon \mathbb{F}_{q^r} \to \mathbb{F}_q$ maps M_{q^r} to M_q and hence induces an isomorphism

$$tr\colon \mathbb{F}_{q^r}/M_{q^r} = \mathbb{Z}/2 \to \mathbb{Z}/2 = \mathbb{F}_q/M_q$$

Since $\alpha \notin M_{q^r}$, it follows that $tr(\alpha) \notin M_q$. In particular this means that $tr(\alpha) \neq 0$.

This means that $B(\sqrt{\alpha},\sqrt{\alpha}) = tr(\sqrt{\alpha}\sqrt{\alpha}) \neq 0$. Since r is even, it follows by 7.2 that we can take g_1 in the orthonormal basis $\{g_i\}_{i=1}^{r}$ of $(\mathbb{F}_{q^r}, B)$ to be

$$g_1 = d\sqrt{\alpha} \qquad d \in \mathbb{F}_q^*$$

where $d = (tr(\alpha))^{-1/2}$. Consequently

$$A(Q) = \Sigma_{i=1}^{r} tr(\alpha g_i^2) tr(g_i^2) = \Sigma_{i=1}^{r} tr(\frac{1}{d^2} g_1^2 g_i^2)$$

$$= \frac{1}{d^2} \Sigma_{i=1}^{r} (tr(g_1 g_i))^2 = \frac{1}{d^2} = tr(\alpha) \notin M_q$$

Consequently by 7.8

$$(\mathbb{F}^2_{q^r}, Q) \cong (\mathbb{F}_q^{2r}, Q_-)$$

Summarizing cases (a)-(c) we have

$$(\mathbb{F}^2_{q^r}, Q) \cong \begin{cases} (\mathbb{F}_q^{2r}, Q_+) & \text{if } \alpha\beta \in M_{q^r} \\ (\mathbb{F}_q^{2r}, Q_-) & \text{if } \alpha\beta \notin M_{q^r} \end{cases}$$

Remark 7.10. When we attempt to assemble the groups $0_{\pm}(n, \mathbb{F}_q)$ into permutative categories, difficulties arise. For instance the most obvious condidate to consider is the category $\tilde{\mathcal{O}}(\mathbb{F}_q)$ whose objects are all nondegenerate quadratic spaces over $\mathbb{F}_q$ and whose morphisms are isometric isomorphisms. This however fails to be permutative category (in fact it is not even symmetric monoidal) since the direct sumof two nondegenerate odd dimensional quadratic spaces is degenerate.

There are two ways to get around this difficulty. One way is to pass to appropriate subcategories of $\tilde{\mathcal{O}}(\mathbb{F}_q)$. There are two obvious choices

(i) $\tilde{\mathcal{O}}^{ev}(\mathbb{F}_q)$ - the full subcategory of $\tilde{\mathcal{O}}(\mathbb{F}_q)$ whose objects are even dimensional quadratic spaces. Direct sum is defined in this subcategory since the direct sum of two nondegenerate even dimensional quadratic spaces is again nondegenerate. It is not difficult to see that $\tilde{\mathcal{O}}^{ev}(\mathbb{F}_q)$ is a symmetric monoidal category. An equivalent permutative category $\bar{\mathcal{O}}^{ev}(\mathbb{F}_q)$ may be constructed as in 4.7.

(ii) $\tilde{\mathcal{O}}^{ev}_{+}(\mathbb{F}_q)$ - the full subcategory of $\tilde{\mathcal{O}}^{ev}(\mathbb{F}_q)$ whose objects are isomorphic to $(\mathbb{F}_q^{2n}, Q_+)$ for some n. This category is also symmetric monoidal. An equivalent permutative category $\mathcal{O}^{ev}(\mathbb{F}_q)$ can be constructed as follows: The objects of $\mathcal{O}^{ev}(\mathbb{F}_q)$ are the even nonnegative integers. The morphisms are

$$\hom(2m,2n) = \begin{cases} \emptyset & \text{if } 2m \neq 2n \\ O_+(2n,\mathbb{F}_q) & \text{if } 2m = 2n \end{cases}$$

The bifunctor $\oplus$: $\mathcal{O}^{ev}(\mathbb{F}_q) \times \mathcal{O}^{ev}(\mathbb{F}_q) \to \mathcal{O}^{ev}(\mathbb{F}_q)$ and the commutativity transformation c are defined as in 2.6.

An alternative procedure is to blow up the category $\tilde{\mathcal{O}}(\mathbb{F}_q)$ into a bigger category $\hat{\mathcal{O}}(\mathbb{F}_q)$.

<u>Definition</u> 7.11. The symmetric monoidal category $(\hat{\mathcal{O}}(\mathbb{F}_q), \oplus, 0, a, \ell, r, c)$ is the category with objects all quadratic spaces over $\mathbb{F}_q$ (degenerate as well as nondegenerate) and with morphisms all isometric epimorphisms between such spaces. The direct sum of two quadratic spaces is defined as in 7.7. The direct sum of two morphisms is defined in the usual way. The 0-module is equipped with the trivial form. The natural isomorphisms a, ℓ, r, c are defined in the usual way (cf. 2.4 or 4.3).

We construct an equivalent permutative category, eg. by the Isbell construction (cf. 2.5) and by abuse of notation we also denote it $\hat{\mathcal{O}}(\mathbb{F}_q)$.

Proposition 7.12. The inclusion of categories $\tilde{\mathcal{O}}(\mathbb{F}_q) \hookrightarrow \hat{\mathcal{O}}(\mathbb{F}_q)$ induces a homotopy equivalence of classifying spaces

$$B\tilde{\mathcal{O}}(\mathbb{F}_q) \xrightarrow{\simeq} B\hat{\mathcal{O}}(\mathbb{F}_q)$$

Proof: In fact $B\tilde{\mathcal{O}}(\mathbb{F}_q)$ is a deformation retract of $B\hat{\mathcal{O}}(\mathbb{F}_q)$. The deformation retraction is supplied by the functor $R\colon \hat{\mathcal{O}}(\mathbb{F}_q) \to \tilde{\mathcal{O}}(\mathbb{F}_q)$ defined as follows:

$$R(V,Q) = (V',Q')$$

where $V' = V/\mathrm{Rad}(V,Q)$ and Q' is the naturally induced form on V'; on morphisms

$$Rf\colon R(V,Q_1) \to R(W,Q_2)$$

is the naturally induced map $f'\colon V' \to W'$ which is defined since any isometric epimorphism sends the radical of the source to the radical of the target. It is obvious that BR is the identity on $B\tilde{\mathcal{O}}(\mathbb{F}_q)$, and there is natural projection $\pi\colon (V,Q) \to R(V,Q)$ which provides a homotopy between $1_{B\hat{\mathcal{O}}(\mathbb{F}_q)}$ and BR.

Remark 7.13. It is not difficult to see that $\pi_0\hat{\mathcal{O}}(\mathbb{F}_q)$ is a commutative monoid on three generators:

δ - corresponding to the object $(\mathbb{F}_q, Q_+)$

ε - corresponding to the object $(\mathbb{F}_q^2, Q_+)$

$\bar{\varepsilon}$ - corresponding to the object $(\mathbb{F}_q^2, Q_-)$

subject to the relations

$$\varepsilon + \delta = \bar{\varepsilon} + \delta$$

$$\delta + \delta = \delta$$

$$\bar{\varepsilon} + \bar{\varepsilon} = \varepsilon + \varepsilon \quad \text{(cf. 7.8)}$$

It is thus clear that $K_0\hat{\mathcal{O}}(\mathbb{F}_q) \cong \mathbb{Z}$ and the natural map

$$\pi_0\hat{\mathcal{O}}(\mathbb{F}_q) \to K_0\hat{\mathcal{O}}(\mathbb{F}_q)$$

sends ε and $\bar{\varepsilon}$ to the same generator and sends δ to 0.

It will be important to note that Prop. 7.12 implies that the inclusion $\tilde{\mathcal{O}}(\mathbb{F}_q) \hookrightarrow \hat{\mathcal{O}}(\mathbb{F}_q)$ induces homotopy equivalences on path components

$$BO_+(2n+1,\mathbb{F}_q) \to B\hat{\mathcal{O}}(\mathbb{F}_q)_{n\varepsilon+\delta}$$

$$BO_+(2n,\mathbb{F}_q) \to B\hat{\mathcal{O}}(\mathbb{F}_q)_{n\varepsilon}$$

$$BO_-(2n,\mathbb{F}_q) \to B\hat{\mathcal{O}}(\mathbb{F}_q)_{\bar{\varepsilon}+(n-1)\varepsilon}$$

(cf. 2.16(vii) regarding notation).

<u>Remark</u> 7.14. It is evident that the same blowing up process could be applied to the various other categories we have already considered. For instance there is a symmetric monoidal category $\hat{\mathcal{S}p}(\mathbb{F}_q)$ whose objects are all symplectic spaces (degenerate as well as nondegenerate) and whose morphisms are all isometric epimorphisms

between such spaces. By a similar argument as in Prop. 7.12 we can construct a deformation retraction R: $\hat{\mathcal{S}p}(\mathbb{F}_q) \to \mathcal{S}p(\mathbb{F}_q)$ given on objects by

$$R(V,B) \to (V/\mathrm{Rad}(V,B),B')$$

where B' is the naturally induced bilinear form, and on morphisms in the obvious way. Hence the inclusion of categories $\mathcal{S}p(\mathbb{F}_q) \to \hat{\mathcal{S}p}(\mathbb{F}_q)$ induces an equivalence of classifying spaces

$$B\,\mathcal{S}p(\mathbb{F}_q) \xrightarrow{\simeq} B\,\hat{\mathcal{S}p}(\mathbb{F}_q)$$

and hence also an equivalence of infinite loop spaces

$$\Gamma B\,\mathcal{S}p(\mathbb{F}_q) \xrightarrow{\simeq} \Gamma B\,\hat{\mathcal{S}p}(\mathbb{F}_q).$$

7.15. <u>The relation of the groups</u> $O_{\pm}(n,\mathbb{F}_q)$ <u>and the groups</u> $Sp(2n,\mathbb{F}_q)$. It follows from the definition of quadratic form that to each quadratic space (V,Q) we can associate a symmetric bilinear form $B_Q\colon V \times V \to \mathbb{F}_q$ defined by

$$B_Q(x,y) = Q(x+y) - Q(x) - Q(y)$$

Since it is clear that $B_Q(x,x) = 0$, we in fact get a symplectic form. Obviously if f: (V,Q) → (W,Q') is an isometric isomorphism (or epimorphism) then so is $f\colon (V,B_Q) \to (W,B_{Q'})$. We thus get an inclusion

$$J\colon O(V,Q) \subseteq Sp(V,B_Q)$$

If we apply this construction to $(\mathbb{F}_q^{2n}, Q_{\pm})$ we obtain the standard symplectic form

$$B_{Q_{\pm}}(x,y) = A(x,y) = \Sigma_{i=1}^{n}(x_{2i-1}y_{2i} - x_{2i}y_{2i-1})$$

and a corresponding inclusion

$$J\colon O_{\pm}(2n,\mathbb{F}_q) \to Sp(2n,\mathbb{F}_q)$$

Of course if we apply this construction to an odd dimensional quadratic space we get a degenerate symplectic form.

However in the odd dimensional case we can perform another construction. Given $T \in Sp(2n,\mathbb{F}_q)$ we can associate a map $\alpha(T) \in O_{+}(2n+1,\mathbb{F}_q)$ as follows:

$$\alpha(T)(x_1,x_2,\ldots,x_{2n},x_{2n+1})$$

$$= (T(x_1,\ldots,x_{2n}), x_{2n+1} + \sqrt{Q_{+}(x_1,\ldots,x_{2n}) - Q_{+}(T(x_1,\ldots,x_{2n}))})$$

This induces an inclusion

$$\alpha\colon Sp(2n,\mathbb{F}_q) \to O_{+}(2n+1,\mathbb{F}_q)$$

This inclusion is in fact an isomorphism (cf. Dickson [16]). It is obvious that the composite

$$Sp(2n,\mathbb{F}_q) \xrightarrow{\alpha} O_+(2n+1,\mathbb{F}_q) \xrightarrow{J} Sp(\mathbb{F}_q^{2n+1},B_{Q_+}) \xrightarrow{R} Sp(2n,\mathbb{F}_q)$$

is the identity where R is the map induced by the retraction $R: \hat{\mathcal{S}p}(\mathbb{F}_q) \to \mathcal{S}p(\mathbb{F}_q)$ of Remark 7.14. Hence the composite

$$O_+(2n+1,\mathbb{F}_q) \xrightarrow{J} Sp(\mathbb{F}_q^{2n+1},B_{Q_+}) \xrightarrow{R} Sp(2n,\mathbb{F}_q)$$

is an isomorphism.

We now recast the remarks of 7.15 in a categorical framework. Obviously there is a symmetric monoidal functor

$$J: \hat{\mathcal{O}}(\mathbb{F}_q) \to \hat{\mathcal{S}p}(\mathbb{F}_q)$$

given on objects by

$$J(V,Q) = (V,B_Q)$$

and on morphisms in the obvious way. This induces an infinite loop map

$$J: \Gamma B\hat{\mathcal{O}}(\mathbb{F}_q) \to \Gamma B\hat{\mathcal{S}p}(\mathbb{F}_q)$$

We now have the following result.

<u>Theorem</u> 7.16. The composite map

$$\Gamma B\hat{\mathcal{O}}(\mathbb{F}_q) \xrightarrow{J} \Gamma B\hat{\mathcal{S}p}(\mathbb{F}_q) \xrightarrow{R} \Gamma B\mathcal{S}p(\mathbb{F}_q)$$

is an equivalence of infinite loop spaces.

Proof: We observe that the composite map

$$R \circ J\colon \pi_0 \hat{\mathcal{O}}(\mathbb{F}_q) \to \pi_0 \mathcal{S}p(\mathbb{F}_q) = 2\mathbb{N}$$

maps the generators of Remark 7.13 as follows.

$$\delta \to 0$$

$$\varepsilon \to 2$$

$$\bar{\varepsilon} \to 2$$

It follows that

$$R \circ J\colon K_0 \hat{\mathcal{O}}(\mathbb{F}_q) \to K_0 \mathcal{S}p(\mathbb{F}_q)$$

is an isomorphism.

Thus it remains to show that

$$R \circ J\colon \Gamma_0 B \hat{\mathcal{O}}(\mathbb{F}_q) \to \Gamma_0 B \mathcal{S}p(\mathbb{F}_q)$$

is an equivalence. However this follows trivially from 2.17(ii), 7.15 and the commutative diagram

$$\begin{array}{ccc} H_*(\Gamma_0 B\,\hat{\mathcal{O}}(\mathbb{F}_q)) & \xrightarrow{\quad R\circ J \quad} & H_*(\Gamma_0 B\,\mathcal{S}p(\mathbb{F}_q)) \\ \| \wr & & \| \wr \\ H_*(\varinjlim_{\alpha\in\pi_0\hat{\mathcal{O}}(\mathbb{F}_q)} B\,\hat{\mathcal{O}}(\mathbb{F}_q)_\alpha) \cong H_*(\varinjlim BO_+(2n+1,\mathbb{F}_q)) & \xrightarrow[\cong]{R\circ J} & H_*(\varinjlim BSp(2n,\mathbb{F}_q)). \end{array}$$

We turn next to identifying the infinite loop spaces associated with the permutative subcategories $\bar{\mathcal{O}}^{ev}(\mathbb{F}_q)$ and $\mathcal{O}^{ev}(\mathbb{F}_q)$ defined in 7.10. An argument similar to that of 4.11 establishes

<u>Proposition</u> 7.17. The infinite loop space $\Gamma B\,\bar{\mathcal{O}}^{ev}(\mathbb{F}_q)$ is homotopy equivalent as a space to

$$\Gamma_0 B\,\mathcal{O}^{ev}(\mathbb{F}_q) \times \mathbb{Z}/2 \times \mathbb{Z} \simeq \Gamma B\,\mathcal{O}^{ev}(\mathbb{F}_q) \times \mathbb{Z}/2$$

The inclusion $\mathcal{O}^{ev}(\mathbb{F}_q) \to \bar{\mathcal{O}}^{ev}(\mathbb{F}_q)$ induces an infinite loop map

$$\Gamma_0 B\,\mathcal{O}^{ev}(\mathbb{F}_q) \to \Gamma_0 B\,\bar{\mathcal{O}}^{ev}(\mathbb{F}_q)$$

which is an equivalence.

<u>Remark</u> 7.18. In the same notation as in 7.13, we see that $\pi_0\,\bar{\mathcal{O}}^{ev}(\mathbb{F}_q)$ is a commutative monoid on two generators $\varepsilon, \bar{\varepsilon}$ subject to the single relation $\bar{\varepsilon} + \bar{\varepsilon} = \varepsilon + \varepsilon$. It follows that $\varepsilon \to 2$, $\bar{\varepsilon} \to \bar{2}$ defines an isomorphism between $\pi_0\,\bar{\mathcal{O}}^{ev}(\mathbb{F}_q)$ and the monoid $\tilde{\mathbb{N}}^{ev}$ of 4.7. We shall also adapt the notation of 4.7 denoting

$$O_+(2n,\mathbb{F}_q) = O(2n,\mathbb{F}_q)$$

$$O_-(2n,\mathbb{F}_q) = O(\overline{2n},\mathbb{F}_q)$$

It follows that $\pi_0 \Gamma B \bar{\mathcal{O}}^{ev}(\mathbb{F}_q) = K_0 \bar{\mathcal{O}}^{ev}(\mathbb{F}_q) = \mathbb{Z} \oplus \mathbb{Z}/2$ on generators $\varepsilon = [2]$ and $\varepsilon - \bar{\varepsilon} = [2] - [\bar{2}]$. As in 4.12 we shall denote

$$\overset{+}{\Gamma_0} B \bar{\mathcal{O}}^{ev}(\mathbb{F}_q) = \Gamma_0 B \bar{\mathcal{O}}^{ev}(\mathbb{F}_q) \cup \Gamma_{\varepsilon-\bar{\varepsilon}} B \bar{\mathcal{O}}^{ev}(\mathbb{F}_q)$$

Thus $\Gamma B \bar{\mathcal{O}}^{ev}(\mathbb{F}_q)$ and $\Gamma B \hat{\mathcal{O}}(\mathbb{F}_q)$ have different zeroth homotopy groups. We shall now show that they also have different fundamental groups.

For on the one hand we have

$$\pi_1 \Gamma B \hat{\mathcal{O}}(\mathbb{F}_q) \cong \pi_1(\Gamma B \mathcal{S}p(\mathbb{F}_q)) = \pi_1(\Gamma_0 B \mathcal{S}p(\mathbb{F}_q)) = 0$$

(cf. Prop. 6.5). On the other hand

$$\pi_1 \Gamma B \bar{\mathcal{O}}^{ev}(\mathbb{F}_q) \cong \pi_1 \Gamma_0 B \mathcal{O}^{ev}(\mathbb{F}_q) \cong H_1(\Gamma_0 B \mathcal{O}^{ev}(\mathbb{F}_q)) \cong \varinjlim_n H_1(BO_+(2n,\mathbb{F}_q))$$

$$\cong \varinjlim_n \frac{O_+(2n,\mathbb{F}_q)}{[O_+(2n,\mathbb{F}_q),O_+(2n,\mathbb{F}_q)]}$$

As we shall now show, this group is nonzero. To see this we shall construct a natural homomorphism

$$O_+(2n,\mathbb{F}_q) \to \mathbb{Z}/2$$

One might be tempted to take this to be the determinant map.

However a moments reflection shows that the determinant of any orthogonal matrix is +1. Thus the determinant map is trivial and useless for this purpose. However we shall now see, there is another invariant which takes the place of determinant for orthogonal groups over a field of characteristic 2.

7.19. <u>The Dickson Invariant</u>. Let (V,Q) be an even dimensional non-degenerate quadratic space over $\mathbb{F}_q$. Then we can consider the corresponding Clifford algebra C(V,Q) defined in the usual way as the universal algebra over $\mathbb{F}_q$ generated by V subject to the relation $v^2 = Q(v)$. It is $\mathbb{Z}/2$-graded: $C(V,Q) = C_0 \oplus C_1$. Moreover the orthogonal group O(V,Q) acts on C(V,Q) and preserves this grading.

It can be shown that there is a unique 2-dimensional subalgebra A of C(V,Q) which is invariant under this action of O(V,Q). In fact A is precisely the center of the subalgebra C_0. The structure of this subalgebra A faithfully reflects the structure of the quadratic space (V,Q): If $(V,Q) \cong (\mathbb{F}_q^{2n}, Q_+)$ then $A \cong \mathbb{F}_q \oplus \mathbb{F}_q$ as an algebra. If $(V,Q) \cong (\mathbb{F}_q^{2n}, Q_-)$ then $A \cong \mathbb{F}_{q^2}$ as an algebra. In either case the automorphism group of the algebra A is

$$\mathrm{Aut}(A) \cong \mathbb{Z}/2$$

The action of O(V,A) on C(V,Q) restricts to an action of O(V,Q) on A. This gives a homomorphism

$$d\colon O(V,Q) \to \mathrm{Aut}(A) \cong \mathbb{Z}/2$$

This homomorphism is called the <u>Dickson invariant</u>.

It is not difficult to see that if $M \in O(V,Q)$ and $N \in O(W,Q')$ then (in additive notation for $\mathbb{Z}/2$)

$$d(M \oplus N) = d(M) + d(N)$$

In particular, the Dickson invariant is compatible with the standard inclusions of orthogonal groups, i.e. if $m < n$ then the following diagram commutes.

$$\begin{array}{ccc} O_{\pm}(2m,\mathbb{F}_q) & \searrow^{d} & \\ \downarrow & & \mathbb{Z}/2 \\ O_{\pm}(2n,\mathbb{F}_q) & \nearrow_{d} & \end{array}$$

Thus there is an induced homomorphism

$$d\colon \lim_{n\to\infty} \frac{O_{+}(2n,\mathbb{F}_q)}{[O_{+}(2n,\mathbb{F}_q),O_{+}(2n,\mathbb{F}_q)]} \to \mathbb{Z}/2$$

which can be shown to be an isomorphism.

It is also clear that we can construct a permutative functor

$$d\colon \bar{\mathcal{O}}^{ev}(\mathbb{F}_q) \to \mathbb{Z}/2$$

along the same lines as in 2.13.

We can also give an explicit formula for the Dickson invariant: if $\{e_i\}_{i=1}^{2n}$ is the standard basis for $(\mathbb{F}_q^{2n},Q_{\pm})$ and

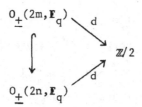

$M \in O_{\pm}(2n, \mathbf{F}_q)$ has matrix form $(a_{ij})_{i,j=1}^{2n}$ with respect to this basis, then (in additive notation)

$$d(M) = \Sigma_{i,j=1}^{n}(\alpha_j w_{ij} y_{ij} + \beta_j x_{ij} z_{ij} + x_{ij} y_{ij})$$

where $\alpha_j = Q_{\pm}(e_{2j-1})$, $\beta_j = Q_{\pm}(e_{2j})$, $w_{ij} = a_{2i-1,2j-1}$, $x_{ij} = a_{2i-1,2j}$, $y_{ij} = a_{2i,2j-1}$, $z_{ij} = a_{2i,2j}$. For more information the reader is referred to Dieudonné [15, p. 65].

We denote by $DO_{\pm}(2n, \mathbf{F}_q)$ the subgroup of $O_{\pm}(2n, \mathbf{F}_q)$ consisting of elements with Dickson invariant 0. As indicated above $DO_{\pm}(2n, \mathbf{F}_q)$ is the commutator subgroup of $O_{\pm}(2n, \mathbf{F}_q)$. We can similarly define $\mathcal{D}\mathcal{O}^{ev}(\mathbf{F}_q)$ to be the subcategory of $\mathcal{O}^{ev}(\mathbf{F}_q)$ with the same objects as $\mathcal{O}^{ev}(\mathbf{F}_q)$ but whose morphisms are $\coprod_{n=0}^{\infty} DO_{+}(2n, \mathbf{F}_q)$. It is easily seen that $\mathcal{D}\mathcal{O}^{ev}(\mathbf{F}_q)$ is a permutative subcategory of $\mathcal{O}^{ev}(\mathbf{F}_q)$.

Remark 7.20. It is interesting to note that the same construction using Clifford algebras can be used in other circumstances to obtain familiar invariants.

(a) If (V,Q) is an odd dimensional nondegenerate quadratic space over a field $\mathbf{F}_q$ of characteristic 2, then there is also a unique 2-dimensional subalgebra A of the Clifford algebra $C(V,Q)$ invariant under the action of $O(V,Q)$. However

$$\mathrm{Aut}(A) = 0$$

so the corresponding homomorphism

$$O(V,Q) \to \mathrm{Aut}(A)$$

is trivial.

(b) If (V,Q) is a nondegenerate quadratic space over a field $\mathbb{F}$ of characteristic $\neq 2$, then there is also a unique 2-dimensional subalgebra A of $C(V,Q)$ invariant under the action of $O(V,Q)$. Moreover

$$\mathrm{Aut}(A) = \mathbb{Z}/2$$

and the corresponding homomorphism

$$O(V,Q) \to \mathrm{Aut}(A) = \mathbb{Z}/2$$

is the determinant map.

(c) If V is an n-dimensional vector space over a field $\mathbb{F}$ of arbitrary characteristic, we may endow it with the zero quadratic form. Then clearly

$$O(V,0) = GL(V),$$

the general linear group of V. It is equally clear that

$$C(V,0) = \Lambda^*(V)$$

the exterior algebra over V. Again there is a unique 2-dimensional subalgebra A of $C(V,0)$ invariant under the action of $O(V,0)$, namely

$$A = \Lambda^0(V) \oplus \Lambda^n(V)$$

Moreover

$$\mathrm{Aut}(A) = \mathbb{F}^*$$

the multiplicative group of units in $\mathbb{F}$. The corresponding homomorphism

$$GL(V) = O(V,0) \to \mathrm{Aut}(A) = \mathbb{F}^*$$

is the standard invariant description of the determinant map.

The following results now complete our description of the infinite loop spaces associated with the permutative categories $\bar{\mathcal{O}}^{ev}(\mathbb{F}_q)$, $\mathcal{O}^{ev}(\mathbb{F}_q)$ and $\mathcal{D}\mathcal{O}^{ev}(\mathbb{F}_q)$.

Theorem 7.21. There are fibrations of infinite loop spaces

$$\Gamma_0 B\mathcal{D}\mathcal{O}^{ev}(\mathbb{F}_q) \to \Gamma_0 B\mathcal{O}^{ev}(\mathbb{F}_q) \xrightarrow{d} B\mathbb{Z}/2$$

$$\Gamma_0 B\mathcal{D}\mathcal{O}^{ev}(\mathbb{F}_q) \to \Gamma B\bar{\mathcal{O}}^{ev}(\mathbb{F}_q) \xrightarrow{(d,\delta)} B\mathbb{Z}/2 \times (\mathbb{Z} \oplus \mathbb{Z}/2)$$

$$\Gamma_0 B\mathcal{D}\mathcal{O}^{ev}(\mathbb{F}_q) \to \Gamma_0^{\pm} B\bar{\mathcal{O}}^{ev}(\mathbb{F}_q) \xrightarrow{(d,\delta)} B\mathbb{Z}/2 \times \mathbb{Z}/2$$

where

$$\delta: \Gamma B\bar{\mathcal{O}}^{ev}(\mathbf{F}_q) \to \pi_0 \Gamma B\bar{\mathcal{O}}^{ev}(\mathbf{F}_q) = \mathbf{Z} \oplus \mathbf{Z}/2$$

is the discretization map which collapses each path component to a point.

Proof: The existence of the first fibration follows by an argument similar to that of Theorem 3.16. The second fibration is then easily derived from the first.

Theorem 7.22. The composite infinite loop map

$$\Gamma B\mathcal{D}\mathcal{O}^{ev}(\mathbf{F}_q) \to \Gamma B\mathcal{O}^{ev}(\mathbf{F}_q) \xrightarrow{J} \Gamma B\mathcal{S}p(\mathbf{F}_q)$$

is an equivalence.

The proof of Theorem 7.22 is postponed until Chapter III §3.

§8. Permutative functors associated with the classical groups.

In this section we introduce some functors between the permutative categories studied in the preceding sections.

8.1. The forgetful functors. The permutative categories of the preceeding sections have as objects vector spaces over $\mathbf{F}_q$ with some additional structure (eg. a bilinear form, or quadratic form) and have as morphisms structure preserving isomorphisms of these vector spaces. By forgetting the additional structure we get a permutative functor to $\mathcal{GL}(\mathbf{F}_q)$:

(a) $\mathcal{O}(\mathbb{F}_q) \to \mathcal{GL}(\mathbb{F}_q)$ q odd

(b) $\mathcal{Sp}(\mathbb{F}_q) \to \mathcal{GL}(\mathbb{F}_q)$

(c) $\mathcal{U}(\mathbb{F}_{q^2}) \to \mathcal{GL}(\mathbb{F}_{q^2})$

(d) $\mathcal{EO}(\mathbb{F}_q) \to \mathcal{GL}(\mathbb{F}_q)$ q even

(e) $\mathcal{O}^{ev}(\mathbb{F}_q) \to \mathcal{GL}(\mathbb{F}_q)$ q even

All of these forgetful functors may be regarded as inclusions of permutative categories. The forgetful functors (a), (d), (c) have extensions to the larger, more elaborate categories $\bar{\mathcal{O}}(\mathbb{F}_q)$, $\mathcal{OBL}(\mathbb{F}_q)$, $\bar{\mathcal{O}}^{ev}(\mathbb{F}_q)$. The targets of these functors are not strictly speaking the category $\mathcal{GL}(\mathbb{F}_q)$ but rather permutative categories equivalent to $\mathcal{GL}(\mathbb{F}_q)$. We thus get permutative functors

(a') $\bar{\mathcal{O}}(\mathbb{F}_q) \to \mathcal{GL}(\mathbb{F}_q)$ q odd

(d') $\mathcal{OBL}(\mathbb{F}_q) \to \mathcal{GL}(\mathbb{F}_q)$ q even

(e') $\bar{\mathcal{O}}^{ev}(\mathbb{F}_q) \to \mathcal{GL}(\mathbb{F}_q)$ q even

which are no longer inclusions.

It should be noted that since symplectic matrices have determinant 1 the functor (b) factors through $\mathcal{SL}^{ev}(\mathbb{F}_q)$.

On morphism sets the forgetful functors (a)-(e) restrict to the standard inclusions of matrix groups

($\bar{a}'$) $O_{\pm}(n,\mathbb{F}_q) \to GL(n,\mathbb{F}_q)$ q odd

($\bar{b}$) $Sp(2n,\mathbb{F}_q) \to GL(2n,\mathbb{F}_q)$

($\bar{c}$) $U(n,\mathbb{F}_{q^2}) \to GL(n,\mathbb{F}_{q^2})$

($\bar{d}$) $EO(n,\mathbb{F}_q) \to GL(n,\mathbb{F}_q)$ q even

($\bar{e}'$) $O_{\pm}(2n,\mathbb{F}_q) \to GL(2n,\mathbb{F}_q)$ q even

8.2. Permutative functors associated with extension of scalars.

If V is a vector space of dimension n over $\mathbb{F}_q$ and $\mathbb{F}_{q^r}$ is any finite extension field over $\mathbb{F}_q$, then

$$\mathbb{F}_{q^r} \otimes_{\mathbb{F}_q} V$$

is naturally a vector space over $\mathbb{F}_{q^r}$ of dimension n. If $B: V \times V \to \mathbb{F}_q$ is a (anti-) symmetric bilinear form on V, then $\overline{B}: (\mathbb{F}_{q^r} \otimes_{\mathbb{F}_q} V) \times (\mathbb{F}_{q^r} \otimes_{\mathbb{F}_q} V) \to \mathbb{F}_{q^r}$ defined by

$$\overline{B}(\Sigma_i \alpha_i \otimes x_i, \Sigma_j \beta_j \otimes y_j) = \Sigma_{i,j} \alpha_i \beta_j B(x_i, y_j)$$

is a (anti-) symmetric bilinear form on $\mathbb{F}_{q^r} \otimes_{\mathbb{F}_q} V$. Similarly if $Q: V \to \mathbb{F}_q$ is a quadratic form on V, then $\overline{Q}: \mathbb{F}_{q^r} \otimes_{\mathbb{F}_q} V \to \mathbb{F}_{q^r}$, defined by

$$\overline{Q}(\Sigma_i \alpha_i \otimes x_i) = \Sigma_i \alpha_i^2 Q(x_i) + \Sigma_{i,j} \alpha_i \alpha_j B_Q(x_i, x_j)$$

where $B_Q: V \times V \to \mathbb{F}_q$ is the symmetric bilinear form associated with Q (cf. 7.14), is a quadratic form on $\mathbb{F}_{q^r} \otimes_{\mathbb{F}_q} V$.

Similarly if $H: V \times V \to \mathbb{F}_{q^2}$ is a Hermitian form on V and $\mathbb{F}_{q^{2r}}$ is an odd dimensional extension of $\mathbb{F}_{q^2}$, then

$$\overline{H}: (\mathbb{F}_{q^{2r}} \otimes_{\mathbb{F}_{q^2}} V) \times (\mathbb{F}_{q^{2r}} \otimes_{\mathbb{F}_{q^2}} V) \to \mathbb{F}_{q^{2r}}$$

defined by

$$\overline{H}(\Sigma_i\alpha_i \otimes x_i, \Sigma_j\beta_j \otimes y_j) = \Sigma_{i,j}\alpha_i\overline{\beta}_j H(x_i,y_j)$$

is a Hermitian form on $\mathbb{F}_{q^{2r}} \otimes_{\mathbb{F}_{q^2}} V$. (The assumption that r be odd is required to make the conjugations on $\mathbb{F}_{q^2}$ and $\mathbb{F}_{q^{2r}}$ agree).

It is obvious that the above constructions send nondegenerate structures to nondegenerate structures. Also if

$$f\colon V \to W$$

is a structure preserving isomorphism of $\mathbb{F}_q$ vector spaces, then

$$\overline{f} = 1_{\mathbb{F}_{q^r}} \otimes f\colon \mathbb{F}_{q^r} \otimes_{\mathbb{F}_q} V \to \mathbb{F}_{q^r} \otimes_{\mathbb{F}_q} W$$

is a structure preserving isomorphism of $\mathbb{F}_{q^r}$ vector spaces.

Thus $V \mapsto \mathbb{F}_{q^r} \otimes_{\mathbb{F}_q} V$, $f \mapsto \overline{f}$ define permutative functors

(a) $\mathcal{GL}(\mathbb{F}_q) \to \mathcal{GL}(\mathbb{F}_{q^r})$

(b) $\mathcal{O}(\mathbb{F}_q) \to \mathcal{O}(\mathbb{F}_{q^r})$ q odd

$\overline{\mathcal{O}}(\mathbb{F}_q) \to \overline{\mathcal{O}}(\mathbb{F}_{q^r})$ q odd

(c) $\mathcal{Sp}(\mathbb{F}_q) \to \mathcal{Sp}(\mathbb{F}_{q^r})$

(d) $\mathcal{U}(\mathbb{F}_{q^2}) \to \mathcal{U}(\mathbb{F}_{q^{2r}})$ r odd

(e) $\mathcal{E}\mathcal{O}(\mathbb{F}_q) \to \mathcal{E}\mathcal{O}(\mathbb{F}_{q^r})$ q even

$\mathcal{O}\mathcal{B}\mathcal{L}(\mathbb{F}_q) \to \mathcal{O}\mathcal{B}\mathcal{L}(\mathbb{F}_{q^r})$ q even

(f) $\mathcal{O}^{ev}(\mathbb{F}_q) \to \mathcal{O}^{ev}(\mathbb{F}_{q^r})$ q even

$$\bar{\mathcal{O}}^{ev}(\mathbb{F}_q) \to \bar{\mathcal{O}}^{ev}(\mathbb{F}_{q^r}) \qquad q \text{ even}$$

All these functors may be regarded as inclusions of permutative categories.

On morphism sets these functors induce the standard inclusions of matrix groups

($\bar{a}$) $GL(n,\mathbb{F}_q) \to GL(n,\mathbb{F}_{q^r})$

($\bar{b}$) $O_{\pm}(n,\mathbb{F}_q) \to O_{\pm}(n,\mathbb{F}_{q^r})$ q odd, r odd

$O_{\pm}(n,\mathbb{F}_q) \to O(n,\mathbb{F}_{q^r})$ q odd, r even

($\bar{c}$) $Sp(2n,\mathbb{F}_q) \to Sp(2n,\mathbb{F}_{q^r})$

($\bar{d}$) $U(n,\mathbb{F}_{q^2}) \to U(n,\mathbb{F}_{q^{2r}})$ r odd

($\bar{e}$) $EO(n,\mathbb{F}_q) \to EO(n,\mathbb{F}_{q^r})$ q even

($\bar{f}$) $O_{\pm}(2n,\mathbb{F}_q) \to O_{\pm}(2n,\mathbb{F}_{q^r})$ q even, r odd

$O_{\pm}(2n,\mathbb{F}_q) \to O(2n,\mathbb{F}_{q^r})$ q even, r even

The dichotomy between r even and odd in case ($\bar{b}$) arises from the fact that if $\mu \in \mathbb{F}_q^*$ is a nonsquare, then μ is a nonsquare in $\mathbb{F}_{q^r}^*$ iff r is odd. Thus the functor (b) transforms the quadratic space $(\mathbb{F}_q^n, Q_-)$ into a space isomorphic to $(\mathbb{F}_{q^r}^n, Q_-)$ if r is odd and to a space isomorphic to $(\mathbb{F}_{q^r}^n, Q_+)$ if r is even. A similar comment applies to ($\bar{f}$).

8.3. <u>Forgetful functors associated with field extensions</u>.

If $\mathbb{F}_{q^r}$ is a finite extension field over $\mathbb{F}_q$ and V is an n-dimensional vector space over $\mathbb{F}_{q^r}$, then V can also be regarded as an nr-dimensional vector space over $\mathbb{F}_q$. If $B: V \times V \to \mathbb{F}_{q^r}$ is a (anti-) symmetric bilinear form on V, then $\overline{B}: V \times V \to \mathbb{F}_q$ given by

$$\overline{B}(x,y) = \text{tr}\cdot B(x,y),$$

where $\text{tr}: \mathbb{F}_{q^r} \to \mathbb{F}_q$ is the trace map $\text{tr}(\alpha) = \alpha + \alpha^q + \cdots + \alpha^{q^{r-1}}$, is also a (anti-) symmetric bilinear form on V. Similarly if $Q: V \to \mathbb{F}_{q^r}$ is a quadratic form on V, then $\overline{Q} = \text{tr}\cdot Q: V \to \mathbb{F}_q$ is a quadratic form on V regarded as a $\mathbb{F}_q$ vector space.

Also if $H: V \times V \to \mathbb{F}_{q^{2r}}$ is a Hermitian form on V where $\mathbb{F}_{q^{2r}}$ is an odd dimensional extension of $\mathbb{F}_{q^2}$, then $\text{tr}\circ H: V \times V \to \mathbb{F}_{q^2}$ is a Hermitian form on V regarded as a vector space over $\mathbb{F}_{q^2}$. (Here again $\text{tr}: \mathbb{F}_{q^{2r}} \to \mathbb{F}_{q^2}$ is the trace map $\text{tr}(\alpha) = \alpha + \alpha^{q^2} + \alpha^{q^4} + \cdots + \alpha^{q^{2(r-1)}}$. The assumption that r be odd is required to insure that the conjugations on $\mathbb{F}_{q^{2r}}$ and $\mathbb{F}_{q^2}$ agree).

It is obvious that the above constructions send nondegenerate bilinear forms to nondegenerate bilinear forms (cf. 4.6). In case $\mathbb{F}_q$ has characteristic 2, this construction sends nondegenerate even dimensional quadratic spaces to nondegenerate quadratic spaces. (However this need not be true if the quadratic space is odd dimensional). Also if $f: V \to V$ is a structure preserving isomorphism of $\mathbb{F}_{q^r}$ vector spaces, then f is also a structure preserving isomorphisms of $\mathbb{F}_q$ vector spaces. Thus we get permutative functors

(a) $\mathcal{GL}(\mathbb{F}_{q^r}) \to \mathcal{GL}(\mathbb{F}_q)$

(b) $\bar{\mathcal{O}}(\mathbb{F}_{q^r}) \to \bar{\mathcal{O}}(\mathbb{F}_q)$ q odd

(c) $\mathcal{Sp}(\mathbb{F}_{q^r}) \to \mathcal{Sp}(\mathbb{F}_q)$

(d) $\mathcal{U}(\mathbb{F}_{q^{2r}}) \to \mathcal{U}(\mathbb{F}_{q^2})$ r odd

(e) $\mathcal{EO}(\mathbb{F}_{q^r}) \to \mathcal{EO}(\mathbb{F}_q)$ q even

$\mathcal{OBL}(\mathbb{F}_{q^r}) \to \mathcal{OBL}(\mathbb{F}_q)$ q even

(f) $\bar{\mathcal{O}}^{ev}(\mathbb{F}_{q^r}) \to \bar{\mathcal{O}}^{ev}(\mathbb{F}_q)$ q even

All these functors may be regarded as inclusions of permutative categories. It should be noted that the corresponding functor $\mathcal{O}(\mathbb{F}_{q^r}) \to \mathcal{O}(\mathbb{F}_q)$ does not exist since the result of applying this construction to $(\mathbb{F}^n_{q^r}, Q_+)$ may yield a quadratic space isomorphic to $(\mathbb{F}^{nr}_q, Q_-)$.

The forgetful functors (a)-(f) restrict on morphism sets to group monomorphisms

($\bar{a}$) $GL(n,\mathbb{F}_{q^r}) \to GL(nr,\mathbb{F}_q)$

($\bar{b}$) $O_\pm(n,\mathbb{F}_{q^r}) \to O_\pm(nr,\mathbb{F}_q)$ q odd, r odd

$O(n,\mathbb{F}_{q^r}) \to O(n\bar{r},\mathbb{F}_q)$

$O(\bar{n},\mathbb{F}_{q^r}) \to O(n\bar{r} + \gamma, \mathbb{F}_q)$

(the two lines above: q odd, r even)

($\bar{c}$) $Sp(2n,\mathbb{F}_{q^r}) \to Sp(2nr,\mathbb{F}_q)$

($\bar{d}$) $U(n,\mathbb{F}_{q^{2r}}) \to U(nr,\mathbb{F}_{q^2})$ r odd

($\bar{e}$) $EO(n,\mathbb{F}_{q^r}) \to EO(nr,\mathbb{F}_q)$ q even

$(\bar{f})$ $O_{\pm}(2n, \mathbb{F}_{q^r}) \to O_{\pm}(nr, \mathbb{F}_q)$ q even

The fact that functors (b) and (f) restrict as shown follows from the calculations in Examples 4.6 and 7.9 respectively.

8.4. <u>The hyperbolic functors</u>.

There are several natural constructions for converting vector spaces into vector spaces with structure. These constructions define permutative functors

(a) $\mathcal{GL}(\mathbb{F}_q) \to \bar{\mathcal{O}}(\mathbb{F}_q)$ q odd

(b) $\mathcal{GL}(\mathbb{F}_q) \to \mathcal{Sp}(\mathbb{F}_q)$

(c) $\mathcal{GL}^{ev}(\mathbb{F}_q) \to \mathcal{Sp}(\mathbb{F}_q)$

(d) $\mathcal{GL}(\mathbb{F}_{q^2}) \to \mathcal{U}(\mathbb{F}_{q^2})$

(e) $\mathcal{GL}(\mathbb{F}_q) \to \mathcal{O}^{ev}(\mathbb{F}_q)$ q even

which can be regarded as inclusions of categories. It should be noted that (c) is <u>not</u> the restriction of (b) but rather a quite different functor.

The constructions proceed as follows: Let

$$E: \mathbb{F}_q^n \times \mathbb{F}_q^n \to \mathbb{F}_q$$

$$A: \mathbb{F}_q^{2n} \times \mathbb{F}_q^{2n} \to \mathbb{F}_q$$

$$H: \mathbb{F}_{q^2}^n \times \mathbb{F}_{q^2}^n \to \mathbb{F}_q$$

denote the standard symmetric, antisymmetric, and Hermitian forms:

$$E(x,y) = \Sigma_{i=1}^{n} x_i y_i$$

$$A(x,y) = \Sigma_{i=1}^{n} (x_{2i} y_{2i-1} - x_{2i-1} y_{2i})$$

$$H(x,y) = \Sigma_{i=1} x_i \bar{y}_i$$

Let $Q_h\colon \mathbb{F}_q^n \oplus \mathbb{F}_q^n \to \mathbb{F}_q$ denote the quadratic form

$$Q_h((x,y)) = E(x,y)$$

Let $A_h\colon (\mathbb{F}_q^n \oplus \mathbb{F}_q^n) \times (\mathbb{F}_q^n \oplus \mathbb{F}_q^n) \to \mathbb{F}_q$ denote the symplectic form

$$A_h((x,y),(z,w)) = E(x,w) - E(y,z).$$

Let $A_h'\colon (\mathbb{F}_q^{2n} \oplus \mathbb{F}_q^{2n}) \times (\mathbb{F}_q^{2n} \oplus \mathbb{F}_q^{2n}) \to \mathbb{F}_q$ denote the symplectic form

$$A_h'((x,y),(z,w)) = A(x,w) + A(y,z)$$

Let $H_h\colon (\mathbb{F}_{q^2}^n \oplus \mathbb{F}_{q^2}^n) \times (\mathbb{F}_{q^2}^n \oplus \mathbb{F}_{q^2}^n) \to \mathbb{F}_{q^2}$ be the Hermitian form

$$H_h((x,y),(z,w)) = H(x,w) + H(y,z)$$

Given a linear isomorphism $f\colon \mathbb{F}_q^n \to \mathbb{F}_q^n$ (resp. $f\colon \mathbb{F}_q^{2n} \to \mathbb{F}_q^{2n}$, resp. $f\colon \mathbb{F}_{q^2}^n \to \mathbb{F}_{q^2}^n$) we denote by f^* the conjugate of f with respect to E (resp. A, resp H). In other words f^* is defined by

$$E(f^*x,y) = E(x,fy)$$

$$A(f^*x,y) = A(x,fy)$$

$$H(f^*x,y) = H(x,fy)$$

Then

$$f \oplus (f^{-1})^*: \mathbb{F}_q^n \oplus \mathbb{F}_q^n \to \mathbb{F}_q^n \oplus \mathbb{F}_q^n$$

$$f \oplus (f^{-1})^*: \mathbb{F}_q^{2n} \oplus \mathbb{F}_q^{2n} \to \mathbb{F}_q^{2n} \oplus \mathbb{F}_q^{2n}$$

$$f \oplus (f^{-1})^*: \mathbb{F}_{q^2}^n \oplus \mathbb{F}_{q^2}^n \to \mathbb{F}_{q^2}^n \oplus \mathbb{F}_{q^2}^n$$

preserves Q_h, A_h (resp. A_h', resp. H_h). Thus

$$\mathbb{F}_q^n \mapsto (\mathbb{F}_q^n \oplus \mathbb{F}_q^n, Q_h) \qquad f \mapsto f \oplus (f^{-1})^*$$

$$\mathbb{F}_q^n \mapsto (\mathbb{F}_q^n \oplus \mathbb{F}_q^n, A_h) \qquad f \mapsto f \oplus (f^{-1})^*$$

$$\mathbb{F}_q^{2n} \mapsto (\mathbb{F}_q^{2n} \oplus \mathbb{F}_q^{2n}, A_h') \qquad f \mapsto f \oplus (f^{-1})^*$$

$$\mathbb{F}_{q^2}^n \mapsto (\mathbb{F}_{q^2}^n \oplus \mathbb{F}_{q^2}^n, H_h) \qquad f \mapsto f \oplus (f^{-1})^*$$

$$\mathbb{F}_q^n \mapsto (\mathbb{F}_q^n \oplus \mathbb{F}_q^n, Q_h) \qquad f \mapsto f \oplus (f^{-1})^*$$

define the permutative functors (a)-(e). (Strictly speaking we should first identify $(\mathbb{F}_q^n \oplus \mathbb{F}_q^n, Q_h)$, $(\mathbb{F}_q^n \oplus \mathbb{F}_q^n, A_h)$, $(\mathbb{F}_q^{2n} \oplus \mathbb{F}_q^{2n}, A_h')$ $(\mathbb{F}_{q^2}^n \oplus \mathbb{F}_{q^2}^n, H_h)$ with $\mathbb{F}_q^{2n}, \mathbb{F}_q^{2n}, \mathbb{F}_q^{4n}, \mathbb{F}_q^{2n}$ equipped with one of the standard quadratic, symplectic, Hermitian forms by using appropriate isomorphisms).

The hyperbolic functors restrict on morphism sets to group

monomorphisms.

($\bar{a}$) $GL(n,\mathbb{F}_q) \to O(2n,\mathbb{F}_q)$ if $q \equiv 1 \pmod 4$

$GL(n,\mathbb{F}_q) \to O(n\bar{2},\mathbb{F}_q)$ if $q \equiv -1 \pmod 4$

($\bar{b}$) $GL(n,\mathbb{F}_q) \to Sp(2n,\mathbb{F}_q)$

($\bar{c}$) $GL(2n,\mathbb{F}_q) \to Sp(4n,\mathbb{F}_q)$

($\bar{d}$) $GL(n,\mathbb{F}_{q^2}) \to U(2n,\mathbb{F}_{q^2})$

($\bar{e}$) $GL(n,\mathbb{F}_q) \to O(2n,\mathbb{F}_q)$ q even

The functor (a) restr' .s as stated, because it is easily seen that the discriminant of the quadratic space $(\mathbb{F}_q^n \oplus \mathbb{F}_q^n, Q_h)$ is $(-1)^n \in \mathbb{F}_q^*/(\mathbb{F}_q^*)^2$ (cf. 4.5), and -1 as a square in $\mathbb{F}_q^*$ q odd iff $q \equiv 1 \pmod 4$.

8.5. Permutative functors associated with the unitary groups

Finally we shall consider permutative functors associated with the unitary groups. These fall into two classes

(a) $\mathcal{S}p(\mathbb{F}_q) \to \mathcal{U}(\mathbb{F}_{q^2})$

(b) $\bar{\mathcal{O}}(\mathbb{F}_q) \to \mathcal{U}(\mathbb{F}_{q^2})$ q odd

(c) $\mathcal{E}\,\mathcal{O}(\mathbb{F}_q) \to \mathcal{U}(\mathbb{F}_{q^2})$ q even

and

(d) $\mathcal{U}(\mathbb{F}_{q^2}) \to \mathcal{S}p(\mathbb{F}_q)$

(e) $\mathcal{U}(\mathbb{F}_{q^2}) \to \bar{\mathcal{O}}(\mathbb{F}_q)$ q odd

(f) $\mathcal{U}(\mathbb{F}_{q^2}) \to \bar{\mathcal{O}}^{ev}(\mathbb{F}_q)$ q even

The functors (a),(b),(c) are defined similarly as those of 8.2. If V is a vector space of dimension n over $\mathbb{F}_q$, then

$$\mathbb{F}_{q^2} \otimes_{\mathbb{F}_q} V$$

is a vector space of dimension n over $\mathbb{F}_{q^2}$. If

$$A: V \times V \to \mathbb{F}_q$$

is a nondegenerate symplectic form on V, then

$$\bar{H}: (\mathbb{F}_{q^2} \otimes_{\mathbb{F}_q} V) \times (\mathbb{F}_{q^2} \otimes_{\mathbb{F}_q} V) \to \mathbb{F}_{q^2}$$

defined by

$$\bar{H}(\Sigma_i \alpha_i \otimes x_i, E_j \beta_j \otimes y_j) = \Sigma_{i,j} \alpha_i \bar{\beta}_j \rho A(x_i, y_j)$$

where $\rho \in \mathbb{F}_{q^2}$ is such that $\bar{\rho} = -\rho$, is a nondegenerate, Hermitian form on $\mathbb{F}_{q^2} \otimes_{\mathbb{F}_q} V$. It is obvious that if $f: V \to V$ is in $Sp(V,A)$ then $1 \otimes f: \mathbb{F}_{q^2} \otimes_{\mathbb{F}_q} V \to \mathbb{F}_{q^2} \otimes_{\mathbb{F}_q} V$ is in $U(\mathbb{F}_{q^2} \otimes_{\mathbb{F}_q} V, \bar{H})$. Then $(V,A) \mapsto (\mathbb{F}_{q^2} \otimes_{\mathbb{F}_q} V, \bar{H})$, $f \mapsto 1 \otimes f$ defines the permutative functor (a).

Similarly if $B: V \times V \to \mathbb{F}_q$ is a nondegenerate symmetric bilinear form, then

$$\overline{H}: (\mathbb{F}_{q^2} \otimes_{\mathbb{F}_q} V) \times (\mathbb{F}_{q^2} \otimes_{\mathbb{F}_q} V) \to \mathbb{F}_{q^2}$$

defined by

$$\overline{H}(\Sigma_i \alpha_i \otimes x_i, \Sigma_j \beta_j \otimes y_j) = \Sigma_{i,j} \alpha_i \overline{\beta}_j B(x_i, y_j)$$

is a nondegenerate Hermitian form on $\mathbb{F}_{q^2} \otimes_{\mathbb{F}_q} V$. Clearly if $f: V \to V$ preserves $B(,)$ then $1 \otimes f: \mathbb{F}_{q^2} \otimes_{\mathbb{F}_q} V \to \mathbb{F}_{q^2} \otimes_{\mathbb{F}_q} V$ preserves $\overline{H}(,)$. Then $(V,B) \mapsto (\mathbb{F}_{q^2} \otimes_{\mathbb{F}_q} V, \overline{H})$, $f \mapsto 1 \otimes f$ defines the permutative functors (b) and (c).

The permutative functors (d), (e), (f) are defined in a manner similar to those of 8.3. If V is an n-dimensional vector space over $\mathbb{F}_{q^2}$ then V can be regarded as a 2n-dimensional vector space over $\mathbb{F}_q$. If $H: V \times V \to \mathbb{F}_{q^2}$ is a nondegenerate Hermitian form on V, then

$$A_H: V \times V \to \mathbb{F}_q$$

defined by

$$A_H(x,y) = \rho[H(x,y) - H(y,x)]$$

where $\rho \in \mathbb{F}_{q^2}$ is such that $\overline{\rho} = -\rho$, is a nondegenerate symplectic form on V. Similarly

$$Q_H : V \to \mathbb{F}_q$$

defined by

$$Q_H(x) = H(x,x)$$

is a nondegenerate quadratic form on V. If $f \in U(V,H)$, clearly $f \in Sp(V,A_H)$ and $f \in O(V,Q_H)$. Then

$$(V,H) \mapsto (V,A_H) \qquad f \mapsto f$$

$$(V,H) \mapsto (V,Q_H) \qquad f \mapsto f$$

define permutative functors (d), (e), (f).

Clearly all the functors (a)-(f) are faithful and all except (b) may be regarded as inclusions of categories.

It should again be noted that since symplectic matrices have have determinant 1 the functor (a) factors through $\mathcal{SU}^{ev}(\mathbb{F}_{q^2})$

(a') $\mathcal{Sp}(\mathbb{F}_q) \to \mathcal{SU}^{ev}(\mathbb{F}_{q^2})$

When restricted to morphism sets, the functors (a)-(f) give group monomorphisms

($\bar{a}'$) $Sp(2n,\mathbb{F}_q) \to SU(2n,\mathbb{F}_{q^2})$

$(\bar{b})$ $O_{\pm}(n,\mathbb{F}_q) \to U(n,\mathbb{F}_{q^2})$ q odd

$(\bar{c})$ $EO(n,\mathbb{F}_q) \to U(n,\mathbb{F}_{q^2})$ q even

$(\bar{d})$ $U(n,\mathbb{F}_{q^2}) \to Sp(2n,\mathbb{F}_q)$

$(\bar{e})$ $U(n,\mathbb{F}_{q^2}) \to O(2n,\mathbb{F}_q)$ if $q \equiv -1 \pmod 4$

$U(n,\mathbb{F}_{q^2}) \to O(n\bar{2},\mathbb{F}_q)$ if $q \equiv 1 \pmod 4$

$(\bar{f})$ $U(n,\mathbb{F}_{q^2}) \to O(n\bar{2},\mathbb{F}_q)$ q even

The fact that (e) and (f) restrict as stated follows by comparing the orders of $U(1,\mathbb{F}_{q^2})$ and $O_{\pm}(2,\mathbb{F}_q)$.

In the case n = 1 in $(\bar{a}')$ above we see from the fact that

$$Sp(2,\mathbb{F}_q) = (q^2-1)q = SU(2,\mathbb{F}_{q^2}) \text{ that}$$

$$(\bar{a}^*) \quad Sp(2,\mathbb{F}_q) \xrightarrow{\cong} SU(2,\mathbb{F}_{q^2})$$

8.6. <u>Relations between permutative functors</u>

We conclude this section by looking at relations between some of the permutative functors we have introduced here. We first note that if $\mathcal{Q}$ is a symmetric monoidal category then the bifunctor

$$\square: \mathcal{Q} \times \mathcal{Q} \to \mathcal{Q}$$

induce functors

$$\square: \mathcal{Q} \times \mathcal{Q} \times \ldots \times \mathcal{Q} \to \mathcal{Q}$$

which are symmetric monoidal functors. Hence so is the composite functor

$$N\colon \mathcal{A} \xrightarrow{\Delta} \underbrace{\mathcal{A} \times \mathcal{A} \times \cdots \times \mathcal{A}}_{N \text{ times}} \xrightarrow{\square} \mathcal{A}$$

Hence it defines an infinite loop map

$$N\colon \Gamma B\mathcal{A} \to \Gamma B\mathcal{A}$$

Since $\square$ gives the H-space structure on $\Gamma B\mathcal{A}$, N induces multiplication by N on homotopy groups.

Theorem 8.6. The composite functors

(1) $\mathcal{GL}(\mathbb{F}_q) \xrightarrow{X} \mathcal{GL}(\mathbb{F}_{q^r}) \xrightarrow{F} \mathcal{GL}(\mathbb{F}_q)$

(2) $\mathcal{S}p(\mathbb{F}_q) \xrightarrow{X} \mathcal{S}p(\mathbb{F}_{q^r}) \xrightarrow{F} \mathcal{S}p(\mathbb{F}_q)$

(3) $\mathcal{U}(\mathbb{F}_{q^2}) \xrightarrow{X} \mathcal{U}(\mathbb{F}_{q^{2r}}) \xrightarrow{F} \mathcal{U}(\mathbb{F}_{q^2})$, r odd

(4) $\bar{\mathcal{O}}(\mathbb{F}_q) \xrightarrow{X} \bar{\mathcal{O}}(\mathbb{F}_{q^r}) \xrightarrow{F} \bar{\mathcal{O}}(\mathbb{F}_q)$, q,r odd

(5) $\bar{\mathcal{O}}^{ev}(\mathbb{F}_q) \xrightarrow{X} \bar{\mathcal{O}}^{ev}(\mathbb{F}_{q^r}) \xrightarrow{F} \bar{\mathcal{O}}^{ev}(\mathbb{F}_q)$ q even

where X denotes the extension of scalars functors of 8.2 and F denotes the forgetful functors of 8.3 are respectively equivalent to

(1') $\mathcal{GL}(\mathbb{F}_q) \xrightarrow{r} \mathcal{GL}(\mathbb{F}_q)$

(2') $\mathcal{S}p(\mathbb{F}_q) \xrightarrow{r} \mathcal{S}p(\mathbb{F}_q)$

(3') $\mathcal{U}(\mathbb{F}_{q^2}) \xrightarrow{r} \mathcal{U}(\mathbb{F}_{q^2})$

(4') $\bar{\mathcal{O}}(\mathbb{F}_q) \xrightarrow{r} \bar{\mathcal{O}}(\mathbb{F}_q)$ q odd

(5') $\bar{\mathcal{O}}^{ev}(\mathbb{F}_q) \xrightarrow{r} \bar{\mathcal{O}}^{ev}(\mathbb{F}_q)$ q even

Proof: We shall confine ourselves to proving (4), (5) are equivalent to (4'), (5'). The other cases are proved similarly and involve fewer complications.

First let us suppose q,r are odd. By 4.6 the quadratic space $(\mathbb{F}_{q^r},T)$ over $\mathbb{F}_q$ with form

$$T(u) = tr(u^2)$$

is isomorphic to $(\mathbb{F}_q^r,Q_+)$. Let $\{e_i\}_{i=1}^r$ be an orthonormal basis for $(\mathbb{F}_{q^r},T)$. Let (V,Q) be an arbitrary nondegenerate quadratic space over $\mathbb{F}_q$. Define

$$\eta: r(V,Q) = (V \oplus V \oplus \cdots \oplus V, Q \oplus Q \oplus \cdots \oplus Q) \to FX(V,Q) = (\bar{V},\bar{Q})$$

by

$$\eta(v_1,v_2,\ldots,v_r) = \Sigma_{i=1}^r e_i \otimes v_i$$

Then

$$\bar{Q}(\eta(v_1,v_2,\ldots,v_r)) = tr(\Sigma_{i=1}^{r} e_i^2 Q(v_i) + \Sigma_{1=i<j<r} e_i e_j B_Q(v_i,v_j))$$

$$= \Sigma_i tr(e_i^2) Q(v_i) + \Sigma_{i<j} tr(e_i e_j) B_Q(v_i,v_j)$$

$$= \Sigma_i Q(v_i)$$

$$= (Q \oplus \cdots \oplus Q)(v_1,v_2,\ldots,v_r)$$

Hence η is a natural equivalence between (4') and (4).

Now suppose q is even. Now consider $\mathbb{F}_{q^r}$ as a vector space over $\mathbb{F}_q$ with nondegenerate symmetric bilinear form

$$T(x,y) = tr(xy)$$

Clearly T is not a symplectic form. Hence $(\mathbb{F}_{q^r},T)$ is isomorphic to $(\mathbb{F}_q^r,E)$. Let $\{e_i\}_{i=1}^r$ be an orthonormal basis for $(\mathbb{F}_{q^r},T)$. Then as above

$$\eta\colon r(V,Q) \to FX(V,Q)$$

defined by

$$\eta(v_1,\ldots,v_r) = \Sigma_{i=1}^{r} e_i \otimes v_i$$

is a natural equivalence between (5') and (5).

Finally we prove a similar result for the hyperbolic functors.

Theorem 8.7. Let q be odd. Then the composite functors

$$(1)\quad \mathcal{O}(\mathbb{F}_q) \xrightarrow{J_1} \mathcal{GL}(\mathbb{F}_q) \xrightarrow{H_1} \bar{\mathcal{O}}(\mathbb{F}_q)$$

$$(2)\quad \mathcal{S}p(\mathbb{F}_q) \xrightarrow{J_2} \mathcal{GL}^{ev}(\mathbb{F}_q) \xrightarrow{H_2} \mathcal{S}p(\mathbb{F}_q)$$

$$(3)\quad \mathcal{U}(\mathbb{F}_{q^2}) \xrightarrow{J_3} \mathcal{GL}(\mathbb{F}_{q^2}) \xrightarrow{H_3} \mathcal{U}(\mathbb{F}_{q^2})$$

where J_1, J_2, J_3 are the forgetful functors of 8.1 and H_1, H_2, H_3 are the hyperbolic functors 8.4(a) (c), (d), are respectively equivalent to

$$(1')\quad \mathcal{O}(\mathbb{F}_q) \xrightarrow{\Delta} \bar{\mathcal{O}}(\mathbb{F}_q) \times \bar{\mathcal{O}}(\mathbb{F}_q) \xrightarrow{1\times\rho} \bar{\mathcal{O}}(\mathbb{F}_q) \times \bar{\mathcal{O}}(\mathbb{F}_q) \xrightarrow{\oplus} \bar{\mathcal{O}}(\mathbb{F}_q)$$

$$(2')\quad \mathcal{S}p(\mathbb{F}_q) \xrightarrow{2} \mathcal{S}p(\mathbb{F}_q)$$

$$(3')\quad \mathcal{U}(\mathbb{F}_{q^2}) \xrightarrow{2} \mathcal{U}(\mathbb{F}_{q^2})$$

where $\rho\colon \bar{\mathcal{O}}(\mathbb{F}_q) \to \bar{\mathcal{O}}(\mathbb{F}_q)$ is the functor induced by

$$\rho(V,Q) = (V,-Q).$$

Proof: We shall confine ourselves to proving that (1) ≅ (1'). The proofs of the other statements are similar.

By definition the composite functor (1) assigns to a quadratic space $(\mathbb{F}_q^n, Q_+)$ the quadratic space $(\mathbb{F}_q^n \oplus \mathbb{F}_q^n, Q_h)$ where

$$Q_h((x,y)) = E(x,y).$$

Now observe that $(\mathbb{F}_q^n \oplus \mathbb{F}_q^n, Q_h)$ can be written as a (vector) direct sum

$$\mathbb{F}_q^n \oplus \mathbb{F}_q^n = V_+ + V_-$$

where V_+, V_- are the subspaces

$$V_+ = \{(x,x) \mid x \in \mathbb{F}_q^n\}$$

$$V_- = \{(x,-x) \mid x \in \mathbb{F}_q^n\}$$

Observe that

$$Q_h((x,x)) = E(x,x) = Q_+(x)$$

$$Q_h((x,-x)) = E(x,-x) = -Q_+(x)$$

Hence there are isomorphisms

$$\alpha: (\mathbb{F}_q^n, Q_+) \cong (V_+, Q_H)$$

$$\beta: \rho(\mathbb{F}_q^n, Q_+) = (\mathbb{F}_q^n, -Q_+) \cong (V_-, Q_h)$$

Observe also that V_+ and V_- are mutually orthogonal:

$$B_h((x,x),(y,-y)) = \frac{1}{2}\left[Q_h((x,x) + (y,-y)) - Q_h((x,x)) - Q_h((y,-y))\right]$$

$$= \frac{1}{2}\left[E(x + y, x - y) - E(x,x) - E(y,-y)\right]$$

$$= \frac{1}{2}\left[E(y,x) + E(x,-y)\right] = 0$$

Hence there is an isomorphism

$$(V_+,Q_h) \oplus (V_-,Q_h) \cong (\mathbb{F}_q^n \oplus \mathbb{F}_q^n, Q_h)$$

Next observe that if

$$f\colon (\mathbb{F}_q^n, Q_+) \to (\mathbb{F}_q^n, Q_+)$$

is an automorphism, then $(f^{-1})^* = f$. Hence we have the commutative diagrams

$$\begin{array}{ccc} (\mathbb{F}_q^n, Q_+) & \xrightarrow[\cong]{\alpha} & (V_+, Q_h) \\ \downarrow{\scriptstyle f} & & \downarrow{\scriptstyle f\oplus(f^{-1})^*} \\ (\mathbb{F}_q^n, Q_+) & \xrightarrow[\cong]{\alpha} & (V_+, Q_h) \end{array}$$

$$\begin{array}{ccc} (\mathbb{F}_q^n, -Q_+) & \xrightarrow[\cong]{\beta} & (V_-, Q_h) \\ \downarrow{\scriptstyle f} & & \downarrow{\scriptstyle f\oplus(f^{-1})^*} \\ (\mathbb{F}_q^n, -Q_+) & \xrightarrow[\cong]{\beta} & (V_-, Q_h) \end{array}$$

Therefore

$$(\mathbb{F}_q^n, Q_+) \oplus \rho(\mathbb{F}_q^n, Q_-) \xrightarrow[\cong]{\alpha\oplus\beta} (V_+,Q_h) \oplus (V_-,Q_h) \cong (\mathbb{F}_q^n \oplus \mathbb{F}_q^n, Q_h)$$

is a natural equivalence $(1') \cong (1)$.

Chapter III

K-Theory of Finite Fields and the ImJ spaces

§1. Introduction

In this chapter we present our main results: the construction of equivalences between the infinite loop spaces of Chapter II and those of Chapter I. The basic construction, known as Brauer lifting, is due to Quillen. In [34,35] he treated the groups $GL(n,\mathbb{F}_q)$ and for q odd, $O(n,\mathbb{F}_q)$ and $Sp(n,\mathbb{F}_q)$. In §2 we extend these results to $U(n,\mathbb{F}_{q^2})$ and for q even, $O(n,\mathbb{F}_q)$ and $Sp(n,\mathbb{F}_q)$. In §3 we state the main results (Theorem 3.1) and outline the proofs. The arguments depend on whether our spaces are localized at characteristic or non-characteristic primes. In §4 we show the triviality of our spaces at the finite field characteristic. In §5,6 we sketch the arguments at non-characteristic primes: first establishing generators and relations for the homology classical groups and then showing Brauer lift is an equivalence of infinite loop spaces. Finally, for completeness, we treat the algebraic closure of finite fields in §7.

§2. The Brauer lift

The construction of maps between the infinite loop spaces of Chapter II and those of Chapter I depends on a result in group representation theory due to Brauer and Green. In [9], Brauer characterized which functions

$$\chi: G \to \mathbb{C}$$

from a finite group to the complex numbers are characters of a virtual representation of the group. In [19] Green applied this result to the following situation: Let $\overline{\mathbb{F}}_p$ denote the algebraic closure of the field of p elements, p any prime. Fix an embedding $\beta: \overline{\mathbb{F}}_p^* \to S^1 \to C^*$ of multiplicative groups. Now let

$$E: G \to GL(n,\overline{\mathbb{F}}_p)$$

be a representation of a finite group G over $\overline{\mathbb{F}}_p$. For any $g \in G$ let $\{\xi_i(g)\}_{i=1}^n$ denote the eigenvalues of the matrix $E(g)$. Then the complex-valued function

$$\chi_E: G \to \mathbb{C}$$

defined by

$$\chi_E(g) = \Sigma_{i=1}^n \beta\xi_i(g)$$

satisfies Brauer's criteria and is therefore the character of a unique virtual complex representation $\overline{\beta}E \in R(G)$ (see also [34]). It is clear that χ_E is additive and multiplicative in E. Thus we obtain a homomorphism of representation rings

$$\overline{\beta}: R_{\overline{\mathbb{F}}_p}(G) \to R(G)$$

In [24], Quillen observed that if p is odd and we start with an orthogonal (resp. symplectic) representation

$$E: G \to O(n,\overline{\mathbb{F}}_p)$$

$$(\text{resp. } E: G \to Sp(2n,\overline{\mathbb{F}}_p))$$

then $\overline{\beta}E$ is a virtual orthogonal (resp. symplectic) representation of G over $\mathbb{C}$. We thus get homomorphisms of representation rings

$$\bar{\beta}: RO_{\bar{\mathbb{F}}_p}(G) \to RO(G)$$

$$\bar{\beta}: RSp_{\bar{\mathbb{F}}_p}(G) \to RSp(G)$$

(where $RO_{\bar{\mathbb{F}}_p}(G)$, $RSp_{\bar{\mathbb{F}}_p}(G)$ are defined in the usual way as the Grothendieck groups of orthogonal; resp. symplectic representations of G over $\bar{\mathbb{F}}_p$). All these homomorphisms $\bar{\beta}$ are called Brauer liftings.

To get corresponding maps of infinite loop spaces we argue as in Quillen [34]. We first observe that passage to classifying maps and then to Grothendieck groups gives ring homomorphisms

$$B: R(G) \to K(BG)$$

$$B: RO(G) \to KO(BG)$$

$$B: RSp(G) \to KSp(BG)$$

Let $r = p^a$ and observe that $\bar{\mathbb{F}}_p = \varinjlim \mathbb{F}_r$. Let

$$E(n,r): GL(n,\mathbb{F}_r) \to GL(n,\bar{\mathbb{F}}_p)$$

$$E(n,r): O(n,\mathbb{F}_r) \to O(n,\bar{\mathbb{F}}_p)$$

$$E(n,r): Sp(2n,\mathbb{F}_r) \to Sp(2n,\bar{\mathbb{F}}_p)$$

denote the standard inclusions. In each case let $[n,r]$ denote the corresponding trivial representation. Then

$$e(n,r) = B\bar{\beta}(E(n,r) - [n,r])$$

represents an element of $K(BGL(n,\mathbb{F}_r))$ (resp. $KO(BO(n,\mathbb{F}_r))$, $KSp(BSp(2n,\mathbb{F}_r))$). These elements $e(n,r)$ are compatible with respect to the standard inclusions of groups as n and r increase. Hence the e(n,r) determine an element of $\varprojlim K(BGL(n,\mathbb{F}_r))$ (resp. $\varprojlim KO(BO(n,\mathbb{F}_r))$, $\varprojlim KSp(BSp(2n,\mathbb{F}_r))$). Now in the Milnor exact sequences

$$0 \to \varprojlim{}^1 K^{-1}(BGL(n,\mathbb{F}_r)) \to K(BGL(\infty,\overline{\mathbb{F}}_p)) \to \varprojlim K(BGL(n,\mathbb{F}_r)) \to 0$$

$$0 \to \varprojlim{}^1 KO^{-1}(BO(n,\mathbb{F}_r)) \to KO(BO(\infty,\overline{\mathbb{F}}_p)) \to \varprojlim KO(BO(n,\mathbb{F}_r)) \to 0$$

$$0 \to \varprojlim{}^1 KSp^{-1}(BSp(2n,\mathbb{F}_r)) \to KSp(BSp(\infty,\overline{\mathbb{F}}_p)) \to \varprojlim KSp(BSp(2n,\mathbb{F}_r)) \to 0$$

the $\varprojlim^1$ terms vanish. For according to Atiyah [4], $K^{-1}(BG) = 0$ for any finite group G, while according to Atiyah and Segal [6], $KO^{-1}(BG)$ and $KSp^{-1}(BG)$ are finite dimensional vector spaces over $\mathbb{F}_2$. Hence the e(n,r) determine a unique element $\tilde{\beta} \in K(BGL(\infty,\overline{\mathbb{F}}_p))$ (resp. $KO(BO(\infty,\overline{\mathbb{F}}_p))$, $KSp(BSp(\infty,\overline{\mathbb{F}}_p))$). Since the maps

$$\bar{1}: BGL(\infty,\overline{\mathbb{F}}_p) \to \Gamma_0 B\mathcal{GL}(\overline{\mathbb{F}}_p)$$

$$\bar{1}: BO(\infty,\overline{\mathbb{F}}_p) \to \Gamma_0 B\mathcal{O}(\overline{\mathbb{F}}_p)$$

$$\bar{1}: BSp(\infty,\overline{\mathbb{F}}_p) \to \Gamma_0 B\mathcal{Sp}(\overline{\mathbb{F}}_p)$$

(cf. II 2.18) induce isomorphisms in integral homology, they also induce isomorphisms in K-theory. Hence there is a unique element $\bar{\beta} \in K(\Gamma_0 B\mathcal{GL}(\overline{\mathbb{F}}_p))$ (resp. $KO(\Gamma_0 B\mathcal{O}(\overline{\mathbb{F}}_p))$, $KSp(\Gamma_0 B\mathcal{Sp}(\overline{\mathbb{F}}_p))$) such that $\bar{1}^*(\bar{\beta}) = \tilde{\beta}$. This element is represented by a homotopy class of maps

(2.1) $\bar{\beta}: \Gamma_0 B\mathcal{GL}(\overline{\mathbb{F}}_p) \to BU$

(2.2) $\bar{\beta}: \Gamma_0 B\mathcal{O}(\overline{\mathbb{F}}_p) \to BO \qquad p \neq 2$

(2.3) $\bar{\beta}: \Gamma_0 B\mathcal{Sp}(\overline{\mathbb{F}}_p) \to BSp \qquad p \neq 2$

These maps are popularly known as Brauer lifts.

It is not clear from the construction that the Brauer lifts are infinite loop maps. However an ingeneous argument of May (cf.[25]) implies that the induced maps

(2.4) $\bar{\beta}: \Gamma_0 B\mathcal{GL}(\bar{\mathbb{F}}_p)^\wedge[\frac{1}{p}] \to BU^\wedge[\frac{1}{p}]$

(2.5) $\bar{\beta}: \Gamma_0 B\mathcal{O}(\bar{\mathbb{F}}_p)^\wedge[\frac{1}{p}] \to BO^\wedge[\frac{1}{p}] \qquad p \neq 2$

(2.6) $\bar{\beta}: \Gamma_0 B\mathcal{S}p(\bar{\mathbb{F}}_p)^\wedge[\frac{1}{p}] \to BSp^\wedge[\frac{1}{p}] \qquad p \neq 2$

are infinite loop maps, where $(\cdot)^\wedge[\frac{1}{p}]$ denotes the Bousfield-Kan completion away from p (cf.[8]). In §7 we show these maps are equivalences of infinite loop spaces.

Moreover May [25] shows that the commutativity of the diagrams

$$\begin{array}{ccc} R_{\bar{\mathbb{F}}_p}(G) & \xrightarrow{\varphi^q} & R_{\bar{\mathbb{F}}_p}(G) \\ \downarrow \bar{\beta} & & \downarrow \bar{\beta} \\ R(G) & \xrightarrow{\psi^q} & R(G) \end{array} \qquad \begin{array}{ccc} RO_{\bar{\mathbb{F}}_p}(G) & \xrightarrow{\varphi^q} & RO_{\bar{\mathbb{F}}_p}(G) \\ \downarrow \bar{\beta} & & \downarrow \bar{\beta} \\ RO(G) & \xrightarrow{\psi^q} & RO(G) \end{array}$$

$$\begin{array}{ccc} RSp_{\bar{\mathbb{F}}_p}(G) & \xrightarrow{\varphi^q} & RSp_{\bar{\mathbb{F}}_p}(G) \\ \downarrow \bar{\beta} & & \downarrow \bar{\beta} \\ RSp(G) & \xrightarrow{\psi^q} & RSp(G) \end{array}$$

where $q = p^a$, φ^q is the map induced by the Frobenius automorphism $x \mapsto x^q$, and ψ^q is the Adams operation, imply the commutativity of the following diagrams of infinite loop maps:

$$(2.7)\qquad \begin{array}{ccc} \Gamma_0 B\mathcal{GL}(\overline{\mathbb{F}}_p)^\wedge[\tfrac{1}{p}] & \xrightarrow{\varphi^q} & \Gamma_0 B\mathcal{GL}(\overline{\mathbb{F}}_p)^\wedge[\tfrac{1}{p}] \\ \downarrow{\scriptstyle\bar{\beta}} & & \downarrow{\scriptstyle\bar{\beta}} \\ BU^\wedge[\tfrac{1}{p}] & \xrightarrow{\psi^q} & BU^\wedge[\tfrac{1}{p}] \end{array}$$

$$(2.8)\qquad \begin{array}{ccc} \Gamma_0 B\mathcal{O}(\overline{\mathbb{F}}_p)^\wedge[\tfrac{1}{p}] & \xrightarrow{\varphi^q} & \Gamma_0 B\mathcal{O}(\overline{\mathbb{F}}_p)^\wedge[\tfrac{1}{p}] \\ \downarrow{\scriptstyle\bar{\beta}} & & \downarrow{\scriptstyle\bar{\beta}} \\ BO^\wedge[\tfrac{1}{p}] & \xrightarrow{\psi^q} & BO^\wedge[\tfrac{1}{p}] \end{array} \qquad p \neq 2$$

$$(2.9)\qquad \begin{array}{ccc} \Gamma_0 B\mathcal{S}p(\overline{\mathbb{F}}_p)^\wedge[\tfrac{1}{p}] & \xrightarrow{\varphi^q} & \Gamma_0 B\mathcal{S}p(\overline{\mathbb{F}}_p)^\wedge[\tfrac{1}{p}] \\ \downarrow{\scriptstyle\bar{\beta}} & & \downarrow{\scriptstyle\bar{\beta}} \\ BSp^\wedge[\tfrac{1}{p}] & \xrightarrow{\psi^q} & BSp^\wedge[\tfrac{1}{p}] \end{array} \qquad p \neq 2$$

Now let β denote the composite infinite loop maps

(2.10) $\beta: \Gamma_0 B\mathcal{GL}(\mathbb{F}_q) \xrightarrow{i} \Gamma_0 B\mathcal{GL}(\overline{\mathbb{F}}_p) \to \Gamma_0 B\mathcal{GL}(\overline{\mathbb{F}}_p)^\wedge[\tfrac{1}{p}] \xrightarrow{\bar{\beta}} BU^\wedge[\tfrac{1}{p}]$

(2.11) $\beta: \Gamma_0^{+} B\bar{\mathcal{O}}(\mathbb{F}_q) \xrightarrow{i} \Gamma_0 B\mathcal{O}(\overline{\mathbb{F}}_p) \to \Gamma_0 B\mathcal{O}(\overline{\mathbb{F}}_p)^\wedge[\tfrac{1}{p}] \xrightarrow{\bar{\beta}} BO^\wedge[\tfrac{1}{p}]$, $p \neq 2$

(2.12) $\beta: \Gamma_0 B\mathcal{S}p(\mathbb{F}_q) \xrightarrow{i} \Gamma_0 B\mathcal{S}p(\overline{\mathbb{F}}_p) \to \Gamma_0 B\mathcal{S}p(\overline{\mathbb{F}}_p)^\wedge[\tfrac{1}{p}] \xrightarrow{\bar{\beta}} BSp^\wedge[\tfrac{1}{p}]$,

$p \neq 2$

where i is the map induced by the inclusion $\mathbb{F}_q \to \overline{\mathbb{F}}_p$. These maps will also be called Brauer lifts. Since $\varphi^q \cdot i = i$ already on the level of permutative categories, it follows that $(\varphi^q - 1)\cdot i$ is trivial as an infinite loop map. From the commutative diagrams above it immediately follows that the composites

$$\Gamma_0 B\mathcal{GL}(\mathbb{F}_q) \xrightarrow{\ \beta\ } BU^\wedge[\tfrac{1}{p}] \xrightarrow{\ \psi^q-1\ } BU^\wedge[\tfrac{1}{p}]$$

$$\Gamma_0^+ B\bar{\mathcal{O}}(\mathbb{F}_q) \xrightarrow{\ \beta\ } BO^\wedge[\tfrac{1}{p}] \xrightarrow{\ \psi^q-1\ } BO^\wedge[\tfrac{1}{p}], \quad p \neq 2$$

$$\Gamma_0 B\mathcal{Sp}(\mathbb{F}_q) \xrightarrow{\ \beta\ } BSp^\wedge[\tfrac{1}{p}] \xrightarrow{\ \psi^q-1\ } BSp^\wedge[\tfrac{1}{p}], \quad p \neq 2$$

are trivial as infinite loop maps.

It follows that there are infinite loop lifts λ completing the diagrams

(2.13)
$$\begin{array}{ccccc} JU(q) & \to & BU^\wedge[\tfrac{1}{p}] & \xrightarrow{\ \psi^q-1\ } & BU^\wedge[\tfrac{1}{p}] \\ & \nwarrow^{\lambda} & \uparrow{\scriptstyle\beta} & & \\ & & \Gamma_0 B\mathcal{GL}(\mathbb{F}_q) & & \end{array}$$

(2.14)
$$\begin{array}{ccccc} J\bar{O}(q) & \to & BO^\wedge[\tfrac{1}{p}] & \xrightarrow{\ \psi^q-1\ } & BO^\wedge[\tfrac{1}{p}] \\ & \nwarrow^{\lambda} & \uparrow{\scriptstyle\beta} & & \\ & & \Gamma_0^+ B\bar{\mathcal{O}}(\mathbb{F}_q) & & \end{array} \qquad p \neq 2$$

(2.15)
$$\begin{array}{ccccc} JSp(q) & \to & BSp^\wedge[\tfrac{1}{p}] & \xrightarrow{\ \psi^q-1\ } & BSp^\wedge[\tfrac{1}{p}] \\ & \nwarrow^{\lambda} & \uparrow{\scriptstyle\beta} & & \\ & & \Gamma_0 B\mathcal{Sp}(\mathbb{F}_q) & & \end{array} \qquad p \neq 2$$

(Strictly speaking the fibers of ψ^q-1 should be $JU(q)^\wedge[\frac{1}{p}]$, $J\bar{O}(q)^\wedge[\frac{1}{p}]$, $JSp(q)^\wedge[\frac{1}{p}]$. However it is easily seen that the homotopy groups of $JU(q)$, $J\bar{O}(q)$, $JSp(q)$ are finite without p-torsion. Hence completion away from p does not change the homotopy type of these spaces.)

Restriction to the zero components of the map

$$\lambda: \overline{\Gamma_0^+ B\bar{\mathcal{O}}}(\mathbb{F}_q) \to J\bar{O}(q)$$

yields an infinite loop map

$$(2.16) \qquad \lambda: \Gamma_0 B\mathcal{O}(\mathbb{F}_q) \to JO(q) \qquad q \text{ odd}$$

Next we turn to the problem of performing a similar construction on the unitary groups. To do this we first observe that the following diagram is commutative

$$\begin{array}{ccc} R_{\bar{\mathbb{F}}_p}(G) & \xrightarrow{\Lambda} & R_{\bar{\mathbb{F}}_p}(G) \\ \downarrow{\scriptstyle \bar{\beta}} & & \downarrow{\scriptstyle \bar{\beta}} \\ R(G) & \xrightarrow{\psi^{-1}} & R(G) \end{array}$$

where Λ is the map induced by the map $GL(n,\bar{\mathbb{F}}_p) \to GL(n,\bar{\mathbb{F}}_p)$ given by

$$\Lambda(A) = (A^t)^{-1}.$$

(Also note that ψ^{-1} is induced by conjugation of complex matrices). By using methods similar to those employed by May [25] in proving commutativity of diagrams (2.7-2.9), one can show that

$$\begin{array}{ccc} \Gamma_0 B\mathcal{GL}(\bar{\mathbb{F}}_p)^\wedge[\frac{1}{p}] & \xrightarrow{\Lambda} & \Gamma_0 B\mathcal{GL}(\bar{\mathbb{F}}_p)^\wedge[\frac{1}{p}] \\ \downarrow{\scriptstyle \bar{\beta}} & & \downarrow{\scriptstyle \bar{\beta}} \\ BU^\wedge[\frac{1}{p}] & \xrightarrow{\psi^{-1}} & BU^\wedge[\frac{1}{p}] \end{array}$$

is a commutative diagram of infinite loop maps. Combining this with diagram (2.7) we get another commutative diagram of infinite loop maps

$$(2.17)\qquad \begin{array}{ccc} \Gamma_0 B\mathcal{GL}(\overline{\mathbb{F}}_p)^\wedge[\frac{1}{p}] & \xrightarrow{\Lambda\cdot\varphi^q} & \Gamma_0 B\mathcal{GL}(\overline{\mathbb{F}}_p)^\wedge[\frac{1}{p}] \\ \downarrow{\scriptstyle\overline{\beta}} & & \downarrow{\scriptstyle\overline{\beta}} \\ BU^\wedge[\frac{1}{p}] & \xrightarrow{\psi^{-q}=\psi^{-1}\cdot\psi^q} & BU^\wedge[\frac{1}{p}] \end{array}$$

Now defining β to be the composite infinite loop map

$$(2.18)\qquad \beta\colon \Gamma_0 B\mathcal{U}(\mathbb{F}_{q^2}) \xrightarrow{i} \Gamma_0 B\mathcal{GL}(\overline{\mathbb{F}}_p) \to \Gamma_0 B\mathcal{GL}(\overline{\mathbb{F}}_p)^\wedge[\tfrac{1}{p}] \xrightarrow{\overline{\beta}} BU^\wedge[\tfrac{1}{p}]$$

and observing that $\Lambda\cdot\varphi^q\cdot i = i$ on the level of permutative categories, we infer from the above diagram that the composite

$$\Gamma_0 B\mathcal{U}(\mathbb{F}_{q^2}) \xrightarrow{\beta} BU^\wedge[\tfrac{1}{p}] \xrightarrow{\psi^{-q}-1} BU^\wedge[\tfrac{1}{p}]$$

is trivial as an infinite loop map. It follows that there is an infinite loop lift λ completing the diagram

$$\begin{array}{ccccc} JU(-q) & \to & BU^\wedge[\frac{1}{p}] & \xrightarrow{\psi^{-q}-1} & BU^\wedge[\frac{1}{p}] \\ & \nwarrow{\scriptstyle\lambda} & \uparrow{\scriptstyle\beta} & & \\ & & \Gamma_0 B\mathcal{U}(\mathbb{F}_{q^2}) & & \end{array}$$

It remains to define a Brauer lift for the orthogonal and symplectic groups over finite fields of characteristic 2. Here one runs into a problem, since if

$$E\colon G \to O(2n,\overline{\mathbb{F}}_2)$$

is an orthogonal representation of a finite group G, it need not follow that the Brauer lift $\overline{\beta}E$ is a orthogonal representation in $R(G)$. Similarly for symplectic representations.

However we can avoid the problem by defining the Brauer lift for symplectic groups to be the composite infinite loop map

$$(2.19) \qquad \beta\colon \Gamma_0 B\mathcal{S}p(\mathbb{F}_q) \xrightarrow{i} \Gamma_0 B\mathcal{GL}(\mathbb{F}_q) \xrightarrow{\beta} BU^\wedge[\tfrac{1}{2}] \xrightarrow{s} BSp^\wedge[\tfrac{1}{2}]$$

where $s\colon BU \to BSp$ is the symplectification map. Similarly for the orthogonal groups we define the Brauer lift to be the composite infinite loop map

$$(2.20) \qquad \beta\colon \Gamma_0^{+} B\mathcal{O}^{ev}(\mathbb{F}_q) \xrightarrow{J} \Gamma_0 B\mathcal{S}p(\mathbb{F}_q) \xrightarrow{\beta} BSp^\wedge[\tfrac{1}{2}] \simeq BO^\wedge[\tfrac{1}{2}]$$

where J is the map induced by the functor defined in II 7.15. (Note: The map corresponding to 2.19 in case q is odd would give 2β where β is the Brauer lift defined in 2.12. In this case however there is no problem since we are completing away from 2, and thus making 2 invertible.)

The commutativity of the diagram

$$\begin{array}{ccc} BU^\wedge[\tfrac{1}{2}] & \xrightarrow{\psi^q - 1} & BU^\wedge[\tfrac{1}{2}] \\ \downarrow s & & \downarrow s \\ BSp^\wedge[\tfrac{1}{2}] & \xrightarrow{\psi^q - 1} & BSp^\wedge[\tfrac{1}{2}] \end{array}$$

together with 2.13 implies that the composite

$$\Gamma_0 B\mathcal{S}p(\mathbb{F}_q) \xrightarrow{\beta} BSp^\wedge[\tfrac{1}{2}] \xrightarrow{\psi^q - 1} BSp^\wedge[\tfrac{1}{2}]$$

is trivial as an infinite loop map. It follows that there is an infinite loop lift λ completing the diagram

$$
(2.21)\qquad
\begin{array}{ccccc}
JSp(q) & \to & BSp^{\wedge}[\frac{1}{2}] & \xrightarrow{\psi^{q}-1} & BSp^{\wedge}[\frac{1}{2}] \\
 & {\scriptstyle\lambda}\nwarrow & \uparrow{\scriptstyle\beta} & & \\
 & & \Gamma_0 B\mathcal{S}p(\bar{\mathbb{F}}_q) & &
\end{array}
\qquad q \text{ even}
$$

Similar reasoning shows that there is an infinite loop lift λ completing the diagram

$$
(2.22)\qquad
\begin{array}{ccccc}
J\bar{O}(q) & \to & BO^{\wedge}[\frac{1}{2}] & \xrightarrow{\psi^{q}-1} & BO^{\wedge}[\frac{1}{2}] \\
 & {\scriptstyle\lambda}\nwarrow & \uparrow{\scriptstyle\beta} & & \\
 & & \overset{+}{\Gamma}_0 B\bar{\mathcal{O}}^{ev}(\bar{\mathbb{F}}_q) & &
\end{array}
\qquad q \text{ even}
$$

We will also denote by λ the composite infinite loop maps

$$(2.23)\qquad \lambda\colon \Gamma_0 B\mathcal{O}^{ev}(\bar{\mathbb{F}}_q) \to \overset{+}{\Gamma}_0 B\bar{\mathcal{O}}^{ev}(\bar{\mathbb{F}}_q) \xrightarrow{\lambda} J\bar{O}(q)$$

and

$$(2.24)\qquad \lambda\colon \Gamma_0 B\mathcal{DO}^{ev}(\bar{\mathbb{F}}_q) \to \overset{+}{\Gamma}_0 B\bar{\mathcal{O}}^{ev}(\bar{\mathbb{F}}_q) \xrightarrow{\lambda} J\bar{O}(q).$$

§3. The main results

We are now in position to state our principal result: the maps λ constructed in the previous section are equivalences of infinite loop spaces (except for the spaces associated with orthogonal groups over fields of characteristic 2).

Theorem 3.1. The infinite loop maps

(a) $\lambda\colon \Gamma_0 B\mathcal{GL}(\bar{\mathbb{F}}_q) \to JU(q)$

(b) $\lambda\colon \Gamma_0 B\mathcal{S}p(\bar{\mathbb{F}}_q) \to JSp(q)$

(c) $\lambda\colon \Gamma_0 B\mathcal{U}(\mathbb{F}_{q^2}) \to JU(-q)$

(d) $\lambda\colon \Gamma_0 B\mathcal{O}(\mathbb{F}_q) \to JO(q)$ q odd

(e) $\lambda\colon \Gamma_0^{+} B\bar{\mathcal{O}}(\mathbb{F}_q) \to J\bar{O}(q)$ q odd

are equivalences. Moreover these maps induce equivalences of infinite loop spaces

(f) $\lambda\colon \Gamma_0 B\mathcal{SL}^{ev}(\mathbb{F}_q) \to JSU(q)$ q odd

(g) $\lambda\colon \Gamma_0 B\mathcal{SU}^{ev}(\mathbb{F}_{q^2}) \to JSU(-q)$ q odd

(h) $\lambda\colon \Gamma_0 B\mathcal{SO}^{ev}(\mathbb{F}_q) \to JSO(q)$ q odd

(i) $\lambda\colon \Gamma_0 B\mathcal{N}^{ev}(\mathbb{F}_q) \simeq \Gamma_0 B\mathcal{ND}^{ev}(\mathbb{F}_q) \to J(q)$ q odd

(j) $\lambda\colon \Gamma_0 B\mathcal{SN}^{ev}(\mathbb{F}_q) \to \tilde{J}(q)$ q odd

(k) $\lambda\colon \Gamma_0 B\mathcal{S}pin^{oct}(\mathbb{F}_q) \to JSpin(q)$ q odd

If q is even, then the following maps are equivalences of infinite loop spaces

(ℓ) $\lambda\colon \Gamma_0 B\mathcal{DO}^{ev}(\mathbb{F}_q) \to J\bar{O}(q)$

(m) $J\colon \Gamma_0 B\mathcal{DO}^{ev}(\mathbb{F}_q) \to \Gamma_0 B\mathcal{S}p(\mathbb{F}_q)$

(n) $(i^{-1}, d)\colon \Gamma_0 B\mathcal{O}^{ev}(\mathbb{F}_q) \to \Gamma_0 B\mathcal{DO}^{ev}(\mathbb{F}_q) \times B\mathbb{Z}/2$

(o) $(i^{-1}, d, \delta)\colon \Gamma_0^{+} B\bar{\mathcal{O}}^{ev}(\mathbb{F}_q) \to \Gamma_0 B\mathcal{DO}^{ev}(\mathbb{F}_q) \times B\mathbb{Z}/2 \times \mathbb{Z}/2$

where the maps J, d, δ are defined as in II 7.21-22 and i is the localization of the inclusion away from 2.

3.3. <u>Outline of proof</u>. We prove that (a)-(e) and (ℓ) are equivalences and derive the other equivalences subsequently. Moreover (e) is derived from (d) by a homological computation.

For the sake of convenience we will denote the maps (a)-(d) and (ℓ) generically by

$$\lambda\colon \Gamma_0 B\mathcal{A} \to F\psi^q$$

To prove that (a)-(d) and (ℓ) are equivalences we observe that $\Gamma_0 B\mathcal{A}$ and $F\psi^q$ are both connected CW complexes. Hence it suffices to show that λ induces isomorphisms on homotopy groups. By Whitehead's theorem this is equivalent to showing that λ induces isomorphisms on integral homology, since $\Gamma_0 B\mathcal{A}$ and $F\psi^q$ are simple spaces (indeed they are infinite loop spaces).

Now a map induces isomorphisms on $H_*(\cdot;\ \mathbb{Z})$ iff it induces isomorphisms on $H_*(\cdot;\ R)$ for $R = \mathbb{Z}/\ell$, ℓ any prime and for $R = \mathbb{Q}$ the rational numbers. The proof therefore breaks up into several steps

Step 1. λ induces an isomorphism in $H_*(\cdot;\ \mathbb{Q})$

This is easy to see. For $F\psi^q$ is defined by the fibration

$$F\psi^q \to X \xrightarrow{\psi^q - 1} Y$$

where X,Y is one of the classical spaces BU, BO, BSO, BSp. According to Chap. I 1.1 we have that $(\psi^q - 1)_*\colon \pi_i X \to \pi_i Y$ is given by multiplication by $q^k - 1$ where $k = [\frac{i+1}{2}]$. From the long exact sequence of homotopy groups associated with this fibration we calculate that the homotopy groups of $F\psi^q$ are either 0, $\mathbb{Z}/2$, $\mathbb{Z}/2 \oplus \mathbb{Z}/2$ or $\mathbb{Z}/q^k - 1$, thus in particular finite. By Serre $\mathcal{C}$-theory, $\tilde{H}_i(F\psi^q;\ \mathbb{Z})$ is finite for all i and hence $\tilde{H}_i(F\psi^q;\ \mathbb{Q}) = \tilde{H}_i(F\psi^q;\ \mathbb{Z}) \otimes \mathbb{Q} = 0$.

Similarly we have

$$\tilde{H}_i(\Gamma_0 B\mathcal{A};\ \mathbb{Q}) = \varinjlim \tilde{H}_i(B\mathcal{A}_n;\ \mathbb{Q})$$

where $\mathcal{A}_n$ denote the finite classical groups making up the category $\mathcal{A}$. But it is a well known fact that the classifying space of a finite group has finite integral homology groups and therefore trivial rational homology. Hence

$$\lambda: \tilde{H}_*(\Gamma_0 B\mathfrak{a};\ \mathbb{Q}) \to \tilde{H}_*(F\psi^q;\ \mathbb{Q})$$

is the trivial isomorphism between trivial groups.

Step 2. λ induces an isomorphism on $H_*(;\ \mathbb{Z}/p)$ where $q = p^n$.

By exactly the same argument as in Step 1, $\pi_i F\psi^q$ is finite without p-torsion for all i. Hence by Serre $\mathcal{C}$ -theory $\tilde{H}_i(F\psi^q;\ \mathbb{Z})$ is finite without p-torsion for all i. Consequently $\tilde{H}_i(F\psi^q;\ \mathbb{Z}/p) = 0$.

Proving that $\tilde{H}_i(\Gamma_0 B\mathfrak{a};\ \mathbb{Z}/p) = 0$ is not that simple. The proof involves group cohomological methods. Details appear in §4. Granting this one gets that

$$\lambda: \tilde{H}_*(\Gamma_0 B\mathfrak{a};\ \mathbb{Z}/p) \to \tilde{H}_*(F\psi^q;\ \mathbb{Z}/p)$$

is also the trivial isomorphism between trivial groups.

Step 3. λ induces an isomorphism on $H_*(\cdot;\ \mathbb{Z}/\ell)$ where ℓ is an odd prime, not dividing q.

We first observe that according to Chapter I, Sections 5 and 6

$$H_*(F\psi^q;\ \mathbb{Z}/\ell) = \mathbb{Z}/\ell[a_1, a_2, \ldots, a_n, \ldots] \otimes E\{b_i \mid i \geq 1\}$$

where

$$\deg a_n = n \times \deg a_1 \text{ and } \deg b_n = \deg a_n - 1$$

In Chapter V we will similarly show that $H_*(\Gamma_0 B\mathfrak{a};\ \mathbb{Z}/\ell)$ is generated by elements $\{\alpha_i, \beta_i \mid i \geq 1\}$, where

$$\deg \alpha_i = \deg a_i,\ \deg \beta_i = \deg b_i$$

In Chapter V we will also show that

$$\lambda_*(\alpha_i) = a_i \text{ mod decomposables}$$

$$\lambda_*(\beta_i) = b_i \text{ mod decomposables.}$$

Since $\{a_i, b_i \mid i \geq 1\}$ is algebraically independent, it follows that so is $\{\alpha_i, \beta_i \mid i \geq 1\}$. Hence

$$H_*(\Gamma_0 B\mathcal{O}\mathcal{C}; \mathbb{Z}/\ell) = \mathbb{Z}/\ell[\alpha_1, \alpha_2, \ldots, \alpha_n, \ldots] \otimes E\{\beta_i \mid i \geq 1\}$$

and

$$\lambda_*: H_*(\Gamma_0 B\mathcal{O}\mathcal{C}; \mathbb{Z}/\ell) \to H_*(F\psi^q; \mathbb{Z}/\ell)$$

is an isomorphism.

Step 4. λ induces an isomorphism on $H_*(; \mathbb{Z}/2)$, q odd.

The proof is similar to that of Step 3. By Chapter I, Sections 3 and 4 we have

$$H_*(F\psi^q; \mathbb{Z}/2) = \mathbb{Z}/2[a_1, a_2, \ldots, a_n, \ldots] \otimes E\{b_i \mid i \geq 1\}$$

where

$$\deg a_n = n \times \deg a_1 \text{ and } \deg b_n = \deg a_n - 1$$

In Chapter IV we will show that $H_*(\Gamma_0 B\mathcal{O}\mathcal{C}; \mathbb{Z}/2)$ is generated by elements $\{\alpha_i, \beta_i \mid i \geq 1\}$ where

$$\deg \alpha_i = \deg a_i, \ \deg \beta_i = \deg b_i$$

In Chapter IV we also show that $\beta_i^2 = 0$ and that

$$\lambda_*(\alpha_i) = a_i \text{ mod decomposables}$$

$$\lambda_*(\beta_i) = b_i \text{ mod decomposables}$$

It follows that the only algebraic relations on $\{\alpha_i, \beta_i | i \geq 1\}$ are $\beta_i^2 = 0$. Hence

$$H_*(\Gamma_0 B\mathcal{O};\ \mathbb{Z}/2) = \mathbb{Z}/2[\alpha_1, \alpha_2, \ldots, \alpha_n, \ldots] \otimes E\{\beta_i | i \geq 1\}$$

and

$$\lambda_*\colon H_*(\Gamma_0 B\mathcal{O};\ \mathbb{Z}/2) \to H_*(F\psi^q;\ \mathbb{Z}/2)$$

is an isomorphism.

Hence modulo results to be proved later, we have demonstrated that (a)-(d) and (ℓ) are equivalences of infinite loop spaces. It remains to derive that (e), (f)-(e) and (m)-(o) are also equivalences.

Since $\Gamma_0^+ B\bar{\mathcal{O}}(\mathbb{F}_q)$ is an infinite loop space with 0-component equivalent to $\Gamma_0 B\mathcal{O}(\mathbb{F}_q)$ and the restriction of (e) is (d) which we have just shown is an equivalence, it remains to check that

$$\lambda\colon \Gamma_0^+ B\bar{\mathcal{O}}(\mathbb{F}_q) \to J\bar{O}(q) \qquad q \text{ odd}$$

is an equivalence on π_0. This is done in Chapter IV, Section 3 by a computation in $H_*(\cdot;\ \mathbb{Z}/2)$.

The maps (f), (g), (j) are obtained from (a), (c), (d) respectively by passing to universal covering spaces and hence are automatically equivalences of infinite loop spaces. Similarly (k) is obtained from (d) by passing to 2-connected covers and is thus an equivalence.

To prove (h) and (i) are equivalences we consider the diagrams of fibrations of infinite loop spaces

$$\begin{array}{ccccc}
B\mathbf{Z}/2 & \xleftarrow{g_1} & \Gamma_0 B\mathcal{O}(\mathbb{F}_q) & \longleftarrow & \Gamma_0 B\mathcal{N}^{ev}(\mathbb{F}_q) \\
\| & & \downarrow \lambda & & \downarrow \lambda_1 \\
\mathbb{R}P^{\infty} & \xleftarrow{g_1} & JO(q) & \longleftarrow & J_1(q)
\end{array}$$

$$\begin{array}{ccccc}
B\mathbf{Z}/2 & \xleftarrow{g_2} & \Gamma_0 B\mathcal{O}(\mathbb{F}_q) & \longleftarrow & \Gamma_0 B\mathcal{N}\mathcal{D}^{ev}(\mathbb{F}_q) \\
\| & & \downarrow \lambda & & \downarrow \lambda_2 \\
\mathbb{R}P^{\infty} & \xleftarrow{g_2} & JO(q) & \longleftarrow & J_2(q)
\end{array}$$

$$\begin{array}{ccccc}
B\mathbf{Z}/2 & \xleftarrow{g_3} & \Gamma_0 B\mathcal{O}(\mathbb{F}_q) & \longleftarrow & \Gamma_0 B\mathcal{S}\mathcal{O}^{ev}(\mathbb{F}_q) \\
\| & & \downarrow \lambda & & \downarrow \lambda_3 \\
\mathbb{R}P^{\infty} & \xleftarrow{g_3} & JO(q) & \longleftarrow & JSO(q)
\end{array}$$

where we use the notation of I 2 and II 3.16. By the results of Chapter IV, Section 2 (cf. IV 2.8 and preceding remark) it follows that the left hand squares commute (even on the infinite loop level). Hence there are infinite loop maps $\lambda_1, \lambda_2, \lambda_3$ completing the diagrams. From the long exact sequences in homotopy and the five lemma we see that $\lambda_1, \lambda_2, \lambda_3$ are equivalences. By II 3.12 and I 2.3 we have

$$\Gamma_0 B\mathcal{N}^{ev}(\mathbb{F}_q) \simeq \Gamma_0 B\mathcal{N}\mathcal{D}^{ev}(\mathbb{F}_q) \text{ and } J_1(q) \simeq J_2(q) \simeq J(q)$$

as infinite loop spaces. Hence there are equivalences of the form (h) and (i).

Finally we turn to the maps associated with orthogonal groups over a field of characteristic 2, i.e. (m), (n), and (o). The fact that (m) is an equivalence follows from the commutative diagram

$$\begin{array}{ccc} \Gamma_0 B\mathcal{D}\mathcal{O}^{ev}(\mathbb{F}_q) & \xrightarrow{\lambda} & J\bar{O}(q) \\ \downarrow J & & \| \\ \Gamma_0 B\mathcal{S}p(\mathbb{F}_q) & \xrightarrow{\lambda} & JSp(q) \end{array}$$

The fact that (n) is an equivalence follows from the fibration

$$\Gamma_0 B\mathcal{D}\mathcal{O}^{ev}(\mathbb{F}_q) \xrightarrow{i} \Gamma_0 B\mathcal{O}^{ev}(\mathbb{F}_q) \xrightarrow{d} B\mathbb{Z}/2$$

cf. II 7.21. Localizing at 2 we get

$$d: \Gamma_0 B\mathcal{O}^{ev}(\mathbb{F}_q)_{(2)} \to (B\mathbb{Z}/2)_{(2)}$$

is an equivalence since by Steps 1 and 2

$$\Gamma_0 B\mathcal{D}\mathcal{O}^{ev}(\mathbb{F}_q)_{(2)} \simeq *$$

Localizing away from 2 we get

$$i: \Gamma_0 B\mathcal{D}\mathcal{O}^{ev}(\mathbb{F}_q)[\tfrac{1}{2}] \to \Gamma_0 B\mathcal{O}^{ev}(\mathbb{F}_q)[\tfrac{1}{2}]$$

is an equivalence since obviously

$$B\mathbb{Z}/2[\tfrac{1}{2}] \simeq *$$

Since $\Gamma_0 B\mathcal{D}\mathcal{O}^{ev}(\mathbb{F}_q)_{(2)} = *$, it follows that

$$\Gamma_0 B \mathcal{DO}^{ev}(\mathbb{F}_q) \to \Gamma_0 B \mathcal{DO}^{ev}(\mathbb{F}_q)[\tfrac{1}{2}]$$

is an equivalence. Thus we get a map

$$i^{-1}: \Gamma_0 B\mathcal{O}^{ev}(\mathbb{F}_q) \to \Gamma_0 B\mathcal{O}^{ev}(\mathbb{F}_q)[\tfrac{1}{2}] \xrightarrow{\cong} \Gamma_0 B \mathcal{DO}^{ev}(\mathbb{F}_q)$$

which induces isomorphisms in $H_*(\cdot; A)$ $A = \mathbb{Q}, \mathbb{Z}/\ell, \ell$ odd prime. Clearly then the map

$$(i^{-1}, d): \Gamma_0 B\mathcal{O}^{ev}(\mathbb{F}_q) \to \Gamma_0 B \mathcal{DO}^{ev}(\mathbb{F}_q) \times B\mathbb{Z}/2$$

induces isomorphisms in $H_*(\cdot; A)$ for $A = \mathbb{Q}, \mathbb{Z}/\ell$ ℓ odd prime and $\mathbb{Z}/2$. Hence (i^{-1}, d) is an equivalence. This proves (n) is an equivalence. The fact that (o) is an equivalence follows by a similar argument.

Remark 3.4. Any space X with finite homotopy groups splits (weakly) into the product of its p-primary components, i.e.

$$X \simeq \Pi_p X_{(p)}$$

Since the spaces $F_\psi{}^q$ have finite homotopy groups, so do the spaces $\Gamma_0 B\mathcal{A}$. Hence all the spaces of Theorem 3.1 split into p-primary components.

We conclude this section with a table of homotopy groups $\pi_i \Gamma_0 B\mathcal{A}$ which by definition are the Quillen K-groups of the category $\mathcal{A}$. In particular $\pi_i \Gamma_0 B\mathcal{GL}(\mathbb{F}_q) = K_i(\mathbb{F}_q)$ $i \geq 1$.

In each case the homotopy groups are computed by means of the equivalences $\Gamma_0 B\mathcal{A} \simeq F_\psi{}^q$ together with the long exact sequence of homotopy groups associated with the fibration

$$F_{\psi}{}^{q} \to X \xrightarrow{\psi^{q}-1} Y$$

(cf. 3.3 Step 1). These calculations are due to Friedlander [18]. We include them here for the sake of completeness and in order to rectify the single error in his table: namely that $\pi_1\Gamma_0 B\mathcal{O}(\mathbb{F}_q)$ is $\mathbb{Z}/2$ and not 0 if q is even.

<u>Theorem</u> 3.5. For $i \geq 1$ the homotopy groups of the indicated spaces are given in the following table

i(mod 8)	$\pi_i(\Gamma_0 B\mathcal{GL}(\mathbb{F}_q))$	q odd $\pi_i\Gamma_0 B\mathcal{O}(\mathbb{F}_q))$	q odd $\pi_i(\Gamma_0 B\mathcal{Sp}(\mathbb{F}_q))$	$\pi_i(\Gamma_0 B\mathcal{U}(\mathbb{F}_{q^2}))$
0	0	$\mathbb{Z}/2$	0	0
1	$\mathbb{Z}/q^{\frac{1}{2}(i+1)}-1$	$\mathbb{Z}/2 \oplus \mathbb{Z}/2$	0	$\mathbb{Z}/q^{\frac{1}{2}(i+1)}+1$
2	0	$\mathbb{Z}_2$	0	0
3	$\mathbb{Z}/q^{\frac{1}{2}(i+1)}-1$	$\mathbb{Z}/q^{\frac{1}{2}(i+1)}-1$	$\mathbb{Z}/q^{\frac{1}{2}(i+1)}-1$	$\mathbb{Z}/q^{\frac{1}{2}(i+1)}-1$
4	0	0	$\mathbb{Z}/2$	0
5	$\mathbb{Z}/q^{\frac{1}{2}(i+1)}-1$	0	$\mathbb{Z}/2 \oplus \mathbb{Z}/2$	$\mathbb{Z}/q^{\frac{1}{2}(i+1)}+1$
6	0	0	$\mathbb{Z}/2$	0
7	$\mathbb{Z}/q^{\frac{1}{2}(i+1)}-1$	$\mathbb{Z}/q^{\frac{1}{2}(i+1)}-1$	$\mathbb{Z}/q^{\frac{1}{2}(i+1)}-1$	$\mathbb{Z}/q^{\frac{1}{2}(i+1)}-1$

while for q even:

i (mod 8)	0	1	2	3	4	5	6	7
$\pi_i\Gamma_0 B\mathcal{O}^{ev}(\mathbb{F}_q)$	0	$\begin{cases}\mathbb{Z}/2 & \text{if } i = 1\\ 0 & \text{if } i > 1\end{cases}$	0	$\mathbb{Z}/q^{\frac{1}{2}(i+1)}-1$	0	0	0	$\mathbb{Z}/q^{\frac{1}{2}(i+1)}-1$
$\pi_i\Gamma_0 B\mathcal{S}p(\mathbb{F}_q)$	0	0	0	$\mathbb{Z}/q^{\frac{1}{2}(i+1)}-1$	0	0	0	$\mathbb{Z}/q^{\frac{1}{2}(i+1)}-1$

§4. Mod p behavior of the classical groups

This section is devoted to showing that the p-primary components of the spaces $\Gamma_0 B\mathcal{A}$ are trivial for the categories $\mathcal{A} = \mathcal{GL}(\mathbb{F}_q)$, $\mathcal{S}p(\mathbb{F}_q)$, $\mathcal{U}(\mathbb{F}_{q^2})$, $\mathcal{O}(\mathbb{F}_q)$ q odd and $\mathcal{DO}^{ev}(\mathbb{F}_q)$ q even, where p is the characteristic of the field. We shall prove this by making a partial computation of the mod p cohomology of the finite groups making up these categories. The arguments employed are virtually identical to those in Friedlander [18]. We include them here for the sake of completeness and because in the case of the orthogonal groups over a field of even characteristic [18] contains an error that we shall rectify.

The key step in our argument is the following result.

Lemma 4.1. Let μ be a generator of $\mathbb{F}_q^*$ and let $\lambda = \mu^b$ act on an $\mathbb{F}_q$-vector space V by multiplication, where b divides p - 1. Then

$$\lambda^*: H^n(BV; \mathbb{Z}/p) \to H^n(BV; \mathbb{Z}/p)$$

has no invariants for $0 < n < d(p-1)/b$, where $q = p^d$.

Proof: If p is odd then

$$H^*(BV; \mathbb{Z}/p) = E\{V\} \otimes_{\mathbb{Z}/p} S[V]$$

where E{V} is the exterior algebra of V over $\mathbb{Z}/p$ in dimension 1, and

S[V] is the symmetric algebra of V over $\mathbb{Z}/p$ in dimension 2. If $p = 2$ then

$$H^*(BV; \mathbb{Z}/p) = S[V]$$

where S[V] is the symmetric algebra of V over $\mathbb{Z}/p$ in dimension 1.

Obviously the minimal polynomial of

$$\mu: V \to V$$

regarded as a map of $\mathbb{F}_p$ vector spaces is equal to $p(x)$ the minimal polynomial (of degree d) of the element μ in $\mathbb{F}_q$ over the base field $\mathbb{F}_p$. The corresponding characteristic polynomial is $p(x)^m$ where m is the dimension of V over $\mathbb{F}_q$. Hence the eigenvalues of

$$\mu: V \to V$$

are $\{\mu, \mu^p, \mu^{p^2}, \ldots, \mu^{p^{d-1}}\}$ each with multiplicity m. Consequently the eigenvalues of

$$\lambda = \mu^b: V \to V$$

are $\{\lambda, \lambda^p, \lambda^{p^2}, \ldots, \lambda^{p^{d-1}}\}$.

Therefore λ_* on $E^i\{V\} \otimes S^j[V]$ has eigenvalues $\lambda^{e_0}(\lambda^p)^{e_1}(\lambda^{p^2})^{e_2} \ldots (\lambda^{p^{d-1}})^{e_{d-1}}$ with $\Sigma_{a=0}^{d-1} e_a = i + j$. If

$$\lambda_*: H^n(BV; \mathbb{Z}/p) \to H^n(BV; \mathbb{Z}/p)$$

has an invariant, then λ_* must have 1 as an eigenvalue on

$$H^n(BV;\ \mathbb{Z}/p) = \oplus_{i+2j=n} E^i[V] \otimes S^j[V] \ (\text{resp.} = S^n[V] \text{ if } p = 2)$$

This implies there is a relation

$$\Sigma_{a=0}^{d-1} e_a p^a = 0 \ (\text{mod } \frac{p^d-1}{b}),\ e_a \geq 0,\ \Sigma_{a=0}^{d-1} e_a \leq n$$

For the smallest such relation (i.e. with $\Sigma_{a=0}^{d-1} e_a$ minimal) we must have $e_a < p$ for all a. But this implies that the relation must be the p-adic expansion of $(p^d-1)/b$:

$$\Sigma_{a=0}^{d-1} \frac{p-1}{b} p^a = \frac{p^d-1}{b} \equiv 0 \ (\text{mod } (p^d-1)/b)$$

Hence the minimal value of $\Sigma_{a=0}^{d-1} e_d$ is $(p-1)d/b$. Consequently

$$n \geq \Sigma_{a=0}^{d-1} e_a \geq (p-1)d/b$$

This completes the proof.

We now employ Lemma 4.1 to prove a vanishing result for the cohomology of the general linear groups.

<u>Lemma</u> 4.2. If $q = p^d$, then

$$H^i(BGL(n,\mathbb{F}_q);\ \mathbb{Z}/p) = 0$$

for $0 < i < d(p-1)$ and all $n > 0$.

<u>Proof</u>: A Sylow p-subgroup of $GL(n,\mathbb{F}_q)$ is $T(n,\mathbb{F}_q)$ the subgroup of upper triangular matrices. Since

$$H^i(BGL(n,\mathbb{F}_q);\ \mathbb{Z}/p) \hookrightarrow H^i(BT(n,\mathbb{F}_q);\ \mathbb{Z}/p)$$

it suffices to prove the same result for $T(n,\mathbb{F}_q)$. We proceed by induction on n. For n = 1, $T(1,\mathbb{F}_q) = \mathbb{F}_q^*$ is group of order prime to p so

$$H^i(BT(1,\mathbb{F}_q);\ \mathbb{Z}/p) = 0 \text{ all } i > 0$$

For n > 1, there is a short exact sequence

$$(4.3)\qquad 1 \to R(n,\mathbb{F}_q) \to T(n,\mathbb{F}_q) \xrightarrow{\pi} T(n-1,\mathbb{F}_q) \to 1$$

where π is projection onto the lower right hand corner and $R(n,\mathbb{F}_q)$ is the "upper row" subgroup of $T(n,\mathbb{F}_q)$.

Now $R(n,\mathbb{F}_q)$ is the semidirect product of $\mathbb{F}_q^*$ and $\mathbb{F}_q^{n-1} = V$, where $\mathbb{F}_q^*$ acts on V by multiplication. We consider the Serre spectral sequence in $H^*(;\ \mathbb{Z}/p)$ of

$$0 \to V \to R(n,\mathbb{F}_q) \to \mathbb{F}_q^* \to 1$$

Since $\mathbb{F}_q^*$ has order prime to p, all the E_2 terms vanish with the possible exception of

$$E_2^{0,i} = H^0(B\mathbb{F}_q^*;\ H^i(BV;\ \mathbb{Z}/p)) = \text{invariants of } H^i(BV;\ \mathbb{Z}/p)$$

By Lemma 4.1, $H^i(BR(n,\mathbb{F}_q);\ \mathbb{Z}/p) = E_\infty^{0,i} = E_2^{0,i} = 0$ for $0 < i < d(p-1)$. Applying this together with the induction hypothesis to the Serre spectral sequence in $H^*(\cdot;\ \mathbb{Z}/p)$ of (4.3) implies that

$$H^i(BT(n,\mathbb{F}_q);\ \mathbb{Z}/p) = 0$$

for $0 < i < d(p-1)$. This completes the induction and proof.

We next turn to the problem of generalizing this result to the other classical groups. In what follows $T(n,\mathbb{F}_q)$ will denote the subgroup of upper triangular matrices in $GL(n,\mathbb{F}_q)$.

Lemma 4.4. There are (semi-split) short exact sequences

(i) $0 \to R(SO(2m,\mathbb{F}_q)) \to \Delta(SO(2m,\mathbb{F}_q)) \xrightarrow{\pi} T(m,\mathbb{F}_q) \to 1$ q odd

(ii) $0 \to R(DO(2m,\mathbb{F}_q)) \to \Delta(DO(2m,\mathbb{F}_q)) \xrightarrow{\pi} T(m,\mathbb{F}_q) \to 1$ q even

(iii) $0 \to R(Sp(2m,\mathbb{F}_q)) \to \Delta(Sp(2m,\mathbb{F}_q)) \xrightarrow{\pi} T(m,\mathbb{F}_q) \to 1$

(iv) $0 \to R(U(2m,\mathbb{F}_{q^2})) \to \Delta(U(2m,\mathbb{F}_{q^2})) \xrightarrow{\pi} T(m,\mathbb{F}_{q^2}) \to 1$

where $\Delta(SO(2m,\mathbb{F}_q))$ (resp. $\Delta(DO(2m,\mathbb{F}_q))$, $\Delta(Sp(2m,\mathbb{F}_q))$, $\Delta(U(2m_{q^2},\mathbb{F}_q))$) contains a Sylow p-subgroup of $SO(2m,\mathbb{F}_q)$ (resp. $DO(2m,\mathbb{F}_q)$), $Sp(2m,\mathbb{F}_q)$, $U(2m,\mathbb{F}_{q^2})$) and $R(SO(2m,\mathbb{F}_q))$ (resp. $R(DO(2m,\mathbb{F}_q))$, $R(Sp(2m,\mathbb{F}_q))$, $R(U(2m,\mathbb{F}_{q^2}))$ is a $\mathbb{F}_q$ vector space on which the scalar matrix λI in $T(m,\mathbb{F}_q)$ (resp. in $T(m,\mathbb{F}_{q^2})$) acts by multiplication by λ^{-2} (resp. λ^{-q-1} in the unitary case).

Proof: First let us recall the hyperbolic forms constructed in II 8.4. If $E\colon \mathbb{F}_q^m \times \mathbb{F}_q^m \to \mathbb{F}_q$, $H\colon \mathbb{F}_{q^2}^m \times \mathbb{F}_{q^2}^m \to \mathbb{F}_{q^2}$ are the standard symmetric and Hermitian forms:

$$E(x,y) = \Sigma_{i=1}^m x_i y_i$$

$$H(x,y) = \Sigma_{i=1}^m x_i \bar{y}_i,$$

then

$$Q_h\colon \mathbb{F}_q^m \oplus \mathbb{F}_q^m \to \mathbb{F}_q \quad Q_h((x,y)) = E(x,y)$$

$$A_h: (\mathbb{F}_q^m \oplus \mathbb{F}_q^m) \times (\mathbb{F}_q^m \oplus \mathbb{F}_q^m) \to \mathbb{F}_q\ ,\ A_h((x,y),(z,w)) = E(x,w) - E(y,z)$$

$$H_h: (\mathbb{F}_{q^2}^m \oplus \mathbb{F}_{q^2}^m) \times (\mathbb{F}_{q^2}^m \oplus \mathbb{F}_{q^2}^m) \to \mathbb{F}_{q^2}\ ,\ H_h((x,y),(z,w)) = H(x,w) + H(y,z)$$

are quadratic, symplectic and Hermitian forms. If $A \in GL(m,\mathbb{F}_q)$ we denote by A^* the conjugate of A with respect to E, i.e. A^* is defined by

$$E(A^*x,y) = E(x,Ay)$$

(In matrix formulation A^* is just the transpose of A). If $A \in GL(m,\mathbb{F}_{q^2})$ denote by A^* the conjugate of A with respect to H, i.e. A^* is defined by

$$H(A^*x,y) = H(x,Ay)$$

(In matrix formulation A^* is just the conjugate transpose of A).

Now define $R(SO(2m,\mathbb{F}_q))$, $R(DO(2m,\mathbb{F}_q))$, $R(Sp(2m,\mathbb{F}_q))$, $R(U(2m,\mathbb{F}_{q^2}))$ to be the additive groups of matrices

$$R(SO(2m,\mathbb{F}_q)) = \{B \mid E(Bx,x) = 0 \text{ for all } x \in \mathbb{F}_q^m\}$$

$$= \text{antisymmetric } m \times m \text{ matrices over } \mathbb{F}_q$$

$$R(DO(2m,\mathbb{F}_q)) = \{B \mid E(Bx,x) = 0 \text{ for all } x \in \mathbb{F}_q^m\}$$

$$= \text{symmetric } m \times m \text{ matrices over } \mathbb{F}_q \text{ with } 0\text{'s on the main diagonal}$$

$$R(Sp(2m,\mathbb{F}_q)) = \{B \mid E(Bx,y) = E(x,By) \text{ for all } x,y \in \mathbb{F}_q^m\}$$

$$= \text{symmetric } m \times m \text{ matrices over } \mathbb{F}_q$$

$$R(U(2m,\mathbb{F}_{q^2}) = \{B \mid H(Bx,y) = -H(x,By) \text{ for all } x,y \in \mathbb{F}_{q^2}^m\}$$

$$= \text{conjugate antisymmetric } m \times m \text{ matrices over } \mathbb{F}_{q^2}$$

Now define $\Delta(SO(2m,\mathbb{F}_q))$, $\Delta(DO(2m,\mathbb{F}_q))$, $\Delta(Sp(2m,\mathbb{F}_q)$, $\Delta(U(2m,\mathbb{F}_{q^2}))$ to be the semidirect product

$$\Delta(SO(2m,\mathbb{F}_q)) = T(m,\mathbb{F}_q) \ltimes R(SO(2m,\mathbb{F}_q))$$

$$\Delta(DO(2m,\mathbb{F}_q)) = T(m,\mathbb{F}_q) \ltimes R(DO(2m,\mathbb{F}_q))$$

$$\Delta(Sp(2m,\mathbb{F}_q)) = T(m,\mathbb{F}_q) \ltimes R(Sp(2m,\mathbb{F}_q))$$

$$\Delta(U(2m,\mathbb{F}_{q^2})) = T(m,\mathbb{F}_{q^2}) \ltimes R(U(2m,\mathbb{F}_{q^2}))$$

with the first factor acting on the second from the right by the formula

$$B \cdot A = A^{-1}B(A^{-1})^*$$

Now define imbeddings

$$\Delta(SO(2m,\mathbb{F}_q)) \to SO(2m,\mathbb{F}_q)$$

$$\Delta(DO(2m,\mathbb{F}_q)) \to DO(2m,\mathbb{F}_q)$$

$$\Delta(Sp(2m,\mathbb{F}_q)) \to Sp(2m,\mathbb{F}_q)$$

$$\Delta(U(2m,\mathbb{F}_{q^2})) \to U(2m,\mathbb{F}_{q^2})$$

by making (A,B) act on $\mathbb{F}_q^m \oplus \mathbb{F}_q^n$ ($\mathbb{F}_{q^2}^m \oplus \mathbb{F}_{q^2}^m$ in the unitary case) according to

$$(A,B)(x,y) = (Ax + ABy, (A^*)^{-1}y).$$

It is easily checked that these imbeddings are group homomorphisms. A straightforward computation of the orders of the groups shows that $\Delta(SO(2m,\mathbb{F}_q))$, $\Delta(DO(2m,\mathbb{F}_q))$, $\Delta(Sp(2m,\mathbb{F}_q))$, $\Delta(U(2m,\mathbb{F}_{q^2}))$ contains a Sylow p-subgroup of $SO(2m,\mathbb{F}_q)$, $DO(2m,\mathbb{F}_q)$, $Sp(2m,\mathbb{F}_q)$, $U(2m,\mathbb{F}_{q^2})$ respectively.

The other statements of the lemma follow immediately.

We can now prove the analog of Lemma 4.2 for the other classical groups

<u>Lemma</u> 4.5. If $q = p^d$ then

(i) $H^i(BSO(2m,\mathbb{F}_q);\ \mathbb{Z}/p) = 0$ q odd $\quad 0 < i < \frac{1}{2}d(p-1)$

(ii) $H^i(BDO(2m,\mathbb{F}_q);\ \mathbb{Z}/p) = 0$ q even $\quad 0 < i < d(p-1)$

(iii) $H^i(BSp(2m,\mathbb{F}_q);\ \mathbb{Z}/p) = 0 \quad 0 < i < \frac{1}{2}d(p-1)$

(iv) $H^i(BU(2m,\mathbb{F}_{q^2});\ \mathbb{Z}/p) = 0 \quad 0 < i < d(p-1)$

for all m.

<u>Proof:</u> We give the proof for $U(2m,\mathbb{F}_{q^2})$; the other cases can be proved by a similar argument.

Since $\Delta(U(2m,\mathbb{F}_{q^2}))$ contains a Sylow p-subgroup of $U(2m,\mathbb{F}_{q^2})$ it suffices to prove that

$$H^i(B\Delta(U(2m,\mathbb{F}_{q^2}));\ \mathbb{Z}/p) = 0 \quad 0 < i < \tfrac{1}{2}d(p-1)$$

We use the map of Serre spectral sequences in $H^*(\ ;\ \mathbb{Z}/p)$ of the diagram

$$\begin{array}{ccccccccc} 1 \to & \pi^{-1}(\mathbb{F}^*_{q^2}) & \to & \Delta(U(2m,\mathbb{F}_{q^2})) & \to & \Delta(U(2m,\mathbb{F}_{q^2}))/\pi^{-1}(\mathbb{F}^*_{q^2}) & \to 1 \\ & \downarrow \pi & & \downarrow \pi & & \approx \downarrow \pi & \\ 1 \to & \mathbb{F}^*_{q^2} & \longrightarrow & T(m,\mathbb{F}_{q^2}) & \longrightarrow & T(m,\mathbb{F}_{q^2})/\mathbb{F}^*_{q^2} & \longrightarrow 1 \end{array}$$

Since $\mathbb{F}^*_{q^2}$ has order prime to p, the spectral sequence of the bottom row collapses and we have

$$E_\infty^{r,s} = E_2^{r,s} = H^r(B(T(m,\mathbb{F}_{q^2})/\mathbb{F}^*_{q^2}); H^s(B\mathbb{F}^*_{q^2}; \mathbb{Z}/p))$$

$$= \begin{cases} 0 & s \neq 0 \\ H^r(B(T(m,\mathbb{F}_{q^2})/\mathbb{F}^*_{q^2}); \mathbb{Z}/p) = H^r(BT(m,\mathbb{F}_{q^2}); \mathbb{Z}/p) & s = 0 \end{cases}$$

From Lemma 4.2 it follows that

$$H^i(B(T(m,\mathbb{F}_{q^2})/\mathbb{F}^*_{q^2}); \mathbb{Z}/p) = 0 \qquad 0 < i < 2d(p-1)$$

Next we observe that $\pi^{-1}(\mathbb{F}^*_{q^2})$ is a semidirect product

$$\pi^{-1}(\mathbb{F}^*_{q^2}) = \mathbb{F}^*_{q^2} \ltimes R(U(2m,\mathbb{F}_{q^2}))$$

with $\lambda \in \mathbb{F}^*_{q^2}$ acting on $R(U(2m,\mathbb{F}_{q^2}))$ by multiplication by $\lambda^{-(q+1)}$. Since $\mathbb{F}^*_{q^2}$ has order prime to p the Serre spectral sequence in $H^*(\ ; \mathbb{Z}/p)$ of

$$0 \to R(U(2m,\mathbb{F}_{q^2})) \to \pi^{-1}(\mathbb{F}^*_{q^2}) \to \mathbb{F}^*_{q^2} \to 1$$

collapses and we have

$$H^i(B\pi^{-1}(\mathbb{F}_{q^2}); \mathbb{Z}/p) = H^0(B\mathbb{F}^*_{q^2}; H^i(BR(U(2m,\mathbb{F}_{q^2})); \mathbb{Z}/p))$$
$$= \text{invariants of } H^i(BR(U(2m,\mathbb{F}_{q^2})); \mathbb{Z}/p)$$
$$\text{under the action of } \mathbb{F}^*_{q^2}$$

Now if μ is a generator of $\mathbb{F}^*_{q^2}$ then μ acts on $R(U(2m,\mathbb{F}_{q^2}))$ by multiplication by $\lambda = \mu^{-(q+1)}$. Since λ is a generator of $\mathbb{F}_q$ and we can regard $R(U(2m,\mathbb{F}_{q^2}))$ as an $\mathbb{F}_q$-vector space, it follows from Lemma 4.1 that

$$H^i(B\pi^{-1}(\mathbb{F}_{q^2}); \mathbb{Z}/p) = 0 \qquad 0 < i < d(p-1)$$

Now returning to the Serre spectral sequence of the top row of 4.6 we see that

$$E_2^{r,s} = H^r(B(\Delta(U(2m,\mathbb{F}_{q^2}))/\pi^{-1}(\mathbb{F}^*_{q^2})); H^s(B\pi^{-1}(\mathbb{F}_{q^2}); \mathbb{Z}/p))$$
$$= H^r(B(T(m,\mathbb{F}_{q^2})/\mathbb{F}^*_{q^2}); H^s(B\pi^{-1}(\mathbb{F}_{q^2}); \mathbb{Z}/p))$$
$$= 0$$

if $0 < s < d(p-1)$ or if $s = 0$ and $0 < r < 2d(p-1)$. It follows that

$$H^i(B\Delta(U(2m,\mathbb{F}_{q^2})); \mathbb{Z}/p) = 0$$

if $0 < i < d(p-1)$.

Putting these results together we are able to show that the p-primary components of $\Gamma_0 B\mathfrak{a}$ are trivial.

<u>Theorem</u> 4.6. Let $q = p^d$, p prime. Then

(a) $\Gamma_0 B\mathcal{GL}(\mathbb{F}_q)_{(p)} \simeq *$

(b) $\Gamma_0 B\mathcal{Sp}(\mathbb{F}_q)_{(p)} \simeq *$

(c) $\Gamma_0 B\mathcal{U}(\mathbb{F}_{q^2})_{(p)} \simeq *$

(d) $\Gamma_0 B\mathcal{O}(\mathbb{F}_q)_{(p)} \simeq *$ p odd

(e) $\Gamma_0 B\mathcal{DO}^{ev}(\mathbb{F}_q)_{(p)} \simeq *$ $p = 2$

<u>Proof</u>: We use Theorem II 8.6. We obtain that

$$\Gamma_0 B\mathcal{GL}(\mathbb{F}_q) \xrightarrow{X} \Gamma_0 B\mathcal{GL}(\mathbb{F}_{q^r}) \xrightarrow{E} \Gamma_0 B\mathcal{GL}(\mathbb{F}_q)$$

$$\Gamma_0 B\mathcal{Sp}(\mathbb{F}_q) \xrightarrow{X} \Gamma_0 B\mathcal{Sp}(\mathbb{F}_{q^r}) \xrightarrow{E} \Gamma_0 B\mathcal{Sp}(\mathbb{F}_q)$$

$$\Gamma_0 B\mathcal{U}(\mathbb{F}_{q^2}) \xrightarrow{X} \Gamma_0 B\mathcal{U}(\mathbb{F}_{q^{2r}}) \xrightarrow{E} \Gamma_0 B\mathcal{U}(\mathbb{F}_{q^2}),\ r \text{ odd}$$

$$\Gamma_0 B\mathcal{O}(\mathbb{F}_q) \xrightarrow{X} \Gamma_0 B\mathcal{O}(\mathbb{F}_{q^r}) \xrightarrow{E} \Gamma_0 B\mathcal{O}(\mathbb{F}_q),\ q,r \text{ odd}$$

$$\Gamma_0 B\mathcal{O}^{ev}(\mathbb{F}_q) \xrightarrow{X} \Gamma_0 B\mathcal{O}^{ev}(\mathbb{F}_{q^r}) \xrightarrow{E} \Gamma_0 B\mathcal{O}^{ev}(\mathbb{F}_q),\ q \text{ even}$$

induces multiplication by r on homotopy groups. By Lemmas 4.2 and 4.5

$$H^i(\Gamma_0 B\mathcal{A};\ Z/p) = \varinjlim_n H^i(B\mathcal{A}_n;\ \mathbb{Z}/p) = 0 \text{ for } 0 < i < \tfrac{1}{2}dr(p-1)$$

for $\mathcal{A} = \mathcal{GL}(\mathbb{F}_{q^r}), \mathcal{Sp}(\mathbb{F}_{q^r}), \mathcal{U}(\mathbb{F}_{q^{2r}}), \mathcal{O}(\mathbb{F}_{q^r}), \mathcal{DO}^{ev}(\mathbb{F}_{q^r})$. Since $\Gamma_0 B\mathcal{A}$ is an infinite loop space, it follows from the universal coefficient theorem and the Hurewicz isomorphism theorem that $\pi_i\Gamma_0 B\mathcal{A}_{(p)} = 0$ for $i < \frac{1}{2}dr(p-1)$. From II 7.21 it also follows that $\pi_i\Gamma_0 B\mathcal{O}^{ev}(\mathbb{F}_{q^r})_{(p)} = 0$ for $1 < i < \frac{1}{2}dr(p-1)$ when q is even.

Taking r prime to p, we conclude that multiplication by r in $\pi_i\Gamma_0 B\mathcal{B}_{(p)}$ is zero for $\mathcal{B} = \mathcal{GL}(\mathbb{F}_q), \mathcal{Sp}(\mathbb{F}_q), \mathcal{U}(\mathbb{F}_{q^2}), \mathcal{O}(\mathbb{F}_q)$

q odd, $\mathcal{O}^{ev}(\mathbb{F}_q)$ q even, and for $i < \frac{1}{2}dr(p-1)$ ($1 < i < \frac{1}{2}dr(p-1)$ in the last case). Since multiplication by r is an isomorphism on the p-primary components, it follows that $\pi_i\Gamma_0 B\mathcal{B}_{(p)} = 0$ for $i < \frac{1}{2}dr(p-1)$ ($1 < i < \frac{1}{2}dr(p-1)$ in the case $\mathcal{B} = \mathcal{O}^{ev}(\mathbb{F}_q)$, q even). Since r can be taken arbitrarily large, we conclude that $\pi_i\Gamma_0 B\mathcal{B}_{(p)} = 0$ for all i except that when q is even $\pi_1\Gamma_0 B\mathcal{O}^{ev}(\mathbb{F}_q)_{(2)} = \mathbb{Z}/2$. From II 7.21 we conclude that $\pi_i\Gamma_0 B\mathcal{D}\mathcal{O}^{ev}(\mathbb{F}_q)_{(2)} = 0$ for all i. This completes the proof.

§5. General pattern of arguments at noncharacteristic primes: determining homology generators and relations for the classical groups.

In this section we outline patterns of arguments used to prove that the Brauer lift induces homology isomorphisms

$$\lambda: H_*(\Gamma_0 B\mathcal{A}; \mathbb{Z}/\ell) \to H_*(F\psi^q; \mathbb{Z}/\ell)$$

for primes $\ell \neq p$ the characteristic of $\mathbb{F}_q$ for any of the categories $\mathcal{A} = \mathcal{GL}(\mathbb{F}_q), \mathcal{O}(\mathbb{F}_q)^*, \mathcal{Sp}(\mathbb{F}_q), \mathcal{U}(\mathbb{F}_{q^2})$. Since complete information about the homology of the spaces $F\psi^q$ was obtained in Chapter I, it remains (a) to obtain similar results about $H_*(\Gamma_0 B\mathcal{A}; \mathbb{Z}/\ell)$ and (b) to relate these results to the information we already possess about $H_*(F\psi^q; \mathbb{Z}/\ell)$. In this section we address ourselves to the first problem. The second problem is treated in the succeeding section.

Our first task will be to determine generators for the homology algebras $H_*(\Gamma_0 B\mathcal{A}, \mathbb{Z}/\ell)$. According to II 2.17

*Note that if q is even then it follows from II 7.21 that $\Gamma_0 B\mathcal{D}\mathcal{O}^{ev}(\mathbb{F}_q)$ is equivalent to $\Gamma_0 B\mathcal{O}^{ev}(\mathbb{F}_q)$ at ℓ. Since there are no substantial difference in the calculations between the cases $\mathcal{O}(\mathbb{F}_q)$ q odd and $\mathcal{O}^{ev}(\mathbb{F}_q)$ q even we shall subsume the second case under the former.

$$H_*(\Gamma B\mathcal{O}\!\mathcal{L}; \mathbb{Z}/\ell) \cong H_*(B\mathcal{O}\!\mathcal{L}; \mathbb{Z}/\ell)[\pi_0\mathcal{O}\!\mathcal{L}]^{-1}$$

$$H_*(\Gamma_0 B\mathcal{O}\!\mathcal{L}; \mathbb{Z}/\ell) \cong \varinjlim_{a\in\pi_0\mathcal{O}\!\mathcal{L}} H_*(B\mathcal{O}\!\mathcal{L}_a; \mathbb{Z}/\ell)$$

Moreover in each case $B\mathcal{O}\!\mathcal{L}_a \simeq BG(a)$ where $G(a)$ is the automorphism group of an object representing $a\in\pi_0\mathcal{O}\!\mathcal{L}$ (cf. II 2.16(vii)). Therefore we begin by looking for generators of the homology algebra

$$H_*(B\mathcal{O}\!\mathcal{L}; \mathbb{Z}/\ell) \cong \bigoplus_{a\in\pi_0\mathcal{O}\!\mathcal{L}} H_*(B\mathcal{O}\!\mathcal{L}_a; \mathbb{Z}/\ell) \cong \bigoplus_{a\in\pi_0\mathcal{O}\!\mathcal{L}} H_*(BG(a); \mathbb{Z}/\ell)$$

(cf. II 2.16(viii)) which for the various categories of interest to us will look like

$$H_*(B\mathcal{GL}(\mathbb{F}_q); \mathbb{Z}/\ell) \cong \oplus_{n\in\mathbb{N}} H_*(BGL(n,\mathbb{F}_q); \mathbb{Z}/\ell)$$

$$H_*(B\mathcal{O}(\mathbb{F}_q); \mathbb{Z}/\ell) \cong \oplus_{n\in\mathbb{N}} H_*(BO(n,\mathbb{F}_q); \mathbb{Z}/\ell) \quad q \text{ odd}$$

$$H_*(B\mathcal{Sp}(\mathbb{F}_q); \mathbb{Z}/\ell) \cong \oplus_{n\in\mathbb{N}} H_*(BSp(2n,\mathbb{F}_q); \mathbb{Z}/\ell)$$

$$H_*(B\mathcal{U}(\mathbb{F}_{q^2}); \mathbb{Z}/\ell) \cong \oplus_{n\in\mathbb{N}} H_*(BU(n,\mathbb{F}_{q^2}); \mathbb{Z}/\ell)$$

$$H_*(B\bar{\mathcal{O}}(\mathbb{F}_q); \mathbb{Z}/\ell) \cong \oplus_{n\in\tilde{\mathbb{N}}} H_*(BO(n,\mathbb{F}_q); \mathbb{Z}/\ell) \quad q \text{ odd}$$

$$H_*(B\bar{\mathcal{O}}^{ev}(\mathbb{F}_q); \mathbb{Z}/\ell) \cong \oplus_{n\in\tilde{\mathbb{N}}^{ev}} H_*(BO(n;\mathbb{F}_q); \mathbb{Z}/\ell) \quad q \text{ even}$$

cf. II 4.7 re notation

(The product in $H_*(B\mathcal{O}\!\mathcal{L}; \mathbb{Z}/\ell)$ is given as in II 2.18)

Since the groups $G(n)$ get fairly large and complicated as n becomes large, our approach will be to compute $H_*(BG(n); \mathbb{Z}/\ell)$ for small n and then relate the homology of the larger groups to that of the smaller groups. We have several means for accomplishing this reduction. One of the simplest is the following well known result:

Lemma 5.1. If H is a subgroup of a finite group G which contains a Sylow ℓ-subgroup of G, then the inclusion maps induces a surjection in homology

$$i_*: H_*(BH; \mathbb{Z}/\ell) \to H_*(BG; \mathbb{Z}/\ell)$$

Proof: We pass to the dual statement in cohomology: is

$$i^*: H^*(BG; \mathbb{Z}/\ell) \to H^*(BH; \mathbb{Z}/\ell)$$

injective? However we know that the composite

$$H^*(BG; \mathbb{Z}/\ell) \xrightarrow{i^*} H^*(BH; \mathbb{Z}/\ell) \xrightarrow{tr} H^*(BG; \mathbb{Z}/\ell)$$

is multiplication by the index [G; H] where tr denotes the transfer homomorphism. Since H contains a Sylow ℓ-subgroup of G, [G; H] is relatively prime to ℓ and $tr \cdot i^*$ is an isomorphism. Therefore i^* is an injection and i_* is a surjection.

In order to exploit this result in our situation we have to introduce the notion of wreath products

Definition 5.2. Let G be a group and let P be a permutation group on m letters. The wreath product of P and G is defined to be the semidirect product

$$P \wr G = P \ltimes G^m$$

where P acts on the m-fold direct product of G by permuting the factors. More precisely the product in $P \wr G$ is given by

$$(\alpha, g_1, g_2, \ldots, g_m) \cdot (\beta, h_1, h_2, \ldots, h_m) = (\alpha\beta, g_{\alpha(1)}h_1, \ldots, g_{\alpha(m)}h_m)$$

If G acts on a vector space V then $P \wr G$ acts on the m-fold direct sum $V \oplus V \oplus \cdots \oplus V$ according to the formula

$$(\alpha, g_1, g_2, \ldots, g_m)(v_1, v_2, \ldots, v_m)$$

$$= (g_{\alpha^{-1}(1)} v_{\alpha^{-1}(1)}, g_{\alpha^{-1}(2)} v_{\alpha^{-1}(2)}, \ldots, g_{\alpha^{-1}(m)} v_{\alpha^{-1}(m)})$$

Thus if $G(n)$ is one of the classical groups, $P \wr G(n)$ can be regarded as a subgroup of $G(mn)$.

5.3. Properties of the wreath product

(i) $P \wr (G \times H) \cong (P \wr G) \times (P \wr H)$

(ii) $(P \times P') \wr G \cong (P \wr G) \times (P' \wr G)$

(iii) $(P \wr P') \wr G \cong P \wr (P' \wr G)$

(iv) $P \wr G$ contains as subgroups $G^m = \{(1, g_1, g_2, \ldots, g_m) \mid g_i \in G\}$ and $P \times G = \{(\alpha, g, g, \ldots, g) \mid \alpha \in P, g \in G\}$.

The relevance of wreath products to computing the homology of classical groups over a finite field lies in the fact that for certain small c, $\mathcal{S}_m \wr G(c)$ contains a Sylow ℓ-subgroup of $G(mc)$. Hence Lemma 5.1 can be immediately applied. More precisely

Proposition 5.4(a) For $G = GL(,\mathbb{F}_q)$, $O(; \mathbb{F}_q)$, $Sp(; \mathbb{F}_q)$ or $U(; \mathbb{F}_{q^2})$ with q odd, the inclusion map

$$i: \mathcal{S}_m \wr G(2) \to G(2m)$$

induces an epimorphism in homology

$$i_*: H_*(B\mathcal{S}_m \wr G(2); \mathbb{Z}/2) \to H_*(BG(2m); \mathbb{Z}/2)$$

(b) If ℓ is odd, let c be the smallest index for which $G(c)$ has order divisible by ℓ. (If $G = O(;\mathbb{F}_q)$ q odd the indices range over the ordered monoid $\tilde{\mathbb{N}}$ defined in II 4.7 rather than over the nonnegative integers. If $G = O(;\mathbb{F}_q)$ q even, then the indices range over $\tilde{\mathbb{N}}^{ev}$). Then the inclusion map

$$i: \mathcal{S}_m \wr G(c) \to G(mc)$$

induces an epimorphism in homology

$$i_*: H_*(B\mathcal{S}_m \wr G(c);\ \mathbb{Z}/\ell) \to H_*(BG(mc);\ \mathbb{Z}/\ell)$$

Proof: One computes the orders of $\mathcal{S}_m \wr G(c)$ and of $G(mc)$ to see that $\mathcal{S}_m \wr G(c)$ contains a Sylow ℓ-subgroup of $G(mc)$ and then applies Lemma 5.1. Detailed computations appear in Chapter VII for the case $\ell = 2$ and in Chapter VIII for the case ℓ odd.

This reduction simplifies our problem considerably, for wreath products of groups have been studied extensively and much is known about their homology and cohomology. In [34] Quillen deduced a result in this area which is basic to the calculation of the $H_*(BG(n);\ \mathbb{Z}/\ell)$. We first need a definition.

Definition 5.5. A collection $\{H_i\}_{i\in I}$ of subgroups of G is said to detect $H^*(BG;\ \mathbb{Z}/\ell)$ or $H_*(BG;\ \mathbb{Z}/\ell)$ if the inclusion maps induce a monomorphism in cohomology

$$H^*(BG;\ \mathbb{Z}/\ell) \to \Pi_{i\in I} H^*(BH_i;\ \mathbb{Z}/\ell)$$

or dually an epimorphism on homology

$$\oplus_{i \in I} H_*(BH_i;\ \mathbb{Z}/\ell) \to H_*(BG;\ \mathbb{Z}/\ell)$$

5.6. <u>Quillen's Lemma</u>. Let $\mathbb{Z}/\ell$ be the cyclic subgroup of $\mathcal{S}_\ell$ generated by the ℓ-cycle $(1,2,\dots,\ell)$. Let G be any finite group. Then the subgroups G^ℓ and $\mathbb{Z}/\ell \times G$ of $\mathbb{Z}/\ell \wr G$ detect $H_*(B\mathbb{Z}/\ell \wr G;\ \mathbb{Z}/\ell)$.

Moreover under favorable circumstances (which occur in the cases of interest to us) we can get a further reduction using the following result.

<u>Lemma</u> 5.7. If the subgroups H_1, H_2 detect $H_*(BG;\ \mathbb{Z}/\ell)$ and H_2 is conjugate in G to a subgroup of H_1, then H_1 alone also detects $H_*(BG;\ \mathbb{Z}/\ell)$.

<u>Proof</u>: Let $g \in G$ be an element such that $gH_2g^{-1} = K \subseteq H_1$. Then we have a commutative diagram

$$\begin{array}{ccccc} H_*(BH_2;\ \mathbb{Z}/\ell) & & \xrightarrow{\quad i_{2*} \quad} & & H_*(BG;\ \mathbb{Z}/\ell) \\ \downarrow{\scriptstyle \overline{c}_{g*}} & & & & \downarrow{\scriptstyle c_{g*}} \\ H_*(BK;\ \mathbb{Z}/\ell) & \xrightarrow{j_*} & H_*(BH_1;\ \mathbb{Z}/\ell) & \xrightarrow{i_{1*}} & H_*(BG;\ \mathbb{Z}/\ell) \end{array}$$

where $c_g, \overline{c}_g$ denotes conjugation by g. But conjugation by an element of a group induces the identity map on the homology of the group. Hence $c_{g*} = 1_{H_*(BG;\ \mathbb{Z}/\ell)}$ so $i_{2*} = i_{1*} j_* \overline{c}_{g*}$ and the following diagram commutes

$$\begin{array}{ccc} H_*(BH_1;\ \mathbb{Z}/\ell) \oplus H_*(BH_2;\ \mathbb{Z}/\ell) & \xrightarrow{i_{1*} + i_{2*}} & \\ \downarrow{\scriptstyle 1_{H_*}(BH;\ \mathbb{Z}/\ell) + j_* \cdot \overline{c}_{g*}} & & H_*(BG;\ \mathbb{Z}/\ell) \\ H_*(BH_1;\ \mathbb{Z}/\ell) & \xrightarrow{i_{1*}} & \end{array}$$

Since $i_{1*} + i_{2*}$ is surjective, so is i_{1*}. Hence H_1 detects $H_*(BG; \mathbb{Z}/\ell)$.

Using Quillen's Lemma and Lemma 5.7 we obtain the following dramatic simplification of Prop. 5.4.

Theorem 5.8. Let c be as in Prop. 5.4. Then the direct sum map

$$G(c)^m \to G(mc)$$

$$(T_1, T_2, \ldots, T_m) \to T_1 \oplus T_2 \oplus \cdots \oplus T_m$$

induces an epimorphism in homology

$$H_*(BG(c); \mathbb{Z}/\ell) \otimes H_*(BG(c); \mathbb{Z}/\ell) \otimes \cdots \otimes H_*(BG(c); \mathbb{Z}/\ell) \to H_*(BG(mc); \mathbb{Z}/\ell)$$

Proof: The direct sum map is just the standard inclusion

$$G(c)^m \to \mathcal{S}_m \wr G(c) \to G(mc)$$

cf. 5.2. Thus the problem amounts to improving Prop. 5.4 so as to show that $G(c)^m$ detects $H_*(BG(mc); \mathbb{Z}/\ell)$. The proof consists of several steps.

Step 1. $G(c)^{\ell^n}$ detects $H_*(BG(\ell^n c); \mathbb{Z}/\ell)$

We proceed by induction on n. For $n = 0$ there is nothing to prove. Assume it is true for n. We first observe by calculating orders that $\mathbb{Z}/\ell \wr G(\ell^n c)$ contains a Sylow ℓ-subgroup of $G(\ell^{n+1}c)$ and hence detects $H_*(BG(\ell^{n+1}c); \mathbb{Z}/\ell)$. By Quillen's Lemma $G(\ell^n c)^{\ell}$ and $\mathbb{Z}/\ell \times G(\ell^n c)$ detect $H_*(BG(\ell^{n+1}c); \mathbb{Z}/\ell)$. We then find an element $g \in G(\ell^{n+1}c)$ and that $g(\mathbb{Z}/\ell \times G(\ell^n c))g^{-1} \subseteq G(\ell^n c)^{\ell}$. Applying Lemma 5.7 we obtain that $G(\ell^n c)^{\ell}$ detects $H_*(BG(\ell^{n+1}c); \mathbb{Z}/\ell)$. Since by induction hypothesis $G(c)^{\ell^n}$ detects $H_*(BG(\ell^n c);\ \mathbb{Z}/\ell)$, it follows that $G(c)^{\ell^{n+1}} = (G(c)^{\ell^n})^{\ell}$ detects $H_*(BG(\ell^{n+1}c);\ \mathbb{Z}/\ell)$. This completes the induction.

Step 2. If $0 \leq k \leq \ell - 1$ then $G(\ell^n c)^k$ detects $H_*(BG(k\ell^n c);\ \mathbb{Z}/\ell)$

Note that this step is vacuous if $\ell = 2$. In any case the proof is by calculating the orders of $G(\ell^n c)^k$ and $G(k\ell^n c)$ and observing that $G(\ell^n c)^k$ contains a Sylow ℓ-subgroup of $G(k\ell^n c)$.

Step 3. $G(c)^m$ detects $H_*(BG(mc);\ \mathbb{Z}/\ell)$ for any m.

We first write m in ℓ-adic form

$$m = \Sigma_{i=0}^{k} a_i \ell^i$$

where $0 \leq a_i \leq \ell - 1$. One then shows by computing orders that the product $\Pi_{i=0}^{k} G(a_i \ell^i c)$ contains a Sylow ℓ-subgroup of $G(mc)$ and hence detects $H_*(BG(mc);\ \mathbb{Z}/\ell)$. But it follows from Steps 1 and 2 that $G(c)^{a_i \ell^i}$ detects $H_*(BG(a_i \ell^i c);\ \mathbb{Z}/\ell)$. Hence $G(c)^m = \Pi_{i=0}^{k} G(c)^{a_i \ell^i}$ detects $H_*(BG(mc);\ \mathbb{Z}/\ell)$.

Details of the arguments appear in Chapter VII for the case $\ell = 2$ and in Chapter VIII for the case ℓ odd. We should also note that in the case ℓ odd the permutative functors defined in Chapter II §8 play a crucial role in simplifying the argument of Step 1.

Theorem 5.8 relates the homology of the groups $G(k)$ to the homology of the smaller groups $G(c)$ if k is a multiple of c. It remains to obtain a similar result in case k is not of that form. This is accomplished by the following

<u>Proposition</u> 5.9.(a) If $\ell = 2$ and q is odd then for $G = GL(\cdot;\ \mathbb{F}_q)$,

$O(\cdot;\ \mathbb{F}_q)$, $U(\cdot;\ \mathbb{F}_{q^2})$ the direct sum map

$$G(2m) \times G(1) \to G(2m+1)$$

induces an epimorphism in homology

$$H_*(BG(2m);\ \mathbb{Z}/\ell) \otimes H_*(BG(1);\ \mathbb{Z}/\ell) \to H_*(BG(2m+1);\ \mathbb{Z}/\ell)$$

(b) If $\ell = 2$ and q is odd then the direct sum maps

$$O(2m,\mathbb{F}_q) \times O(\overline{1},\mathbb{F}_q) \to O(\overline{2m+1},\mathbb{F}_q)$$

$$O(2m-1,\mathbb{F}_q) \times O(\overline{1},\mathbb{F}_q) \to O(\overline{2m},\mathbb{F}_q)$$

induce epimorphisms in homology

$$H_*(BO(2m,\mathbb{F}_q);\ \mathbb{Z}/\ell) \otimes H_*(BO(\overline{1},\mathbb{F}_q);\ \mathbb{Z}/\ell) \to H_*(BO(\overline{2m+1},\mathbb{F}_q);\ \mathbb{Z}/\ell)$$

$$H_*(BO(2m-1,\mathbb{F}_q);\ \mathbb{Z}/\ell) \otimes H_*(BO(\overline{1},\mathbb{F}_q);\ \mathbb{Z}/\ell) \to H_*(BO(\overline{2m},\mathbb{F}_q);\ \mathbb{Z}/\ell)$$

(c) If ℓ is odd, let c be as in Prop. 5.4 and let mc be the largest multiple of c such that $mc \leq k$ (In case $G = O(\cdot,\mathbb{F}_q)$ q even or odd then $c,k \in \tilde{\mathbb{N}}$ and we use the notation of II 4.7). Then the standard inclusion $G(mc) \to G(k)$ induces an epimorphism in homology

$$H_*(BG(mc);\ \mathbb{Z}/\ell) \to H_*(BG(k);\ \mathbb{Z}/\ell)$$

Proof: In each case one computes the orders of the groups and sees that the subgroup contains a Sylow ℓ-subgroup and hence detects the homology of the containing group.

Having thus reduced our problem, we turn to calculating

$H_*(BG(n); \mathbb{Z}/\ell)$ for small n. The pattern of argument here differs markedly between the cases $\ell = 2$ and ℓ odd.

The problem of computing $H_*(BG(n); \mathbb{Z}/2)$ for n = 1 and n = 2 is more complicated, requiring a variety of ad hoc arguments. We summarize some of the results and arguments.

5.10 <u>Calculations of</u> $H_*(BG(n); \mathbb{Z}/2)$ <u>for</u> n = 1 <u>and</u> 2

(a) $H_*(BO(n; \mathbb{F}_q); \mathbb{Z}/2)$

We first observe that $O(1,\mathbb{F}_q) \cong O(\bar{1},\mathbb{F}_q) \cong \mathbb{Z}/2$. Hence $H_i(BO(n,\mathbb{F}_q); \mathbb{Z}/2) = \mathbb{Z}/2$, n = 1 or $\bar{1}$ and all i. We denote by $v_i \in H_i(BO(1,\mathbb{F}_q); \mathbb{Z}/2)$ and $y_i \in H_i(BO(\bar{1},\mathbb{F}_q); \mathbb{Z}/2)$ the unique nonzero elements.

Next we observe that $O(2,\mathbb{F}_q)$ is a dihedral group, i.e. it has two generators s,b with relations $s^k = 1$, $b^2 = 1$, $bsb^{-1} = s^{-1}$. Using the Serre-Leray spectral sequence and the standard representation of the dihedral group in $O(2,\mathbb{R})$, we compute the cohomology ring of the dihedral group and show that the subgroups $O(1,\mathbb{F}_q) \times O(1,\mathbb{F}_q)$ and $O(\bar{1},\mathbb{F}_q) \times O(\bar{1},\mathbb{F}_q)$ detect $H^*(BO(2,\mathbb{F}_q); \mathbb{Z}/2)$ and hence also $H_*(BO(2,\mathbb{F}_q); \mathbb{Z}/2)$.

This implies that $H_*(BO(2,\mathbb{F}_q); \mathbb{Z}/2)$ is generated by monomials of the form $v_i v_j$ and $y_i y_j$. A simple computation establishes the relations

$$v_i^2 = y_i^2.$$

Detailed calculations appear in Chapter VI §3-4.

(b) $H_*(BSp(2,\mathbb{F}_q); \mathbb{Z}/2)$

Here we observe that the Sylow 2-subgroup of $Sp(2,\mathbb{F}_q)$ is a generalized quaternion group Q_t, i.e. it has two generators x,y with relations $x^{2^{t-1}} = y^2$, $x^{2^t} = 1$, $yxy^{-1} = x^{-1}$. Using a result of Swan, that any group with a generalized quaternionic Sylow 2-subgroup has

4-periodic cohomology, together with the Bockstein spectral sequence we determine that

$$H^*(BSp(2,\mathbb{F}_q);\ \mathbb{Z}/2) = \mathbb{Z}/2[P] \otimes E[\tilde{x}]$$

where degree $P = 4$ and degree $\tilde{x} = 3$.

It follows therefore that $H_i(BSp(2,\mathbb{F}_q);\ \mathbb{Z}/2) = 0,0,\mathbb{Z}/2,\mathbb{Z}/2$ for $i \equiv 1,2,3,4 \bmod 4$. Let $\sigma_j \in H_{4j}(BSp(2,\mathbb{F}_q);\ \mathbb{Z}/2)$ $j \geq 0$ and $\tau_k \in H_{4k-1}(BSp(2,\mathbb{F}_q);\ \mathbb{Z}/2)$, $k \geq 1$ denote the generators. Finally we show that $\tau_k^2 = 0$ in $H_*(BSp(4,\mathbb{F}_q);\ \mathbb{Z}/2)$ by using homology operations. Detailed computations appear in Chapter VI §5.

(c) $H_*(BGL(n,\mathbb{F}_q);\ \mathbb{Z}/2)$ $n = 1,2$.

We observe that $GL(1,\mathbb{F}_q) \cong \mathbb{F}_q^* \cong \mathbb{Z}/q-1$. Hence $H_i(BGL(1,\mathbb{F}_q);\ \mathbb{Z}/2) = \mathbb{Z}/2$ for all i. We denote by α_i, β_i the generators of $H_*(BGL(1,\mathbb{F}_q);\ \mathbb{Z}/2)$ in degrees $2i, 2i-1$ respectively.

Using homology operations and some homological calculations we determine that $\beta_i^2 = 0$ in $H_*(BGL(2,\mathbb{F}_q);\ \mathbb{Z}/2)$. Next using the Brauer lift

$$BGL(2,\mathbb{F}_q) \to BGL(\infty,\mathbb{F}_q) \to \Gamma_0 B\,\mathcal{GL}(\mathbb{F}_q) \xrightarrow{\lambda} JU(q)$$

we determine that $\{\alpha_0\beta_1, \alpha_0\alpha_1, \alpha_0\beta_2, \alpha_1\beta_1, \alpha_0\alpha_2, \alpha_1\alpha_1, \beta_1\beta_2\}$ is a linearly independent set in $H_*(BGL(2,\mathbb{F}_q);\ \mathbb{Z}/2)$.

Using these facts together with the Serre-Leray spectral sequence of

$$1 \to SL(2,\mathbb{F}_q) \to GL(2,\mathbb{F}_q) \xrightarrow{\det} \mathbb{F}_q^* \to 1$$

and using the fact that we computed the cohomology ring of $SL(2,\mathbb{F}_q) = Sp(2,\mathbb{F}_q)$ in part (b), we are able to compute the cohomology ring $H^*(BGL(2,\mathbb{F}_q);\ \mathbb{Z}/2)$. Moreover the same argument shows that

$GL(1,\mathbb{F}_q) \times GL(1,\mathbb{F}_q)$ detects $H^*(BGL(2,\mathbb{F}_q); \mathbb{Z}/2)$ and hence also $H_*(BGL(2,\mathbb{F}_q); \mathbb{Z}/2)$. This implies that $H_*(BGL(2,\mathbb{F}_q); \mathbb{Z}/2)$ is generated by monomials of the form $\alpha_i\alpha_j$, $i \leq j$; $\beta_i\beta_j$, $i < j$, and $\alpha_i\beta_j$. Detailed calculations appear in Chapter VI §6.

(d) $H_*(BU(n,\mathbb{F}_{q^2}); \mathbb{Z}/2)$, $n = 1,2$

The computations here are virtually identical to those of part (c). We first observe that

$$U(1,\mathbb{F}_{q^2}) = \{x \in \mathbb{F}_{q^2} \mid x\bar{x} = 1\} \cong \mathbb{Z}/q + 1$$

Hence $H_i(BU(1,\mathbb{F}_{q^2}); \mathbb{Z}/2) = \mathbb{Z}/2$ for all i. We denote by ξ_i, η_i the generators of $H_*(BU(1,\mathbb{F}_{q^2}); \mathbb{Z}/2)$ in degrees $2i, 2i-1$ respectively.

We then proceed as in (c) to show that $U(1,\mathbb{F}_{q^2}) \times U(1,\mathbb{F}_{q^2})$ detects $H_*(BU(2,\mathbb{F}_{q^2}); \mathbb{Z}/2)$ and to show that $\eta_i^2 = 0$ in $H_*(BU(2,\mathbb{F}_q); \mathbb{Z}/2)$. The argument uses the fact that the permutative functor $\mathcal{S}p(\mathbb{F}_q) \to \mathcal{U}(\mathbb{F}_{q^2})$ defined in II 8.5a induces an isomorphism $SU(2,\mathbb{F}_{q^2}) \cong Sp(2,\mathbb{F}_q)$.

All this implies $H_*(BU(2,\mathbb{F}_{q^2}); \mathbb{Z}/2)$ is generated by monomials of the form $\xi_i\xi_j$, $i \leq j$; $\eta_i\eta_j$, $i < j$; and $\xi_i\eta_j$. Detailed calculations appear in Chapter VI §6.

Next we turn to computing $H_*(BG(n); \mathbb{Z}/\ell)$ for small n when ℓ is odd. This is comparatively simpler than the case $\ell = 2$.

<u>Proposition</u> 5.11. Let ℓ be odd. As in Prop. 5.4 let c be the smallest index for which $G(c)$ has order divisible by ℓ ($G = GL(\cdot,\mathbb{F}_q)$, $O(\cdot,\mathbb{F}_q)$ q even or odd, $Sp(\cdot,\mathbb{F}_q)$ or $U(\cdot,\mathbb{F}_{q^2})$). Then

(a) The Sylow ℓ-subgroup of $G(c)$ is cyclic

(b)

$$H_i(BG(c);\ \mathbb{Z}/\ell) = \begin{cases} \text{cyclic on generator } \gamma_k \text{ if } i = 2kc \\ \text{cyclic on generator } \delta_k \text{ if } i = 2kc - 1 \\ 0 \quad \text{otherwise} \end{cases}$$

<u>Sketch of Proof</u>: To prove (a) one uses the functors of Chapter II §8 to derive the Sylow ℓ-subgroups of $G(c)$ from one of the cyclic groups $GL(1,\mathbb{F}_{q^r})$ or $U(1,\mathbb{F}_{q^{2r}})$ for some appropriately chosen r.

Let H be the Sylow ℓ-subgroup of $G(c)$. Since H is cyclic $H_i(BH;\ \mathbb{Z}/\ell) = \mathbb{Z}/\ell$ for all i. Let us pick a generator $a_i \in H_i(BH;\ \mathbb{Z}/\ell)$. Let $j\colon H \to G(c)$ denote the inclusion. We have to show that $j_*(a_i) = 0$ in $H_i(BG(c);\ \mathbb{Z}/\ell)$ unless i of the form $2kc$ or $2kc - 1$.

We do this by showing that for certain n the n-th power μ^n of the generator $\mu \in H$ is conjugate in $G(c)$ to μ, i.e. $g\mu g^{-1} = \mu^n$. This implies a commutative diagram

$$\begin{array}{ccc} H & \xrightarrow{j} & G(c) \\ \Big\downarrow (\cdot)^n & & \Big\downarrow g(\cdot)g^{-1} \\ H & \xrightarrow{j} & G(c) \end{array}$$

Since conjugation induces the identity on homology, we obtain that in $H_i(BG(c);\ \mathbb{Z}/\ell)$ $i = 2m, 2m-1$ there is an equation

$$j_*(a_i) = n^m j_*(a_i)$$

$$(n^m - 1)\, j_*(a_i) = 0$$

Hence $j_*(a_i) = 0$ unless ℓ divides $n^m - 1$. It turns out that for these values of n, ℓ divides $n^m - 1$ iff m is a multiple of c.

Details of the proof appear in Chapter VIII.

Combining Theorem 5.8, Prop. 5.9, 5.10 and Prop. 5.11 we obtain generators for the homology algebras $H_*(B\mathcal{A};\ \mathbb{Z}/\ell)$:

Theorem 5.12. The homology algebra

$$H_*(B\mathcal{A};\ \mathbb{Z}/\ell) = \oplus_{a\in\pi_0\mathcal{A}} H_*(BG(a);\ \mathbb{Z}/\ell)$$

is graded commutative generated as follows:

(i) If $\ell = 2$, $\mathcal{A} = \bar{\mathcal{O}}(\mathbb{F}_q)$ q odd, then the elements $v_i \in H_i(BO(1,\mathbb{F}_q);\ \mathbb{Z}/2)$, $y_i \in H_i(BO(\bar{1},\mathbb{F}_q);\ \mathbb{Z}/2)$ defined in 5.10(a) generate $H_*(B\bar{\mathcal{O}}(\mathbb{F}_q);\ \mathbb{Z}/2)$. There are relations $v_i^2 = y_i^2$.

(ii) If $\ell = 2$, $\mathcal{A} = \mathcal{S}p(\mathbb{F}_q)$ q odd, then the elements $\alpha_i \in H_{4i}(BSp(2,\mathbb{F}_q);\ \mathbb{Z}/2)$, $\tau_i \in H_{4i-1}(BSp(2,\mathbb{F}_q);\ \mathbb{Z}/2)$ defined in 5.10(b) generate $H_*(B\mathcal{S}p(\mathbb{F}_q);\ \mathbb{Z}/2)$. There are relations $\tau_i^2 = 0$.

(iii) If $\ell = 2$, $\mathcal{A} = \mathcal{GL}(\mathbb{F}_q)$ q odd, then the elements $\alpha_i \in H_{2i}(BGL(1,\mathbb{F}_q);\ \mathbb{Z}/2)$, $\beta_i \in H_{2i-1}(BGL(1,\mathbb{F}_q);\ \mathbb{Z}/2)$ defined in 5.10(c) generate $H_*(B\mathcal{GL}(\mathbb{F}_q);\ \mathbb{Z}/2)$. There are relations $\beta_i^2 = 0$.

(iv) If $\ell = 2$, $\mathcal{A} = \mathcal{U}(\mathbb{F}_{q^2})$ q odd, then the elements $\xi_i \in H_{2i}(BU(1,\mathbb{F}_{q^2});\ \mathbb{Z}/2)$, $\eta_i \in H_{2i-1}(BU(1,\mathbb{F}_{q^2});\ \mathbb{Z}/2)$ defined in 5.10(d) generate $H_*(B\mathcal{U}(\mathbb{F}_{q^2});\ \mathbb{Z}/2)$. There are relations $\eta_i^2 = 0$.

(v) If ℓ is odd, $\mathcal{A} = \mathcal{GL}(\mathbb{F}_q)$, $\mathcal{S}p(\mathbb{F}_q)$, $\mathcal{U}(\mathbb{F}_{q^2})$, $\bar{\mathcal{O}}(\mathbb{F}_q)$ q odd, or $\bar{\mathcal{O}}^{ev}(\mathbb{F}_q)$ q even, then the homology algebra is generated by the elements $\gamma_i \in H_{2ic}(BG(c);\ \mathbb{Z}/\ell)$, $\delta_i \in H_{2ic-1}(BG(c);\ \mathbb{Z}/\ell)$ defined in Prop. 5.11 together with the elements (cf. II 2.16(viii) regarding notation)

(a) $[1] \in H_0(BGL(1,\mathbb{F}_q);\ \mathbb{Z}/\ell)$ if $\mathcal{A} = \mathcal{GL}(\mathbb{F}_q)$
(b) $[2] \in H_0(BSp(2,\mathbb{F}_q);\ \mathbb{Z}/\ell)$ if $\mathcal{A} = \mathcal{S}p(\mathbb{F}_q)$
(c) $[1] \in H_0(BU(1,\mathbb{F}_{q^2});\ \mathbb{Z}/\ell)$ if $\mathcal{A} = \mathcal{U}(\mathbb{F}_{q^2})$
(d) $[1] \in H_0(BO(1,\mathbb{F}_q);\ \mathbb{Z}/\ell)$ and $[\bar{1}] \in H_0(BO(\bar{1},\mathbb{F}_q);\ \mathbb{Z}/\ell)$ if

$\mathcal{O} = \bar{\mathcal{O}}(\mathbb{F}_q)$ q odd with relation $[1]^2 = [\bar{1}]^2$

(e) $[2] \in H_0(BO(2,\mathbb{F}_q); \mathbb{Z}/\ell)$ and $[\bar{2}] \in H_0(BO(\bar{2},\mathbb{F}_q); \mathbb{Z}/\ell)$ if $\mathcal{O} = \bar{\mathcal{O}}^{ev}(\mathbb{F}_q)$ q even with relation $[2]^2 = [\bar{2}]^2$

(The relations $\delta_i^2 = 0$ are implicit in the commutativity of $H_*(B\mathcal{O}; \mathbb{Z}/\ell)$.)

Theorem 5.12 enables us to immediately obtain generators for the homology of the corresponding infinite loop spaces $H_*(\Gamma_0 B\mathcal{O}; \mathbb{Z}/\ell)$.

<u>Corollary</u> 5.13. The algebra $H_*(\Gamma_0 B\mathcal{O}; \mathbb{Z}/\ell)$ is generated by two families of elements $\{\bar{\alpha}_i\}$ $i \geq 1$ and $\{\bar{\beta}_i\}$ $i \geq 1$. There are relations $\bar{\beta}_i^2 = 0$. More precisely

(i) If $\ell = 2$ and $\mathcal{O} = \bar{\mathcal{O}}(\mathbb{F}_q)$ q odd, then by Theorem 5.12 $\bar{\alpha}_i = \bar{v}_i = v_i * [-1]$ and $\bar{y}_i = y_i * [-\bar{1}] \in H_i(\Gamma_0 B\bar{\mathcal{O}}(\mathbb{F}_q); \mathbb{Z}/2)$ generate the homology algebra. We define a new set of generators by $\bar{\beta}_i = \bar{u}_i = \Sigma_{j=0}^{j} \bar{y}_j \chi(\bar{v}_{c-j})$ where χ: $\Gamma_0 B\bar{\mathcal{O}}(\mathbb{F}_q) \to \Gamma_0 B\bar{\mathcal{O}}(\mathbb{F}_q)$ is the negative of the identity map (so that $\bar{y}_i = \Sigma_{j=0}^{i} \bar{u}_j \bar{v}_{i-j}$ cf. Prop IV 2.8). The relations $\bar{\beta}_i^2 = \bar{u}_i^2 = 0$ are equivalent to the relations $v_i^2 = y_i^2$.

(ii) If $\ell = 2$ and $\mathcal{O} = \mathcal{S}p(\mathbb{F}_q)$ q odd, then

$\bar{\alpha}_i = \bar{\sigma}_i = \sigma_i * [-2] \in H_{4i}(\Gamma_0 B\, \mathcal{S}p(\mathbb{F}_q); \mathbb{Z}/2)$ and

$\bar{\beta}_i = \bar{\tau}_i = \tau_i * [-2] \in H_{4i-1}(\Gamma_0 B\, \mathcal{S}p(\mathbb{F}_q); \mathbb{Z}/2)$

(iii) If $\ell = 2$ and $\mathcal{O} = \mathcal{GL}(\mathbb{F}_q)$ q odd, then

$\bar{\alpha}_i = \alpha_i * [-1] \in H_{2i}(\Gamma_0 B\, \mathcal{GL}(\mathbb{F}_q); \mathbb{Z}/2)$ and

$\bar{\beta}_i = \beta_i * [-1] \in H_{2i-1}(\Gamma_0 B\, \mathcal{GL}(\mathbb{F}_q); \mathbb{Z}/2)$

(iv) If $\ell = 2$ and $\mathcal{O} = \mathcal{U}(\mathbb{F}_{q^2})$ q odd, then

$\bar{\alpha}_i = \bar{\xi}_i = \xi_i * [-1] \in H_{2i}(\Gamma_0 B\, \mathcal{U}(\mathbb{F}_{q^2}); \mathbb{Z}/2)$ and

$\bar{\beta}_i = \bar{\eta}_i = \eta_i * [-1] \in H_{2i-1}(\Gamma_0 B\, \mathcal{U}(\mathbb{F}_{q^2}); \mathbb{Z}/2)$

(v) If ℓ is odd and $\mathcal{O} = \mathcal{GL}(\mathbb{F}_q)$, $\mathcal{S}p(\mathbb{F}_q)$, $\mathcal{U}(\mathbb{F}_{q^2})$, $\mathcal{O}(\mathbb{F}_q)$ q odd or $\mathcal{O}^{ev}(\mathbb{F}_q)$ q even, then $\bar{\alpha}_i = \gamma_i * [-c] \in H_{2ic}(\Gamma_0 B\mathcal{O}; \mathbb{Z}/\ell)$ and $\bar{\beta}_i = \delta_i * [-c] \in H_{2ic-1}(\Gamma_0 B\mathcal{O}; \mathbb{Z}/\ell)$.

§6. General pattern of arguments at noncharacteristic primes: the Brauer lift.

In this section we relate the results of Corollary 5.13 to those of Chapter I on $H_*(F\psi^q;\ \mathbb{Z}/\ell)$ and show that λ is an equivalence at ℓ. Recall that the latter results were obtained by computing with the Serre spectral sequence of the fibration sequence

$$H \xrightarrow{i} F\psi^q \xrightarrow{\pi} BH \xrightarrow{\psi^q - 1} BH$$

In each case we get

$$H_*(F\psi^q;\ \mathbb{Z}/\ell) = \mathbb{Z}/\ell[a_i \mid i \geq 1] \otimes E\{b_i \mid i \geq 1\}$$

where

$$\pi_*(a_m) = p_{ms} \qquad i_*(r_{ms}) = b_m$$

where the p_i and r_i are standard generators of $H_*(BH;\ \mathbb{Z}/\ell) = \mathbb{Z}/\ell[p_i \mid i \geq 1]$ and $H_*(H;\ \mathbb{Z}/\ell) = E\{r_i \mid i \geq 1\}$ and s is an integer depending on q and ℓ. (In each case $\deg a_m = \deg \bar{\alpha}_m$, $\deg b_m = \deg \bar{\beta}_m$ where $\bar{\alpha}_m, \bar{\beta}_m \in H_*(\Gamma_0 B\ ;\ \mathbb{Z}/\ell)$ are the generators of Cor. 5.13.)

Our proof that $\lambda\colon \Gamma_0 B \to F\psi^q$ is an equivalence at ℓ will be completed once we show that

$$\lambda_*(\bar{\alpha}_i) = a_i \qquad \lambda_*(\bar{\beta}_i) = b_i$$

modulo decomposable elements, where the $\bar{\alpha}_i, \bar{\beta}_i$ are the generators of $H_*(\Gamma_0 B\mathfrak{a};\ \mathbb{Z}/\ell)$ chosen in Corollary 5.13. Thus λ_* is an epimorphism. Also since $H_*(F\psi^q;\ \mathbb{Z}/\ell)$ is a free algebra on the generators $\{a_i, b_i\}$ modulo the relations $b_i^2 = 0$, it will follow that $H_*(\Gamma_0 B\mathfrak{a};\ \mathbb{Z}/\ell)$ is free on the generators $\{\bar{\alpha}_i, \bar{\beta}_i\}$ modulo the relations $\bar{\beta}_i^2 = 0$. It follows then that λ_* is an isomorphism and λ is an equivalence at ℓ.

We first will show that

$$\lambda_*(\bar{a}_i) = a_i + \text{decomposables}$$

To do this we will use the diagram

$$\begin{array}{ccccc} H & \xrightarrow{i} & F_\psi^q & \xrightarrow{\pi} & BH \\ & & \uparrow{\scriptstyle\lambda} & \nearrow{\scriptstyle\beta} & \\ & & \Gamma_0 B\mathcal{A} & & \end{array}$$

From the way the a_i's are defined it suffices to show that

$$\beta(\bar{a}\ ) = p_{is} + \text{decomposables}$$

We will prove this using character arguments and the following well-known lemma.

<u>Lemma</u> 6.1. Let A be a cyclic ℓ-subgroup of S^1. Then the inclusion $i\colon A \to S^1$ induces isomorphisms

$$i_*\colon H_{2m}(BA;\ \mathbb{Z}/\ell) \xrightarrow{\cong} H_{2m}(BS^1;\ \mathbb{Z}/\ell)$$

for all m.

<u>Proof</u>: There is a short exact sequence

$$1 \to A \xrightarrow{i} S^1 \xrightarrow{f} S^1 \to 1$$

where $f(z) = z^\ell$, which induces a fibration sequence

$$S^1 \to BA \to BS^1 \to BS^1$$

We observe that by dimensional considerations the Serre spectral sequence in homology of $S^1 \to BA \to BS^1$ collapses, which proves the lemma.

<u>Theorem</u> 6.2. $\lambda_*(\bar{a}_i) = a_i$ + decomposables

<u>Sketch of Proof</u>: As mentioned above it is enough to show that

$$\beta_*(\bar{a}_i) = p_{is} + \text{decomposables}$$

Our strategy will be as follows: From the calculations of 5.10-5.13, we have that $\bar{a}_i$ is the image of the standard generator in $H_{is}(BA; \mathbb{Z}/\ell)$ under the inclusion map $BA \to BG(c) \to BG(\infty) \to \Gamma_0 B\mathfrak{A}$ where A is some cyclic ℓ-subgroup of $G(c)$. Using character arguments we shall relate the composite

$$BA \to BG(c) \to BG(\infty) \to \Gamma_0 B\mathfrak{A} \xrightarrow{\beta} BH$$

to the composite

$$BA \to BS^1 \to BH$$

where $S^1 \to H$ is one of the classical inclusions. We then use Lemma 6.1. We now look at the various cases separately

(a) $\ell = 2$, $\mathfrak{A} = \mathcal{O}(\mathbb{F}_q)$ q odd

In this case we see that the Brauer lift of the inclusion

$$\mathbb{Z}/2 = O(1,\mathbb{F}_q) \to O(\infty,\mathbb{F}_q)$$

is the inclusion

$$\mathbb{Z}/2 = O(1,\mathbb{R}) \to O(\infty,\mathbb{R}) = O$$

Hence we have a commutative diagram

$$\begin{array}{ccc} B\mathbb{Z}/2 = BO(1,\mathbb{F}_q) \to BO(\infty,\mathbb{F}_q) & \to & \Gamma_0 B\mathcal{O}(\mathbb{F}_q) \\ \Vert & & \downarrow \beta \\ B\mathbb{Z}/2 = BO(1,\mathbb{R}) & \longrightarrow & BO \end{array}$$

which shows that

$$\beta_*(\overline{v}_i) = \overline{e}_i$$

(cf. Cor. 5.13 and I §3 regarding notation). Details appear in Chapter IV §2.

(b) $\ell = 2$, $\mathcal{A} = \mathcal{GL}(\mathbb{F}_q)$ q odd

In this case the Brauer lift of the inclusion

$$\mathbb{Z}/q-1 = \mathbb{F}_q^* = GL(1,\mathbb{F}_q) \to GL(\infty,\mathbb{F}_q)$$

is the inclusion

$$\mathbb{Z}/q-1 \to S^1 = U(1) \to U$$

Hence we have a connected diagram

$$\begin{array}{ccc} B\mathbb{Z}/q\text{-}1 = BGL(1,\mathbb{F}_q) \to BGL(\infty,\mathbb{F}_q) & \to & \Gamma_0 B\mathcal{GL}(\mathbb{F}_q) \\ \| & & \downarrow \beta \\ B\mathbb{Z}/q\text{-}1 \to BS^1 = BU(1) & \longrightarrow & BU \end{array}$$

Using Lemma 6.1, we obtain

$$\beta_*(\overline{a}_i) = a_i$$

(cf. Corollary 5.13 and I §4 regarding notation). Details appear in Chapter IV §7.

(c) $\ell = 2$, $\mathcal{A} = \mathcal{U}(\mathbb{F}_{q^2})$ q odd

The case is virtually identical to that of (b)

(d) $\ell = 2$, $\mathcal{A} = \mathcal{Sp}(\mathbb{F}_q)$ q odd

In this case the generator σ_i is the image of the generator of $H_{4i}(B\mathbb{Z}/2, \mathbb{Z}/2)$ under the chain of inclusions

$$B\mathbb{Z}/2 \to BQ_t \to BSp(2,\mathbb{F}_q) \to BSp(\infty,\mathbb{F}_q) \to \Gamma_0 B\mathcal{Sp}(\mathbb{F}_q)$$

where $Q_t = \{x,y \mid x^{2^t} = 1,\ x^{2^{t-1}} = y^2,\ yxy^{-1} = x^{-1}\}$ is the Sylow 2-sub group of $Sp(2,\mathbb{F}_q)$ (cf. 5.10). We find that the Brauer lift of

$$\mathbb{Z}/2 \to Q_t \to Sp(2,\mathbb{F}_q) \to Sp(\infty,\mathbb{F}_q)$$

is

$$\mathbb{Z}/2 \to S^1 \to S^3 = Sp(1) \to Sp$$

Hence we get a commutative diagram

$$\begin{array}{ccccccccc}
B\mathbb{Z}/2 & \to & BQ_t & \to & BSp(2,\mathbb{F}_q) & \to & BSp(\infty,\mathbb{F}_q) & \to & \Gamma_0 B\,\mathcal{S}p(\mathbb{F}_q) \\
\| & & & & & & & & \downarrow \beta \\
B\mathbb{Z}/2 & \to & BS^1 & \to & BS^3 = BSp(1) & & \longrightarrow & & BSp
\end{array}$$

Using Lemma 6.1, we obtain

$$\beta_*(\bar{\sigma}_i) = g_i$$

(cf. Corollary 5.13 and I §4 regarding notation). Details appear in Chapter IV §5.

(e) ℓ odd, $\mathcal{A} = \mathcal{GL}(\mathbb{F}_q)$, $\mathcal{U}(\mathbb{F}_{q^2})$, $\mathcal{S}p(\mathbb{F}_q)$, $\mathcal{O}(\mathbb{F}_q)$ q odd, or $\mathcal{O}^{ev}(\mathbb{F}_q)$ q even

The generator $\bar{\gamma}_i \in H_{2ic}(\Gamma_0 B\mathcal{A};\ \mathbb{Z}/\ell)$ is the image of a generator in $H_{2ic}(BA;\ \mathbb{Z}/\ell)$ under the chain of inclusions

$$BA \to BG(c) \to BG(\infty) \to \Gamma_0 B\mathcal{A}$$

where A is the cyclic Sylow ℓ-subgroup of $G(c)$

A character argument shows that the Brauer lift of the inclusion

$$A \to G(c) \to G(\infty)$$

is the inclusion

$$A \xrightarrow{f} \Pi_{i=1}^{d} A \to \Pi_{i=1}^{d} S^1 \to \Pi_{i=1}^{d} H \xrightarrow{\oplus} H$$

where f is the map $f(a) = (a, a^q, a^{q^d}, \ldots, a^{q^{d-1}})$ where d is an integer which depends on $\mathcal{A}$ and ℓ. We thus get a commutative diagram

$$\begin{array}{ccccccc} BA \to BG(c) & \longrightarrow & BG(\infty) & \longrightarrow & \Gamma_0 B\mathcal{A} \\ \Big\| & & & & \downarrow \beta \\ BA \to \Pi_{i=1}^{d} BA \to \Pi_{i=1}^{d} BS^1 \to \Pi_{i=1}^{d} BH & \xrightarrow{B\oplus} & BH \end{array}$$

Evaluating this diagram in $H_*(\,;\,\mathbb{Z}/\ell)$ using Lemma 6.1 we obtain

$$\beta_*(\overline{\gamma}_i) = p_{2ic} + \text{decomposables}$$

(cf. Corollary 5.13, regarding notation). Details appear in Chapter V.

It remains to show that

$$\lambda_*(\overline{\beta}_i) = b_i + \text{decomposables}$$

We shall do this in two stages. We will first show that this holds for small i. This is done by means of the Bockstein spectral sequence

Remark 6.3. Recall that if X is a space, then the mod ℓ Bockstein spectral sequence in homology has E^1 term $E^1_m = H_m(X;\,\mathbb{Z}/\ell)$. The differentials in the spectral sequence all have degree -1, $d_n\colon E^n_m \to E^n_{m-1}$. The first differential d_1 is the ordinary Bockstein homomorphism. An element in $H_*(X;\,\mathbb{Z}/\ell)$ is an infinite cycle in E^*_* iff it is in the image of the reduction homomorphism $\rho\colon H_*(X;\,\mathbb{Z}) \to H_*(X;\,\mathbb{Z}/\ell)$. Moreover an element $x \in H_*(X;\,\mathbb{Z}/\ell)$ is a d_n-boundary iff $x = \rho(y)$ where y has order 2^n. Thus an element in $H_*(X;\,\mathbb{Z}/\ell)$ which is an infinite cycle determines a nonzero element in E^∞ iff it is a reduction of a nontorsion element in $H_*(X;\,\mathbb{Z})$. If X is an H-space, then the Bockstein spectral sequence is a spectral sequence of algebras and the differentials are algebra derivations.

There is a dual Bockstein spectral sequence in cohomology with similar properties. Further details may be found in Browder [10].

Definition 6.4. If n is an integer $\nu_\ell(n)$ will denote the largest nonnegative integer ν such that ℓ^ν divides n.

Proposition 6.5. For small i, $\lambda_*(\overline{\beta}_i) = b_i$ + decomposables.

Sketch of Proof: We use the notation of Chapter I and Corollary 5.13.

(a) If $\ell = 2$, $\mathcal{O} = \overline{\mathcal{O}}(\mathbb{F}_q)$ q odd, then $\lambda_*(\overline{u}_1) = \hat{u}_1$.

If not we would have $\lambda_*(\overline{u}_1) = 0$ and by Lemma I 3.2 $\lambda_*(\overline{u}_i) = 0$ for all $i \geq 1$. By Corollary 5.13 and Theorem 6.2 we would have im $\lambda_* \subseteq \mathbb{Z}/2[\hat{v}_1, \hat{v}_2, \ldots]$. On the other hand by Lemma I 7.4

$$d_\nu(\hat{v}_2^2) = \hat{u}_1\hat{u}_2$$

where $\nu = \nu(\frac{1}{2}(q^2-1))$ in the largest integer for which 2^ν divides $\frac{1}{2}(q^2-1)$. By Lemma VI 4.5, $d_\nu(\overline{v}_2^2)$ is defined. Hence by Theorem 6.2

$$\lambda_* d_\nu(\overline{v}_2^2) = d_\nu \lambda_*(\overline{v}_2^2) = d_\nu(\hat{v}_2^2) = \hat{u}_1\hat{u}_2 \notin \mathbb{Z}/2[\hat{v}_1, \hat{v}_2, \ldots]$$

This contradiction establishes $\lambda_*(\overline{u}_1) = \hat{u}_1$. Details appear in IV 2.13

(b) If $\ell = 2$, $\mathcal{O} = \mathcal{Sp}(\mathbb{F}_q)$ q odd, then $\lambda_*(\overline{\tau}_i) = \hat{h}_i$ + decomposables $i = 1,2$.

This follows immediately from Lemma I 8.2 which states that $d_\nu(\hat{g}_i) = \hat{h}_i$ + decomposables $i = 1,2$ where $\nu = \nu(q^2-1)$ is the largest integer for which 2^ν divides q^2-1; from Prop. VI 5.7 which implies that $d_\nu(\overline{\sigma}_i) = \overline{\tau}_i$ for all i; and from Theorem 6.2:

$$\lambda_*(\overline{\tau}_i) = \lambda_* d_\nu(\overline{\sigma}_i) = d_\nu \lambda_*(\overline{\sigma}_i) = d_\nu(\hat{g}_i) = \hat{h}_i \quad i = 1,2$$

modulo decomposables. Details appear in Chapter IV, §6.

(c) If $\ell = 2$, $\mathcal{A} = \mathcal{GL}(\mathbb{F}_q)$ q odd, then $\lambda_*(\overline{\beta}_i) = \hat{b}_i$ + decomposables, i = 1,2

The proof that $\lambda_*(\overline{\beta}_1) = \hat{b}_1$ is the same as that of (b) except that we use Prop. VI 2.4 instead of Prop. VI 5.7. To prove that $\lambda_*(\overline{\beta}_2) = \hat{b}_2$ + decomposables one uses Lemma I 4.6 and the fact that λ_* is an infinite loop map:

$$\lambda_* Q^2(\overline{\beta}_1) = Q^2\lambda_*(\overline{\beta}_1) = Q^2(\hat{b}_1) = \hat{b}_2$$

while by Theorem IX 4.4, $Q^2(\overline{\beta}_1) = \overline{\beta}_2$ + decomposables. It follows that

$$\lambda_*(\overline{\beta}_2) = \hat{b}_2 + \text{decomposables}$$

Details appear in Chapter IV §7.

(d) If $\ell = 2$, $\mathcal{A} = \mathcal{U}(\mathbb{F}_{q^2})$ q odd, then $\lambda_*(\overline{\eta}_i) = \hat{b}_i$ i = 1,2.

The proof is identical to that of case (c)

(e) If $\ell > 2$, $\mathcal{A} = \mathcal{GL}(\mathbb{F}_q)$, $\mathcal{U}(\mathbb{F}_{q^2})$, $\mathcal{Sp}(\mathbb{F}_q)$, $\overline{\mathcal{O}}(\mathbb{F}_q)$ q odd or $\overline{\mathcal{O}}^{ev}(\mathbb{F}_q)$ q even, then $\lambda_*(\overline{\beta}_i) = b_i$ + decomposables, $i = 1,2,\ldots,\ell$.

The proof is identical to that case (b) except that we use Lemmas I 7.5 and 8.3 and Prop. VI 2.4 instead of Lemma I 8.2 and Prop. VI 5.7 respectively.

It remains to show that $\lambda_*(\overline{\beta}_i) = b_i$ + decomposables for all i. We do this by using Steenrod operations. First we introduce a new concept.

<u>Definition</u> 6.6. Let $\mathcal{A}_*$ denote the dual mod ℓ Steenrod algebra. Let M_* be an $\mathcal{A}_*$ module and let N_* be an $\mathcal{A}_*$ submodule. We say that an element $x \in M_i$ is <u>Steenrod related</u> to N_* if there is a $y \in N_*$ and $P_*^I \in \mathcal{A}_*$ such that

(a) $P_*^I x = y$

(b) $P_*^I: M_i \to M_{i-|I|}$ is a monomorphism of groups.

The Steenrod span of N_* denoted $Sp(N_*)$ is defined to be the $\mathcal{A}_*$ module generated by all x which are Steenrod related to N_*. Clearly $N_* \subseteq Sp(N_*)$. We denote by $Sp^{(i)}(N_*)$ the i-fold iterated Steenrod span of N_*, $Sp(Sp(\ldots Sp(N_*)))$.

Clearly $\{Sp^{(i)}(N_*)\}$ is an ascending chain of $\mathcal{A}_*$ submodules of M_*. We denote $CL(N_*) = \bigcup_{i=1}^{\infty} Sp^{(i)}(N_*)$ and call it the Steenrod closure of N_*. It is clear that $CL(CL(N_*)) = CL(N_*)$.

Lemma 6.7. Let M_*, $\overline{M}_*$ be $\mathcal{A}_*$-modules. Let $N_* \subseteq M_*$, $\overline{N}_* \subseteq \overline{M}_*$ be $\mathcal{A}_*$-submodules. Suppose we are given two homomorphisms of $\mathcal{A}_*$-module pairs

$$f,g\colon (\overline{M}_*,\overline{N}_*) \to (M_*,N_*)$$

Assume also that

(a) $CL(\overline{N}_*) = \overline{M}_*$

(b) g is an isomorphism

(c) $f|\overline{N}_* = g|\overline{N}_*$

Then $f = g$.

Proof: We first show $f|Sp(\overline{N}_*) = g|Sp(\overline{N}_*)$. To do this, it suffices to show that $f|\overline{S}_* = g|\overline{S}_*$ where $\overline{S}_*$ is the set of x in $\overline{M}_*$ which are Steenrod related to $\overline{N}_*$.

Suppose $x \in \overline{S}_i$. Then we can find $z \in \overline{N}_*$ and $P_*^I \in \mathcal{A}_*$ such that $P_*^I x = z$ and $P_*^I: \overline{M}_i \to \overline{M}_{i-|I|}$ is a monomorphism. Since g is an isomorphism, $P_*^I: M_i \to M_{i-|I|}$ is also a monomorphism. We now have

$$P_*^I f(x) = fP_*^I(x) = f(z) = g(z) = gP_*^I(x) = P_*^I g(x)$$

Hence $f(x) = g(x)$ and $f|\overline{S}_* = g|\overline{S}_*$. Consequently $f|Sp(\overline{N}_*) = g|Sp(\overline{N}_*)$.

Since g is an isomorphism,

$$f(Sp(\overline{N}_*)) = g(Sp(\overline{N}_*)) = Sp(N_*)$$

Hence we can apply the same argument to the pair of maps

$$f,g\colon (\overline{M}_*, Sp(\overline{N}_*)) \to (M_*, Sp(N_*))$$

We conclude that $f|Sp^{(2)}(\overline{N}_*) = g|Sp^{(2)}(N_*)$.

Iterating the argument we conclude that $f|Sp^{(i)}(\overline{N}_*) = g|Sp^{(i)}(\overline{N}_*)$ for all i. Consequently $f|CL(\overline{N}_*) = g|CL(\overline{N}_*)$. Since $CL(\overline{N}_*) = \overline{M}_*$, $f = g$.

We now complete the argument that $\lambda_*(\overline{\beta}_i) = b_i$ + decomposables. We will use Prop. 6.5 and Lemma 6.7 except in the case $\mathcal{A} = \mathcal{O}(\mathbb{F}_q)$ q odd, d = 2 where a slight modification is required.

Theorem 6.8. $\lambda_*(\overline{\beta}_i) = b_i$ + decomposables for all i.

Sketch of Proof: (a) All cases except $\ell = 2$, $\mathcal{A} = \mathcal{O}(\mathbb{F}_q)$ q odd.

We denote by M_* the graded $\mathbb{Z}/\ell$ module generated by $\{b_i\}_{i=1}^{\infty}$. We denote by $\overline{M}_*$ the graded $\mathbb{Z}/\ell$ module generated by $\{\overline{\beta}_i\}_{i=1}^{\infty}$. Let $N_* \subseteq M_*$, $\overline{N}_* \subseteq \overline{M}_*$ denote the submodules generated by $\{b_i | i \text{ small}\}$, $\{\overline{\beta}_i | i \text{ small}\}$ respectively, where "small" is taken in the sense of Prop. 6.5.

It is easily checked that (M_*, N_*) and $(\overline{M}_*, \overline{N}_*)$ are $\mathcal{A}_*$-module pairs (where the $\mathcal{A}_*$-module structure comes from the inclusions $\overline{M}_* \subseteq H_*(\Gamma_0 B\mathcal{A};\ \mathbb{Z}/\ell)$, $M_* \subseteq H_*(F\psi^q;\ \mathbb{Z}/\ell)$). Some simple calculations using Prop. VI 2.5, 2.6 or 5.8 show that $CL_*(\overline{N}_*) = \overline{M}_*$.

It is easily seen that $g\colon (\overline{M}_*, \overline{N}_*) \to (M_*, N_*)$ given by $g(\overline{\beta}_i) = b_i$

is an $\mathcal{O}_*$-module isomorphism. Also if we mod out by decomposable elements, we see that $\lambda_*: H_*(\Gamma_0 B\mathcal{O}; \mathbb{Z}/\ell) \to H_*(F\psi^q; \mathbb{Z}/\ell)$ induces a map $\lambda_*: (\overline{M}_*, \overline{N}_*) \to (M_*, N_*)$. Since Steenrod operations send decomposable elements to decomposable elements, $\lambda_*: (\overline{M}_*, \overline{N}_*) \to (M_*, N_*)$ is an $\mathcal{O}_*$-map.

Now according to Prop. 6.5 $\lambda_* | \overline{N}_* = g | N_*$. We are now in the situation of Lemma 6.7 and can conclude that $\lambda_* = g$ on $\overline{M}_*$. Consequently for all i

$$\lambda_*(\overline{\beta}_i) = g(\overline{\beta}_i) = b_i \text{ modulo decomposables.}$$

(b) $\ell = 2$ $\mathcal{O} = \mathcal{O}(\mathbb{F}_q)$ q odd

In this case we use Lemma I 3.2 which gives that $\lambda_*(\overline{u}_1) = \hat{u}_1$ implies $\lambda_*(\overline{u}_i) = \hat{u}_i$ for all i.

As mentioned at the beginning of the section an immediate corollary of Theorem 6.2 and 6.8 is the following

<u>Corollary</u> 6.9. $\lambda: \Gamma_0 B\mathcal{O} \to F\psi^q$ is an equivalence at ℓ.

Another consequence is the following

<u>Corollary</u> 6.10. The generators and relations of Theorem 5.12 and Corollary 5.13 are a complete list of generators and relations in $H_*(B\mathcal{O}; \mathbb{Z}/\ell)$ and $H_*(\Gamma_0 B\mathcal{O}; \mathbb{Z}/\ell)$ respectively.

<u>Sketch of Proof</u>: The result for $H_*(\Gamma_0 B\mathcal{O}; \mathbb{Z}/\ell)$ is immediate from Cor. 6.9 and the calculation of $H_*(F\psi^q; \mathbb{Z}/\ell)$. The result for $H_*(B\mathcal{O}; \mathbb{Z}/\ell)$ follows by examining the image of $H_*(BG(n); \mathbb{Z}/\ell)$ in $H_*(F\psi^q; \mathbb{Z}/\ell)$ under the chain of maps

$$BG(n) \to BG(\infty) \to \Gamma_0 B\mathcal{Q} \xrightarrow{\lambda} F\psi^q$$

seeing that the only relations in the image are those relations specified in Theorem 5.12. Details appear in Chapters IV and V.

Finally by dualizing Cor. 6.10 we can get a complete description of the cohomology rings $H^*(BG(n); \mathbb{Z}/\ell)$.

§7. The algebraic closures of finite fields

In this section we carry out the computations of the mod-ℓ homology and cohomology of the classical groups over the algebraic closure $\overline{\mathbb{F}}_p$ of the field $\mathbb{F}_p$, where ℓ is a prime different from p. We also show that the Brauer lift

$$\overline{\beta}: \Gamma_0 B\mathcal{GL}(\overline{\mathbb{F}}_p) \to BU$$

$$\overline{\beta}: \Gamma_0 B\mathcal{O}(\overline{\mathbb{F}}_p) \to BO$$

$$\overline{\beta}: \Gamma_0 B\mathcal{Sp}(\overline{\mathbb{F}}_p) \to BSp$$

is an equivalence when completed away from p. We also use the results of §3 to compute the homotopy groups of these spaces, which are the ordinary, orthogonal and symplectic K-groups of $\overline{\mathbb{F}}_p$.

These results are essentially due to Quillen [34]. We include them here for the sake of completeness and because the computations illustrate in a simplified way the corresponding calculations for finite fields which were sketched out in §5 and §6 and carried out in the succeeding chapters.

Remark 7.1. Before proceeding further we clarify what we mean by the notation $\mathcal{GL}(\overline{\mathbb{F}}_p)$, $\mathcal{O}(\overline{\mathbb{F}}_p)$ and $\mathcal{Sp}(\overline{\mathbb{F}}_p)$.

The category $\mathcal{GL}(\mathbb{F})$ was defined for an arbitrary field $\mathbb{F}$ in

II 2.6. Thus $\mathcal{GL}(\overline{\mathbb{F}}_p)$ has as objects the nonnegative integers and as morphisms

$$\hom(m,n) = \begin{cases} \emptyset & \text{if } m \neq n \\ GL_n(\overline{\mathbb{F}}_p) & \text{if } m = n \end{cases}$$

If V is an m-dimensional vector space over $\overline{\mathbb{F}}_p$ and $A\colon V \times V \to \overline{\mathbb{F}}_p$ is a nondegenerate symplectic form on V, then m must be even say $m = 2n$ and (V,A) is isomorphic to the standard symplectic space $(\overline{\mathbb{F}}_p^{2n}, A)$

$$A(x,y) = \Sigma_{i=1}^{n}(x_{2i}y_{2i-1} - x_{2i-1}y_{2i})$$

We denote the group of automorphisms of $(\overline{\mathbb{F}}_p^{2n}, A)$ by $Sp(2n,\overline{\mathbb{F}}_p)$. The category $\mathcal{Sp}(\overline{\mathbb{F}}_p)$ is then defined to have as objects the even nonnegative integers and as morphisms

$$\hom(2m,2n) = \begin{cases} \emptyset & \text{if } m \neq n \\ Sp(2n,\mathbb{F}_q) & \text{if } m = n \end{cases}$$

(compare II 6.4).

If p is odd, V is an n-dimensional vector space over $\overline{\mathbb{F}}_p$ and $Q\colon V \to \overline{\mathbb{F}}_p$ is a nondegenerate quadratic form, then (V,Q) is isomorphic to $(\overline{\mathbb{F}}_p^n, Q_+)$ with

$$Q_+(x) = \Sigma_{i=1}^{n} x_i^2$$

We denote by $O(n,\overline{\mathbb{F}}_p)$ the group of automorphisms of $(\overline{\mathbb{F}}_p^n, Q_+)$. The category $\mathcal{O}(\overline{\mathbb{F}}_p)$ has as objects the nonnegative integers and as morphisms

$$\hom(m,n) = \begin{cases} \emptyset & \text{if } m \neq n \\ O(n,\overline{\mathbb{F}}_p) & \text{if } m = n \end{cases}$$

(cf. II 2.8).

If $p = 2$, V is a 2n-dimensional vector space over $\overline{\mathbb{F}}_2$ and $Q\colon V \to \overline{\mathbb{F}}_2$ is a nondegenerate quadratic form, then (V,Q) is isomorphic to the standard quadratic space $(\overline{\mathbb{F}}_2^{2n}, Q_+)$

$$Q_+(x) = \Sigma_{i=1}^n x_{2i-1}x_{2i}$$

We denote by $O(2n,\overline{\mathbb{F}}_2)$ the group of automorphisms of $(\overline{\mathbb{F}}_2^{2n}, Q_+)$. The category $\mathcal{O}(\overline{\mathbb{F}}_2)$ has as objects the even nonnegative integers and as morphisms

$$\hom(2m,2n) = \begin{cases} \emptyset & \text{if } m \neq n \\ O(2n,\overline{\mathbb{F}}_2) & \text{if } m = n \end{cases}$$

(compare II 7.10(ii)).

Since the Galois group of $\overline{\mathbb{F}}_p$ is the profinite completion $\hat{\mathbb{Z}}$, which has no torsion elements, it follows that $\overline{\mathbb{F}}_p$ has no involution. Consequently the concept of Hermitian space over $\overline{\mathbb{F}}_p$ is meaningless and there is no such category as $\mathcal{U}(\overline{\mathbb{F}}_p)$.

Our first step will be to determine generators for the homology algebras

$$H_*(B\mathcal{GL}(\overline{\mathbb{F}}_p);\ \mathbb{Z}/\ell) = \oplus_{n=0}^{\infty} H_*(BGL(n,\overline{\mathbb{F}}_p);\ \mathbb{Z}/\ell)$$

$$H_*(B\mathcal{O}(\overline{\mathbb{F}}_p);\ \mathbb{Z}/\ell) = \oplus_{n=0}^{\infty} H_*(BO(n,\overline{\mathbb{F}}_p);\ \mathbb{Z}/\ell) \qquad p \neq 2$$

$$H_*(B\mathcal{O}(\overline{\mathbb{F}}_2);\ \mathbb{Z}/\ell) = \oplus_{n=0}^{\infty} H_*(BO(2n,\overline{\mathbb{F}}_2);\ \mathbb{Z}/\ell)$$

$$H_*(B\mathcal{Sp}(\overline{\mathbb{F}}_p);\ \mathbb{Z}/\ell) = \oplus_{n=0}^{\infty} H_*(BSp(2n,\overline{\mathbb{F}}_p);\ \mathbb{Z}/\ell)$$

where ℓ is a prime such that $\ell \neq p$.

We begin by observing that

$$GL(1,\overline{\mathbb{F}}_p) = \overline{\mathbb{F}}_p^* = \lim_{\vec{r}} \mathbb{F}^*_{p^r} = \lim_{\vec{r}} \mathbb{Z}/(p^r-1) = \mathbb{Z}_{(p)}/\mathbb{Z} = \oplus_{\ell\neq p}\mathbb{Z}/\ell^\infty$$

Consequently according to VI 2.9, we have

$$(7.2)\qquad H_i(BGL(1,\overline{\mathbb{F}}_p);\ \mathbb{Z}/\ell) = \begin{cases}\mathbb{Z}/\ell \text{ on generator } \alpha_j \text{ if } i = 2j\\ 0 \text{ if } i \text{ is odd}\end{cases}$$

Since $O(1,\overline{\mathbb{F}}_p) = \mathbb{Z}/2$ when $p > 2$, we have

$$(7.3)\qquad H_i(BO(1,\overline{\mathbb{F}}_p);\ \mathbb{Z}/2) = \mathbb{Z}/2 \text{ on generator } \epsilon_i \text{ for all } i \geq 0$$

Also according to VI 5.10, we have for $p > 2$

$$(7.4)\qquad H_i(BSp(2,\overline{\mathbb{F}}_p);\ \mathbb{Z}/2) = \begin{cases}\mathbb{Z}/2 \text{ on generator } \gamma_j \text{ if } i = 4j\\ 0 \text{ if } i \not\equiv 0 \pmod 4\end{cases}$$

Now suppose ℓ is an odd prime different from p. We observe that the hyperbolic inclusion maps II 8.4($\overline{a}$), ($\overline{b}$), ($\overline{c}$) yield inclusions

$$(7.5)\qquad \begin{aligned} GL(1,\overline{\mathbb{F}}_p) &= \lim_{\vec{r}} GL(1,\mathbb{F}_{p^r}) \to \lim_{\vec{r}} O(2,\mathbb{F}_{p^r}) = O(2,\overline{\mathbb{F}}_p)\\ GL(1,\overline{\mathbb{F}}_p) &= \lim_{\vec{r}} GL(1,\mathbb{F}_{p^r}) \to \lim_{\vec{r}} Sp(2,\mathbb{F}_{p^r}) = Sp(2,\overline{\mathbb{F}}_p)\end{aligned}$$

<u>Lemma</u> 7.6. For ℓ an odd prime $\neq p$

$$H_i(BO(2,\overline{\mathbb{F}}_p);\ \mathbb{Z}/\ell) = \begin{cases}\text{cyclic on generator } \sigma_j \text{ if } i = 4j\\ 0 \text{ if } i \not\equiv 0 \pmod 4\end{cases}$$

$$H_i(BSp(2,\overline{\mathbb{F}}_p);\ \mathbb{Z}/\ell) = \begin{cases} \text{cyclic on generator } \sigma_j & \text{if } i = 4j \\ 0 \text{ if } & i \not\equiv 0 \pmod 4 \end{cases}$$

where in each case σ_j is the image of the generator α_{2j} of (7.2) under the inclusions (7.5).

Proof: Pick r large enough so that $p^r \equiv 1 \pmod{\ell}$. If p is odd, also take r to be even. It is then easy to see from the group orders

$$|GL(1,\mathbb{F}_{p^r})| = p^r-1,\quad |O(2,\mathbb{F}_{p^r})| = 2(p^r-1),\quad |Sp(2,\mathbb{F}_{p^r})| = (p^r-1)(p^r+1)$$

that under the hyperbolic inclusions

$$(*) \qquad GL(1,\mathbb{F}_{p^r}) \to O(2,\mathbb{F}_{p^r}) \qquad GL(1,\mathbb{F}_{p^r}) \to Sp(2,\mathbb{F}_{p^r})$$

$GL(1,\mathbb{F}_{p^r})$ contains a Sylow ℓ-subgroup. Hence in each case (*) induces an epimorphism in mod-ℓ homology. Passing to direct limits we see that (7.5) induces epimorphisms in mod-ℓ homology.

In view of (7.2) all that remains to be done is to show that under the inclusions (7.5) the element $\alpha_j \in H_{2j}(BGL(1,\mathbb{F}_q);\ \mathbb{Z}/\ell)$ maps to 0 if j is odd.

We note that the inclusion

$$GL(1,\overline{\mathbb{F}}_p) \to O(2,\overline{\mathbb{F}}_p)$$

is given by endowing $\overline{\mathbb{F}}_p \oplus \overline{\mathbb{F}}_p$ with the quadratic form

$$Q((x,y)) = xy$$

and regarding $a \in \overline{\mathbb{F}}_q^* = GL(1,\overline{\mathbb{F}}_p)$ as acting on $\overline{\mathbb{F}}_p \oplus \overline{\mathbb{F}}_p$ by

$$a(x,y) = (ax, a^{-1}y)$$

Let $\beta\colon \overline{\mathbb{F}}_p \oplus \overline{\mathbb{F}}_p \to \overline{\mathbb{F}}_p \oplus \overline{\mathbb{F}}_p$ denote the map

$$\beta(x,y) = (y,x)$$

Clearly β is an orthogonal map. Note that

$$\beta a \beta^{-1}(x,y) = \beta a(y,x) = \beta(ay, a^{-1}x) = (a^{-1}x, ay) = a^{-1}\cdot(x,y)$$

Hence $\beta a \beta^{-1} = a^{-1}$ for all $a \in GL(1,\overline{\mathbb{F}}_p)$ and the following diagram commutes

$$\begin{array}{ccc}
H_*(B\mathbb{Z}/\ell^\infty;\ \mathbb{Z}/\ell) & \xrightarrow{(-1)_*} & H_*(B\mathbb{Z}/\ell^\infty;\ \mathbb{Z}/\ell) \\
\downarrow \cong & & \downarrow \cong \\
H_*(BGL(1,\overline{\mathbb{F}}_p);\ \mathbb{Z}/\ell) & \xrightarrow{((\cdot)^{-1})_*} & H_*(BGL(1,\overline{\mathbb{F}}_p);\ \mathbb{Z}/\ell) \\
\downarrow h_* & & \downarrow h_* \\
H_*(BO(2,\overline{\mathbb{F}}_p);\ \mathbb{Z}/\ell) & \xrightarrow{(\beta(\cdot)\beta^{-1})_*} & H_*(BO(2,\overline{\mathbb{F}}_q);\ \mathbb{Z}/\ell)
\end{array}$$

Since conjugation always induces the identity map in homology, we have

$$(-1)^j h_*(\alpha_j) = h_*(\alpha_j)$$

or

$$(1-(-1)^j) h_*(\alpha_j) = 0$$

If j is odd, it follows that $h_*(\alpha_j) = 0$. This completes the proof in the orthogonal case. The symplectic case is handled similarly.

We now have the following result about generators for the various homology algebras. We assume that ℓ is a prime $\neq p$.

Theorem 7.7 (a). The homology algebra

$$H_*(B\mathcal{GL}(\overline{\mathbb{F}}_p);\ \mathbb{Z}/\ell) = \oplus_{n=0}^{\infty} H_*(BGL(n,\overline{\mathbb{F}}_p);\ \mathbb{Z}/\ell)$$

is generated by the elements $\{\alpha_j \mid j \geq 0\}$ of (7.2).

(b) If $p > 2$, the homology algebra

$$H_*(B\mathcal{O}(\overline{\mathbb{F}}_p);\ \mathbb{Z}/2) = \oplus_{n=0}^{\infty} H_*(BO(n,\overline{\mathbb{F}}_p);\ \mathbb{Z}/2)$$

is generated by the elements $\{\epsilon_j \mid j \geq 0\}$ of (7.3)

(c) If $p > 2$, the homology algebra

$$H_*(B\mathcal{S}p(\overline{\mathbb{F}}_p);\ \mathbb{Z}/2) = \oplus_{n=0}^{\infty} H_*(BSp(2n,\overline{\mathbb{F}}_p);\ \mathbb{Z}/2)$$

is generated by the elements $\{\gamma_j \mid j \geq 0\}$ of (7.4)

(d) If $\ell > 2$ and $p > 2$, the homology algebra

$$H_*(B\mathcal{O}(\overline{\mathbb{F}}_p);\ \mathbb{Z}/\ell) = \oplus_{n=0}^{\infty} H_*(BO(n,\overline{\mathbb{F}}_p);\ \mathbb{Z}/\ell)$$

is generated by the elements $\{\sigma_j \mid j \geq 0\}$ of 7.6 together with the generator $[1] \in H_0(BO(1,\overline{\mathbb{F}}_p);\ \mathbb{Z}/\ell) = \mathbb{Z}/\ell$.

(e) If $\ell > 2$, the homology algebra

$$H_*(B\mathcal{O}(\overline{\mathbb{F}}_2);\ \mathbb{Z}/\ell) = \oplus_{n=0}^{\infty} H_*(BO(2n,\overline{\mathbb{F}}_2);\ \mathbb{Z}/\ell)$$

is generated by the elements $\{\sigma_j \mid j \geq 0\}$ of 7.6.

(f) If $\ell > 2$, the homology algebra

$$H_*(B\mathcal{Sp}(\overline{\mathbb{F}}_p);\ \mathbb{Z}/\ell) = \oplus_{n=0}^{\infty} H_*(BSp(2n,\overline{\mathbb{F}}_p);\ \mathbb{Z}/\ell)$$

is generated by the elements $\{\sigma_j | j \geq 0\}$ of 7.6.

Let us denote by $\overline{\alpha}_j, \overline{\epsilon}_j, \overline{\gamma}_j, \overline{\sigma}_j$ the images of $\alpha_j, \epsilon_j, \gamma_j, \sigma_j$ under the standard inclusions

$$BGL(1,\overline{\mathbb{F}}_p) \to BGL(\infty,\overline{\mathbb{F}}_p) \to \Gamma_0 B\mathcal{GL}(\overline{\mathbb{F}}_p)$$

$$BO(1,\overline{\mathbb{F}}_p) \to BO(\infty,\overline{\mathbb{F}}_p) \to \Gamma_0 B\mathcal{O}(\overline{\mathbb{F}}_p)$$

$$BSp(2,\overline{\mathbb{F}}_p) \to BSp(\infty,\overline{\mathbb{F}}_p) \to \Gamma_0 B\mathcal{Sp}(\overline{\mathbb{F}}_p)$$

$$BO(2,\overline{\mathbb{F}}_p) \to BO(\infty,\overline{\mathbb{F}}_p) \to \Gamma_0 B\mathcal{O}(\overline{\mathbb{F}}_p)$$

$$BSp(2,\overline{\mathbb{F}}_p) \to BSp(\infty,\overline{\mathbb{F}}_p) \to \Gamma_0 B\mathcal{Sp}(\overline{\mathbb{F}}_p)$$

An immediate consequence of Theorem 7.7 is

Corollary 7.8 (a) $H_*(\Gamma_0 B\mathcal{GL}(\overline{\mathbb{F}}_p);\ \mathbb{Z}/\ell)$ is generated by $\{\overline{\alpha}_j | j \geq 1\}$

(b) If $p > 2$, $H_*(\Gamma_0 B\mathcal{O}(\overline{\mathbb{F}}_p);\ Z/2)$ is generated by $\{\overline{\epsilon}_j | j \geq 1\}$

(c) If $p > 2$, $H_*(\Gamma_0 B\mathcal{Sp}(\overline{\mathbb{F}}_p);\ Z/2)$ is generated by $\{\overline{\gamma}_j | j \geq 1\}$

(d) If $\ell > 2$, $H_*(\Gamma_0 B\mathcal{O}(\overline{\mathbb{F}}_p);\ \mathbb{Z}/\ell)$ is generated by $\{\overline{\sigma}_j | j \geq 1\}$

(e) If $\ell > 2$, $H_*(\Gamma_0 B\mathcal{Sp}(\overline{\mathbb{F}}_p);\ \mathbb{Z}/\ell)$ is generated by $\{\overline{\sigma}_j | j \geq 1\}$.

We shall prove Theorem 7.7 by a series of lemmas beginning with the general linear case.

Lemma 7.9. If $q = p^r \equiv 1 \pmod{\ell}$ when ℓ is odd or if $q = p^r \equiv 1 \pmod 4$ when $\ell = 2$, then the direct sum homomorphism

$$GL(1,\overline{\mathbb{F}}_q)^n \xrightarrow{\oplus} GL(n,\overline{\mathbb{F}}_q)$$

induces an epimorphism in mod-ℓ homology.

Proof: If ℓ is odd then this is proved explicitly in Chapter VIII §3 using Quillen's detection methods as outlined in §6. If $\ell = 2$, then it is proved similarly in Chapter VII §4 that the direct sum homomorphisms

$$GL(2,\mathbb{F}_q)^n \xrightarrow{\oplus} GL(2n,\mathbb{F}_q)$$

$$GL(2n,\mathbb{F}_q) \times GL(1,\mathbb{F}_q) \xrightarrow{\oplus} GL(2n+1,\mathbb{F}_q)$$

induce epimorphisms on mod-2 homology. Thus the proof will be complete once we show that

$$GL(1,\mathbb{F}_q) \times GL(1,\mathbb{F}_q) \xrightarrow{\oplus} GL(2,\mathbb{F}_q)$$

induces an epimorphism in mod-2 homology.

We begin by noting that the wreath product $\mathcal{S}_2 \wr GL(1,\mathbb{F}_q)$ contains a Sylow 2-subgroup of $GL(2,\mathbb{F}_q)$. For we have

$$|\mathcal{S}_2 \wr GL(1,\mathbb{F}_q)| = 2(q-1)^2 \qquad |GL(2,\mathbb{F}_q)| = (q^2-1)q(q-1)$$

Hence the index of $\mathcal{S}_2 \wr GL(1,\mathbb{F}_q)$ in $GL(2,\mathbb{F}_q)$ is

$$[GL(2,\mathbb{F}_q);\ \mathcal{S}_2 \wr GL(1,\mathbb{F}_q)] = q(\tfrac{q+1}{2})$$

which is odd since $q \equiv 1 \pmod 4$. Hence

$$\mathcal{S}_2 \wr GL(1,\mathbb{F}_q) \to GL(2,\mathbb{F}_q)$$

induces an epimorphism on mod-2 homology. By Lemma 5.6 the subgroups $\mathcal{S}_2 \times GL(1,\mathbb{F}_q)$ and $GL(1,\mathbb{F}_q)^2$ detect $H_*(BGL(2,\mathbb{F}_q);\ \mathbb{Z}/2)$. Since

$$\begin{pmatrix} -1 & 1 \\ 1 & 1 \end{pmatrix}(\mathcal{S}_2 \times GL(1,\mathbb{F}_q))\begin{pmatrix} -1 & 1 \\ 1 & 1 \end{pmatrix}^{-1} \subseteq GL(1,\mathbb{F}_q)^2$$

it follows from Lemma 5.7, that

$$GL(1,\mathbb{F}_q)^2 \xrightarrow{\oplus} GL(2,\mathbb{F}_q)$$

induces an epimorphism in mod-2 homology.

Corollary 7.10. The direct sum homomorphism

$$GL(1,\overline{\mathbb{F}}_p)^n \xrightarrow{\oplus} GL(n,\overline{\mathbb{F}}_p)$$

induces an epimorphism in mod-ℓ homology ($\ell \neq p$).

Proof: We have $\overline{\mathbb{F}}_p = \varinjlim_q \mathbb{F}_q$ where q are as in Lemma 7.9. Hence the result follows from Lemma 7.9 by taking direct limits.

Lemma 7.11. If $p > 2$, the direct sum homomorphisms

(i) $Sp(2,\overline{\mathbb{F}}_p)^n \xrightarrow{\oplus} Sp(2n,\overline{\mathbb{F}}_p)$

(ii) $O(1,\overline{\mathbb{F}}_p)^n \xrightarrow{\oplus} O(n,\overline{\mathbb{F}}_p)$

induce epimorphisms in mod-2 homology.

Proof: According to Prop. VII 3.1 the direct sum homomorphism

$$Sp(2,\mathbb{F}_{p^r})^n \xrightarrow{\oplus} Sp(2n,\mathbb{F}_{p^r})$$

induces epimorphisms in mod-2 homology for all r. Taking direct limits we obtain that (i) induces an epimorphism in mod-2 homology.

According to Prop. VII 2.1 and Prop. VI 4.3(b) the subgroups

$\{O(1,\mathbb{F}_{p^r})^i \times O(\overline{1},\mathbb{F}_{p^r})^{2j} \mid i + 2j = n\}$ detect $H_*(BO(n,\mathbb{F}_{p^r}); \mathbb{Z}/2)$. Since under the permutative functor $\mathcal{O}(\mathbb{F}_{p^r}) \to \mathcal{O}(\overline{\mathbb{F}}_p)$, $O(1,\mathbb{F}_{p^r})$ and $O(\overline{1},\mathbb{F}_{p^r})$ are both mapped into $O(1,\overline{\mathbb{F}}_p)$ it follows that $O(1,\overline{\mathbb{F}}_p)^n$ detects the image of $H_*(BO(n,\mathbb{F}_{p^r}); \mathbb{Z}/2)$ in $H_*(BO(n,\overline{\mathbb{F}}_p); \mathbb{Z}/2)$. Taking direct limits we get that (ii) induces an epimorphism in mod-2 homology.

Lemma 7.12. If ℓ is odd, r is even and $p^r \equiv 1 \pmod{\ell}$, then the direct sum homomorphisms

$$O(2,\mathbb{F}_{p^r})^n \xrightarrow{\oplus} O(2n,\mathbb{F}_{p^r})$$

$$Sp(2,\mathbb{F}_{p^r})^n \xrightarrow{\oplus} Sp(2n,\mathbb{F}_{p^r})$$

induce epimorphisms in mod-ℓ homology. If in addition $p > 2$, then the direct sum homorphism

$$O(2n,\mathbb{F}_{p^r}) \times O(1,\mathbb{F}_{p^r}) \to O(2n+1,\mathbb{F}_{p^r})$$

also induces an epimorphism in mod-ℓ homology.

Proof: This is a special case of Theorems VIII 4.3 and 5.3.

Taking direct limits we obtain

Corollary 7.13. If ℓ is an odd prime $\neq p$, then the direct sum homomorphisms

$$O(2,\overline{\mathbb{F}}_p)^n \xrightarrow{\oplus} O(2n,\overline{\mathbb{F}}_p)$$

$$O(2n,\overline{\mathbb{F}}_p) \times O(1,\overline{\mathbb{F}}_p) \xrightarrow{\oplus} O(2n+1,\overline{\mathbb{F}}_p) \quad (p > 2)$$

$$Sp(2,\overline{\mathbb{F}}_p)^n \xrightarrow{\oplus} Sp(2n,\overline{\mathbb{F}}_p)$$

induce epimorphisms in mod-ℓ homology.

7.14. Proof of Theorem 7.7. This is immediate from Cor. 7.10, Lemma 7.11 and Cor. 7.13.

We can now prove our main result

Theorem 7.15. The Brauer lift maps

$$\bar{\beta}: \Gamma_0 B\mathcal{GL}(\bar{\mathbb{F}}_p) \to BU$$

$$\bar{\beta}: \Gamma_0 B\mathcal{O}(\bar{\mathbb{F}}_p) \to BO$$

$$\bar{\beta}: \Gamma_0 B\mathcal{Sp}(\bar{\mathbb{F}}_q) \to BSp$$

induce isomorphisms in mod-ℓ homology for any prime $\ell \neq p$. Consequently the maps β are equivalences when completed away from p.

Proof: (a) The general linear case. From the definition of Brauer lift it is easy to see that the following diagram commutes

$$\begin{array}{ccccccc} & \overset{i}{\nearrow} & BGL(1,\bar{\mathbb{F}}_p) & \to & BGL(\infty,\bar{\mathbb{F}}_p) & \to & \Gamma_0 B\mathcal{GL}(\bar{\mathbb{F}}_p) \\ B\mathbb{Z}/\ell & & & & & & \downarrow \beta \\ & \underset{i}{\searrow} & B\bar{\mathbb{F}}_q^{*} & \xrightarrow{B\beta} & BS^1 = BU(1) & \longrightarrow & BU \end{array}$$

where $\beta: \bar{\mathbb{F}}_p^* \to S^1$ is the imbedding chosen in the definition of Brauer lift (cf. §2). According to VI 2.9, $i_*(x_{2i}) = \alpha_i \in H_*(BGL(1,\bar{\mathbb{F}}_p); \mathbb{Z}/\ell)$. Chasing the element x_{2i} around the top of the diagram in mod-ℓ homology we obtain $\beta_*(\bar{\alpha}_i)$. Pursuing the element x_{2i} around the bottom of the diagram and applying Lemma 6.11 we obtain a_i where $\{a_i \mid i \geq 1\}$ are the standard generators of $H_*(BU; \mathbb{Z}/\ell) = \mathbb{Z}/\ell[a_1, a_2, \ldots]$.

Consequently $\overline{\beta}_*(\overline{\alpha}_i) = a_i$ and in the chain of maps

$$\mathbb{Z}/\ell[\overline{\alpha}_1,\overline{\alpha}_2,\ldots] \xrightarrow{i} H_*(\Gamma_0 B\,\mathcal{GL}(\overline{\mathbb{F}}_p);\ \mathbb{Z}/\ell)$$

$$\xrightarrow{\overline{\beta}_*} H_*(BU;\ \mathbb{Z}/\ell) = \mathbb{Z}/\ell[a_1,a_2,\ldots]$$

the composite $\overline{\beta}_* \cdot i$ is an isomorphism. But by Corollary 7.8, i is an epimorphism. Hence i and $\overline{\beta}_*$ must both be isomorphisms.

(b) The orthogonal case, $\ell = 2$, $p \neq 2$. From the definition of Brauer lift it is easy to see that the following diagram commutes

$$\begin{array}{ccccc} BO(1,\overline{\mathbb{F}}_p) & \rightarrow & BO(\infty,\overline{\mathbb{F}}_p) & \rightarrow & \Gamma_0 B\mathcal{O}(\overline{\mathbb{F}}_p) \\ \| & & & & \downarrow{\scriptstyle \beta} \\ B\mathbb{Z}/2 & = & BO(1,\mathbb{R}) & \rightarrow & BO \end{array}$$

Applying mod-2 homology and chasing the element $\epsilon_i \in H_i(BO(1,\overline{\mathbb{F}}_p);\ \mathbb{Z}/2)$ around the diagram we obtain $\overline{\beta}_*(\overline{\epsilon}_i) = \overline{e}_i$ where $\{\overline{e}_i \mid i \geq 1\}$ are the standard generators of $H_*(BO;\ \mathbb{Z}/2) = \mathbb{Z}/2[\overline{e}_1,\overline{e}_2,\ldots]$. The rest of the proof now proceeds as in case (a).

(c) The orthogonal case $\ell \neq 2$. In this case the generators $\sigma_j \in H_{4j}(BO(2,\mathbb{F}_q);\ \mathbb{Z}/\ell)$ are the images of the generators a_{2j} under the hyperbolic inclusion 7.5

$$\overline{\mathbb{F}}_p^* = GL(1,\overline{\mathbb{F}}_p) \rightarrow O(2,\overline{\mathbb{F}}_p)$$

obtained by endowing $\overline{\mathbb{F}}_p \oplus \overline{\mathbb{F}}_p$ with the quadratic form $Q((x,y)) = xy$ and regarding $a \in \overline{\mathbb{F}}_p^*$ as acting on $\overline{\mathbb{F}}_p \oplus \overline{\mathbb{F}}_p$ by

$$a(x,y) = (ax, a^{-1}y)$$

Now consider the representation

$$E\colon \mathbb{Z}/\ell \to \overline{\mathbb{F}}_p^* \to O(2,\overline{\mathbb{F}}_p)$$

If $\mu \in \mathbb{Z}/\ell$ is a generator, the eigenvalues of μ^r under this representation are $\{\mu^r, (\mu^r)^{-1}\}$. The Brauer character of this element is then

$$\chi_E(\mu^r) = \beta(\mu^r) + \beta(\mu^r)^{-1} = 2\cos(2\pi r/\ell)$$

But this is precisely the character of the standard representation

$$\rho\colon \mathbb{Z}/\ell \to S^1 = SO(2,\mathbb{R}) \subset O(2,\mathbb{R})$$

Hence the following diagram commutes

$$\begin{array}{ccccccccc} & \nearrow & B\overline{\mathbb{F}}_p^* = BGL(1,\overline{\mathbb{F}}_p) & \to & BO(2,\overline{\mathbb{F}}_p) & \to & BO(\infty,\overline{\mathbb{F}}_p) & \to & \Gamma_0 B\mathcal{O}(\overline{\mathbb{F}}_p) \\ B\mathbb{Z}/\ell & & & & & & & & \downarrow \overline{\beta} \\ & \searrow & BS^1 = BSO(2,\mathbb{R}) & \to & BO(2,\mathbb{R}) & & \longrightarrow & & BO \end{array}$$

Applying mod-ℓ homology to the above diagram and chasing the element $x_{4i} \in H_{4i}(B\mathbb{Z}/\ell;\ \mathbb{Z}/\ell)$ around the diagram, using Lemma 6.1 we obtain $\overline{\beta}_*(\overline{\sigma}_j) = p_j$ where $\{p_j \mid j \geq 1\}$ are the standard generators of $H_*(BO;\ \mathbb{Z}/\ell) = \mathbb{Z}/\ell[p_1, p_2, \ldots]$. The rest of the proof now proceeds as in case (a).

(d) The symplectic case $\ell = 2$, $p \neq 2$. Let

$$E\colon \mathbb{Z}/2 \to Sp(2,\overline{\mathbb{F}}_p)$$

be the natural inclusion. Then the Brauer character of E is given by

$$\chi_E(1) = \beta(1) + \beta(1) = 2 \qquad \chi_E(-1) = \beta(-1) + \beta(-1) = -2$$

But this is the character of the standard representation

$$\rho\colon \mathbb{Z}/2 \to S^1 \to S^3 = Sp(1) \to U(2,\mathbf{C})$$

Hence the following diagram commutes

$$\begin{array}{ccccccc} & \nearrow & BSp(2,\overline{\mathbb{F}}_p) & \to & BSp(\infty,\overline{\mathbb{F}}_p) & \to & \Gamma_0 B\mathcal{S}p(\overline{\mathbb{F}}_p) \\ B\mathbb{Z}/2 & & & & & & \downarrow \overline{\beta} \\ & \searrow & BS^1 \to BS^3 & = & BSp(1) & \longrightarrow & BSp \end{array}$$

According to VI 5.5, $i_*(x_{4j}) = \gamma_j$. Chasing the element x_{4j} around the above diagram in mod-2 homology, using Lemma 6.1, we obtain $\overline{\beta}_*(\overline{\gamma}_j) = g_j$ where $\{g_j \mid j \geq 1\}$ are the standard generators of $H_*(BSp;\ \mathbb{Z}/2) = \mathbb{Z}/2[g_1, g_2, \ldots]$. The rest of the argument now proceeds as in case (a).

(e) The symplectic case $\ell \neq 2$. This is virtually identical to the corresponding orthogonal case (c).

The following results were also proved during the course of the proof of Theorem 7.15. Again we assume ℓ is a prime $\neq p$ and use the notation of Cor. 7.8.

<u>Theorem</u> 7.16 (a) $H_*(\Gamma_0 B\mathcal{GL}(\overline{\mathbb{F}}_p);\ \mathbb{Z}/\ell) = \mathbb{Z}/\ell[\overline{\alpha}_1, \overline{\alpha}_2, \ldots]$

(b) If $p > 2$, $H_*(\Gamma_0 B\mathcal{O}(\overline{\mathbb{F}}_p);\ \mathbb{Z}/2) = \mathbb{Z}/2[\overline{\epsilon}_1, \overline{\epsilon}_2, \ldots]$

(c) If $p > 2$, $H_*(\Gamma_0 B\mathcal{S}p(\overline{\mathbb{F}}_p);\ \mathbb{Z}/2) = \mathbb{Z}/2[\overline{\gamma}_1, \overline{\gamma}_2, \ldots]$

(d) If $\ell > 2$, $H_*(\Gamma_0 B\mathcal{O}(\overline{\mathbb{F}}_p);\ \mathbb{Z}/\ell) = \mathbb{Z}/\ell[\overline{\sigma}_1, \overline{\sigma}_2, \ldots]$

(e) If $\ell > 2$, $H_*(\Gamma_0 B\mathcal{S}p(\overline{\mathbb{F}}_p);\ \mathbb{Z}/\ell) = \mathbb{Z}/\ell[\overline{\sigma}_1, \overline{\sigma}_2, \ldots]$

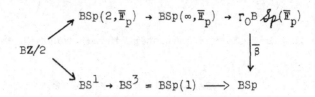

We then obtain the following results about the mod-ℓ homology of the classical groups over $\overline{\mathbb{F}}_p$ assuming $\ell \neq p$. We use the notation of Theorem 7.7.

Theorem 7.17 (a). The homology algebra $H_*(B\mathcal{GL}(\overline{\mathbb{F}}_p); \mathbb{Z}/\ell) = \oplus_{n=0}^{\infty} H_*(BGL(n,\overline{\mathbb{F}}_p); \mathbb{Z}/\ell)$ is free commutative on the generators $\{\alpha_j | j \geq 0\}$, i.e.

$$H_*(B\mathcal{GL}(\overline{\mathbb{F}}_p); \mathbb{Z}/\ell) = \mathbb{Z}/\ell[\alpha_0, \alpha_1, \alpha_2, \ldots]$$

(b) If $p > 2$, the homology algebra $H_*(B\mathcal{O}(\overline{\mathbb{F}}_p); \mathbb{Z}/2) = \oplus_{n=0}^{\infty} H_*(BO(n,\overline{\mathbb{F}}_p); \mathbb{Z}/2)$ is free commutative on the generators $\{\epsilon_j | j \geq 0\}$, i.e.

$$H_*(B\mathcal{O}(\overline{\mathbb{F}}_p); \mathbb{Z}/2) = \mathbb{Z}/2[\epsilon_0, \epsilon_1, \epsilon_2, \ldots]$$

(c) If $p > 2$, the homology algebra $H_*(B\mathcal{Sp}(\overline{\mathbb{F}}_p); \mathbb{Z}/2) = \oplus_{n=0}^{\infty} H_*(BSp(2n,\overline{\mathbb{F}}_p); \mathbb{Z}/2)$ is free commutative on the generators $\{\gamma_j | j \geq 0\}$, i.e.

$$H_*(B\mathcal{Sp}(\overline{\mathbb{F}}_p); \mathbb{Z}/2) = \mathbb{Z}/2[\gamma_0, \gamma_1, \gamma_2, \ldots]$$

(d) If $\ell > 2$ and $p > 2$, the homology algebra $H_*(B\mathcal{O}(\overline{\mathbb{F}}_p); \mathbb{Z}/\ell) = \oplus_{n=0}^{\infty} H_*(BO(n,\overline{\mathbb{F}}_p); \mathbb{Z}/\ell)$ is free commutative on the generators $\{[1], \sigma_j | j \geq 0\}$, i.e.

$$H_*(B\mathcal{O}(\overline{\mathbb{F}}_p); \mathbb{Z}/\ell) = \mathbb{Z}/\ell[[1], \sigma_0, \sigma_1, \sigma_2, \ldots]$$

(e) If $\ell > 2$, the homology algebra $H_*(B\mathcal{O}(\overline{\mathbb{F}}_2); \mathbb{Z}/\ell) = \oplus_{n=0}^{\infty} H_*(BO(2n,\overline{\mathbb{F}}_2); \mathbb{Z}/\ell)$ is free commutative on the generators $\{\sigma_j | j \geq 0\}$, i.e.

$$H_*(B\mathcal{O}(\overline{\mathbb{F}}_2);\ \mathbb{Z}/\ell) = \mathbb{Z}/\ell[\sigma_0,\sigma_1,\sigma_2,\ldots]$$

(f) If $\ell > 2$, the homology algebra $H_*(B\mathcal{S}p(\overline{\mathbb{F}}_p);\ \mathbb{Z}/\ell) = \oplus_{n=0}^{\infty} H_*(BSp(2n,\overline{\mathbb{F}}_p);\ \mathbb{Z}/\ell)$ is free commutative on the generators $\{\sigma_j \mid j \geq 0\}$, i.e.

$$H_*(B\mathcal{S}p(\overline{\mathbb{F}}_p);\ \mathbb{Z}/\ell);\ \mathbb{Z}/\ell) = \mathbb{Z}/\ell[\sigma_0,\sigma_1,\sigma_2,\ldots]$$

Proof: We confine ourselves to the general linear case. The other cases are proved similarly. Consider the infinite component version

$$\Gamma B\mathcal{GL}(\overline{\mathbb{F}}_p) \simeq \Gamma_0 B\mathcal{GL}(\overline{\mathbb{F}}_p) \times \mathbb{Z}$$

It follows from Theorem 7.16 that

$$H_*(\Gamma B\mathcal{GL}(\overline{\mathbb{F}}_p);\ \mathbb{Z}/\ell) = \mathbb{Z}/\ell[[-1],[1],\overline{\alpha}_1,\overline{\alpha}_2,\ldots]/([-1][1] = [0])$$

Let us denote by α_j the image of α_j under the natural map

$$i_*: H_*(B\mathcal{GL}(\overline{\mathbb{F}}_p);\ \mathbb{Z}/\ell) \to H_*(\Gamma B\mathcal{GL}(\overline{\mathbb{F}}_p);\ \mathbb{Z}/\ell).$$

Then $\alpha_0 = [1]$ and $\alpha_j = \overline{\alpha}_j[1]$ for $j \geq 1$. It is therefore clear that $H_*(\Gamma B\mathcal{GL}(\overline{\mathbb{F}}_p);\ \mathbb{Z}/\ell)$ may be equally well described as

$$H_*(\Gamma B\mathcal{GL}(\overline{\mathbb{F}}_p);\ \mathbb{Z}/\ell) = \mathbb{Z}/\ell[[-1],\alpha_0,\alpha_1,\alpha_2,\ldots]/([-1]\alpha_0 = [0])$$

Now consider the chain of maps

$$\mathbb{Z}/\ell[\alpha_0,\alpha_1,\alpha_2,\ldots] \xrightarrow{\pi_*} H_*(B\mathcal{GL}(\overline{\mathbb{F}}_p);\ \mathbb{Z}/\ell) \xrightarrow{i_*} H_*(\Gamma B\mathcal{GL}(\overline{\mathbb{F}}_p);\ \mathbb{Z}/\ell)$$

Clearly the composite $i_* \circ \pi_*$ is a monomorphism. Hence π_* is a monomorphism. But by Theorem 7.7, π_* is an epimorphism. Consequently π_* is an isomorphism.

We now dualize the results of Theorem 7.17 to obtain the structure of the mod-ℓ cohomology rings of the classical groups over $\bar{\mathbb{F}}_p$ when $\ell \neq p$. We begin by computing the cohomology rings of the lowest dimensional groups.

<u>Lemma</u> 7.18. (a) $H^*(BGL(1,\bar{\mathbb{F}}_p);\ \mathbb{Z}/\ell) = \mathbb{Z}/\ell[y]$ where degree $y = 2$.

(b) If $p > 2$, $H^*(BO(1,\bar{\mathbb{F}}_p);\ \mathbb{Z}/2) = \mathbb{Z}/2[x]$ where degree $x = 1$

(c) If $p > 2$, $H^*(BSp(2,\bar{\mathbb{F}}_p);\ \mathbb{Z}/2) = \mathbb{Z}/2[P]$ where degree $P = 4$

(d) If $\ell > 2$, $H^*(BO(2,\bar{\mathbb{F}}_p);\ \mathbb{Z}/\ell) = \mathbb{Z}/\ell[z]$ where degree $z = 4$

(e) If $\ell > 2$, $H^*(BSp(2,\bar{\mathbb{F}}_p);\ \mathbb{Z}/\ell) = \mathbb{Z}/\ell[z]$ where degree $z = 4$.

<u>Proof</u>: Since $GL(1,\bar{\mathbb{F}}_p) = \bar{\mathbb{F}}_p^* = \oplus_{\ell\neq p}\mathbb{Z}/\ell^\infty$, (a) follows immediately from VI 2.9 . Since $O(1,\bar{\mathbb{F}}_p) = \mathbb{Z}/2$, (b) is immediate from VI 2.1. Case (c) is explicitly proved in VI 5.10.

To prove (d) we note that by Lemma 7.6, the hyperbolic inclusion

$$h\colon GL(1,\bar{\mathbb{F}}_p) \to O(2,\bar{\mathbb{F}}_p)$$

induces an epimorphism in mod-ℓ homology and hence a monomorphism in mod-ℓ cohomology. Since by Theorem 7.17

$$H^i(BO(2,\bar{\mathbb{F}}_p);\ \mathbb{Z}/\ell) = \begin{cases} \mathbb{Z}/\ell \text{ if } i \equiv 0 \pmod 4,\ i \geq 0 \\ 0 \quad \text{otherwise} \end{cases}$$

$H^*(BO(2,\bar{\mathbb{F}}_p);\ \mathbb{Z}/\ell) \cong \operatorname{im} h^* = \mathbb{Z}/\ell[y^2]$. Case (d) now follows. Case (c) is proved similarly.

Lemma 7.18 can be described concisely as follows: Let G = GL, O or Sp and let k = 1 or 2 depending on the case considered. Then

$$H^*(BG(k);\ \mathbb{Z}/\ell) = \mathbb{Z}/\ell[u]$$

According to Theorem 7.17 the direct sum homomorphisms

$$G(k)^n \xrightarrow{\oplus} G(kn)$$

induce epimorphisms on mod-ℓ homology and hence monomorphisms on mod-ℓ cohomology. Moreover 7.17 also implies that $H_*(BG(kn);\ \mathbb{Z}/\ell)$ is obtained from $H_*(BG(k)^n;\ \mathbb{Z}/\ell)$ by dividing out by all the relations implied by commutativity, in other words by dividing out by the action of the symmetric group $\mathcal{S}_n$ on $G(k)^n$. Dually we must have that $H^*(BG(kn);\ \mathbb{Z}/\ell)$ is the symmetric subring of $H^*(BG(k)^n;\ \mathbb{Z}/\ell) = \mathbb{Z}/\ell[u_1,u_2,\ldots,u_n]$. Hence

$$H^*(BG(kn);\ \mathbb{Z}/\ell) = \mathbb{Z}/\ell[s_1,s_2,\ldots,s_n]$$

where s_i is the i-th elementary symmetric polynomial in $u_1,u_2,\ldots,u_n$, with degree $s_i = i \times$ degree a. Thus we have the following results

<u>Theorem</u> 7.19 (a) $H^*(BGL(n,\overline{\mathbb{F}}_p);\ \mathbb{Z}/\ell) = \mathbb{Z}/\ell[c_1,c_2,\ldots,c_n]$ as a subring of $H^*(BGL(1,\overline{\mathbb{F}}_p)^n;\ \mathbb{Z}/\ell) = \mathbb{Z}/\ell[y_1,y_2,\ldots,y_n]$ where

$$c_i = \sigma_i(y_1,y_2,\ldots,y_n) \in H^{2i}(BGL(1,\overline{\mathbb{F}}_p)^n;\ \mathbb{Z}/\ell)$$

is the i-th elementary symmetric polynomial

(b) If $p > 2$, $H^*(BO(n,\overline{\mathbb{F}}_p);\ \mathbb{Z}/2) = \mathbb{Z}/2[w_1,w_2,\ldots,w_n]$ as a subring of $H^*(BO(1,\overline{\mathbb{F}}_p)^n;\ \mathbb{Z}/2) = \mathbb{Z}/2[x_1,x_2,\ldots,x_n]$ where

$$w_i = \sigma_i(x_1, x_2, \ldots, x_n) \in H^i(BO(1, \overline{\mathbb{F}}_p)^n; \mathbb{Z}/2)$$

is the i-th elementary symmetric polynomial

(c) If $p > 2$, $H^*(BSp(2n, \overline{\mathbb{F}}_p); \mathbb{Z}/2) = \mathbb{Z}/2[\overline{g}_1, \overline{g}_2, \ldots, \overline{g}_n]$ as a subring of $H^*(BSp(2, \overline{\mathbb{F}}_p)^n; \mathbb{Z}/\ell) = \mathbb{Z}/\ell[P_1, P_2, \ldots, P_n]$ where

$$\overline{g}_i = \sigma_i(P_1, P_2, \ldots, P_n) \in H^{4i}(BSp(2, \overline{\mathbb{F}}_p)^n; \mathbb{Z}/2)$$

is the i-th elementary symmetric polynomial

(d) If $\ell > 2$, $H^*(BO(2n, \overline{\mathbb{F}}_p); \mathbb{Z}/\ell) = \mathbb{Z}/\ell[\overline{p}_1, \overline{p}_2, \ldots, \overline{p}_n]$ as a subring of $H^*(BO(2, \overline{\mathbb{F}}_p)^n; \mathbb{Z}/\ell) = \mathbb{Z}/\ell[z_1, z_2, \ldots, z_n]$ where

$$\overline{p}_i = \sigma_i(z_1, z_2, \ldots, z_n) \in H^{4i}(BO(2, \overline{\mathbb{F}}_p)^n; \mathbb{Z}/\ell)$$

is the i-th elementary symmetric polynomial

(e) If $\ell > 2$, $H^*(BSp(2n, \overline{\mathbb{F}}_p); \mathbb{Z}/\ell) = \mathbb{Z}/\ell[\overline{p}_1, \overline{p}_2, \ldots, \overline{p}_n]$ as a subring of $H^*(BSp(2, \overline{\mathbb{F}}_p)^n; \mathbb{Z}/\ell) = \mathbb{Z}/\ell[z_1, z_2, \ldots, z_n]$ where

$$\overline{p}_i = \sigma_i(z_1, z_2, \ldots, z_n) \in H^{4i}(BSp(2, \overline{\mathbb{F}}_p)^n; \mathbb{Z}/\ell)$$

is the i-th elementary symmetric polynomial.

(f) If $\ell > 2$ and $p > 2$, then the inclusion

$$O(2n, \overline{\mathbb{F}}_p) \to O(2n+1, \overline{\mathbb{F}}_p) \qquad n \geq 0$$

induces an isomorphism of mod-ℓ cohomology rings.

Next we turn to the rational and mod-p homology and cohomology of the classical groups over $\overline{\mathbb{F}}_p$.

<u>Proposition</u> 7.20 (a) The reduced rational homology and cohomology of

the groups $GL(n,\overline{\mathbb{F}}_p)$, $O(n,\overline{\mathbb{F}}_p)$, $Sp(2n,\overline{\mathbb{F}}_p)$ is zero in all degrees

(b) The reduced mod-p homology and cohomology of the groups $GL(n,\overline{\mathbb{F}}_p)$ and $Sp(2n,\overline{\mathbb{F}}_p)$ is zero in all degrees

(c) If $p > 2$, the reduced mod-p homology and cohomology of the groups $O(n,\overline{\mathbb{F}}_p)$ is zero in all degrees

(d) The Dickson invariant homomorphism

$$d: O(2n,\overline{\mathbb{F}}_2) \to \mathbb{Z}/2$$

induces isomorphisms in mod-2 homology. The direct sum homomorphism

$$\mu: O(2m,\overline{\mathbb{F}}_2) \times O(2n,\overline{\mathbb{F}}_2) \xrightarrow{\oplus} O(2m+2n,\overline{\mathbb{F}}_2)$$

induces on mod-2 homology the map given by

$$\mu_*(x_{mi}, x_{nj}) = (i,j)\, x_{m+n,i+j}$$

where $x_{nj} \in H_{4j}(BO(2n,\overline{\mathbb{F}}_2); \mathbb{Z}/2) \cong H_{4j}(B\mathbb{Z}/2; \mathbb{Z}/2) = \mathbb{Z}/2$ denotes the generator.

<u>Proof</u>: Since $GL(n,\overline{\mathbb{F}}_p)$, $O(n,\overline{\mathbb{F}}_p)$, $Sp(2n,\overline{\mathbb{F}}_p)$ are direct limits of the finite groups $GL(n,\mathbb{F}_{p^r})$, $O(n,\mathbb{F}_{p^r})$, $Sp(2n,\mathbb{F}_{p^r})$ respectively, (a) follows immediately. Similarly (b) and (c) follow by taking direct limits in Lemma 4.2 and 4.5.

In case (d) we first note that the Dickson invariant homomorphisms

$$d: O(2n,\mathbb{F}_{2^r}) \to \mathbb{Z}/2$$

defined in II 7.19 are consistent with respect to extension of scalars and hence we can pass to direct limits to define

$$d\colon O(2n,\overline{\mathbb{F}}_2) \to \mathbb{Z}/2$$

Let $DO(2n,\overline{\mathbb{F}}_2)$ denote the kernel of d. Then by taking direct limits in Lemma 4.5, we find that $\tilde{H}_*(BDO(2n,\overline{\mathbb{F}}_2);\, \mathbb{Z}/2) = \tilde{H}^*(BDO(2n,\overline{\mathbb{F}}_2);\, \mathbb{Z}/2) = 0$. The first part of part (d) now follows from the Serre spectral sequence of the fibration

$$BDO(2n,\overline{\mathbb{F}}_2) \to BO(2n,\overline{\mathbb{F}}_2) \to B\mathbb{Z}/2$$

The second part of (d) follows from the commutative diagram

$$\begin{array}{ccc} O(2m,\overline{\mathbb{F}}_2) \times O(2n,\overline{\mathbb{F}}_2) & \to & O(2m+2n,\overline{\mathbb{F}}_2) \\ \downarrow{\scriptstyle d\times d} & & \downarrow{\scriptstyle d} \\ \mathbb{Z}/2 \times \mathbb{Z}/2 & \xrightarrow{\mu} & \mathbb{Z}/2 \end{array}$$

where the bottom arrow denotes multiplication, which is evaluated in homology in VI 2.3(e).

As a consequence we get the following results about the rational and p-homotopy type of the spaces $\Gamma_0 B\mathcal{GL}(\overline{\mathbb{F}}_p)$, $\Gamma_0 B\mathcal{O}(\overline{\mathbb{F}}_p)$, $\Gamma_0 B\mathcal{Sp}(\overline{\mathbb{F}}_p)$.

<u>Corollary</u> 7.21 (a). The spaces $\Gamma_0 B\mathcal{GL}(\overline{\mathbb{F}}_p)$, $\Gamma_0 B\mathcal{O}(\overline{\mathbb{F}}_p)$, $\Gamma_0 B\mathcal{Sp}(\overline{\mathbb{F}}_p)$ all have trivial rational homotopy type.

(b) The spaces $\Gamma_0 B\mathcal{GL}(\overline{\mathbb{F}}_p)$, $\Gamma_0 B\mathcal{O}(\overline{\mathbb{F}}_p)$ $p > 2$, $\Gamma_0 B\mathcal{Sp}(\overline{\mathbb{F}}_p)$ all have trivial p-homotopy type

(c) The Dickson invariant induces an equivalence

$$d\colon \Gamma_0 B\mathcal{O}(\overline{\mathbb{F}}_2) \to B\mathbb{Z}/2$$

of infinite loop spaces at 2.

Proof: By Prop. 7.20 the space of (a) and (b) have trivial rational and mod-p homology, from which (a) and (b) follow. Similarly Prop. 7.20(d) implies that d induces an isomorphism in mod-2 homology, from which (c) follows.

Note. Corollary 7.21 implies that the Brauer lift maps of Theorem 7.15 are not equivalences rationally or at p. Hence they must be completed away from p in order to become equivalences.

We conclude by a calculation of homotopy groups.

Theorem 7.22. For $i \geq 1$ the homotopy groups of the indicated spaces are given in the following table

i(mod 8)	$\pi_i \Gamma_0 B\mathcal{GL}(\overline{\mathbb{F}}_p)$	p > 2 $\pi_i \Gamma_0 B\mathcal{O}(\overline{\mathbb{F}}_p)$	p > 2 $\pi_i \Gamma_0 B\mathcal{Sp}(\overline{\mathbb{F}}_p)$
0	0	0	0
1	$\overline{\mathbb{F}}_p^* = \oplus_{\ell \neq p} \mathbb{Z}/\ell^\infty$	$\mathbb{Z}/2$	0
2	0	$\mathbb{Z}/2$	0
3	$\overline{\mathbb{F}}_p^* = \oplus_{\ell \neq p} \mathbb{Z}/\ell^\infty$	$\overline{\mathbb{F}}_p^* = \oplus_{\ell \neq p} \mathbb{Z}/\ell^\infty$	$\overline{\mathbb{F}}_p^* = \oplus_{\ell \neq p} \mathbb{Z}/\ell^\infty$
4	0	0	0
5	$\overline{\mathbb{F}}_p^* = \oplus_{\ell \neq p} \mathbb{Z}/\ell^\infty$	0	$\mathbb{Z}/2$
6	0	0	$\mathbb{Z}/2$
7	$\overline{\mathbb{F}}_p^* = \oplus_{\ell \neq p} \mathbb{Z}/\ell^\infty$	$\overline{\mathbb{F}}_p^* = \oplus_{\ell \neq p} \mathbb{Z}/\ell^\infty$	$\overline{\mathbb{F}}_p^* = \oplus_{\ell \neq p} \mathbb{Z}/\ell^\infty$

i(mod 8)	$\pi_i \Gamma_0 B\mathcal{O}(\overline{\mathbb{F}}_2)$	$\pi_i \Gamma_0 B\mathcal{Sp}(\overline{\mathbb{F}}_2)$
0	0	0
1	$\begin{cases} \mathbb{Z}/2 & \text{if } i=1 \\ 0 & \text{if } i>1 \end{cases}$	0
2	0	0
3	$\overline{\mathbb{F}}_2^* = \oplus_{\ell \neq 2} \mathbb{Z}/\ell^\infty$	$\overline{\mathbb{F}}_2^* = \oplus_{\ell \neq 2} \mathbb{Z}/\ell^\infty$
4	0	0
5	0	0
6	0	0
7	$\overline{\mathbb{F}}_2^* = \oplus_{\ell \neq 2} \mathbb{Z}/\ell^\infty$	$\overline{\mathbb{F}}_2^* = \oplus_{\ell \neq 2} \mathbb{Z}/\ell^\infty$

Proof: We illustrate by computing the homotopy groups $\pi_i\Gamma_0 B\mathcal{O}(\overline{\mathbb{F}}_p)$ $p > 2$. The other computations are similar.

Our starting point is Theorem 7.15 which implies that

$$(7.23) \qquad \pi_i\Gamma_0 B\mathcal{O}(\overline{\mathbb{F}}_p)[\tfrac{1}{p}]^\wedge \cong \pi_i BO[\tfrac{1}{p}]^\wedge .$$

According to Bousfield-Kan [8], for any simple space X there is a short exact sequence

$$(7.24) \quad 0 \to \Pi_{\ell\neq p}\mathrm{Ext}(\mathbb{Z}/\ell^\infty,\pi_i X) \to \pi_i X[\tfrac{1}{p}]^\wedge \to \Pi_{\ell\neq p}\mathrm{Hom}(\mathbb{Z}/\ell^\infty,\pi_{i-1}X) \to 0$$

and if X has finitely generated homotopy groups, then $\pi_i X[\frac{1}{p}]^\wedge$ is just the profinite completion of $\pi_i X$ away from p.

Since for $i \geq 1$, $\pi_i BO = \mathbb{Z}$, $\mathbb{Z}/2$, $\mathbb{Z}/2$, 0, $\mathbb{Z}$, 0, 0, 0 according as $i \equiv 0,1,2,3,4,5,6,7 \pmod 8$ we must have $\pi_i BO = \Pi_{\ell\neq p}\hat{\mathbb{Z}}_{(\ell)}$, $\mathbb{Z}/2$, $\mathbb{Z}/2$, 0, $\Pi_{\ell\neq p}\hat{\mathbb{Z}}_{(\ell)}$, 0, 0, 0 according as $i \equiv 0,1,2,3,4,5,6,7 \pmod 8$. It follows from 3.5 that $\pi_i\Gamma_0 B\mathcal{O}(\overline{\mathbb{F}}_p) = \varinjlim_r \pi_i\Gamma_0 B\mathcal{O}(\mathbb{F}_{p^r})$ is 0 for $i = 4,5,6 \pmod 8$, is a $\mathbb{Z}/2$-vector space for $i = 0,1,2 \pmod 8$, and is a direct limit of finite $\frac{1}{p}$ -local groups and hence a torsion $\frac{1}{p}$ -local group for $i \equiv 3,7 \pmod 8$.

For $i \equiv 1,2 \pmod 8$, (7.23) and (7.24) give us

$$(7.25) \quad 0 \to \Pi_{\ell\neq p}\mathrm{Ext}(\mathbb{Z}/\ell^\infty,\pi_i\Gamma_0 B\mathcal{O}(\overline{\mathbb{F}}_p)) \to \mathbb{Z}/2$$

$$\to \Pi_{\ell\neq p}\mathrm{Hom}(\mathbb{Z}/\ell^\infty,\pi_{i-1}\Gamma_0 B\mathcal{O}(\overline{\mathbb{F}}_p)) \to 0$$

Since $\pi_j\Gamma_0 B\mathcal{O}(\overline{\mathbb{F}}_p)$ is a $\mathbb{Z}/2$-vector space for $j \equiv 0,1,2 \pmod 8$

$$(7.26)\quad \begin{aligned} &\mathrm{Hom}(\mathbb{Z}/\ell^\infty, \pi_j\Gamma_0 B\mathcal{O}(\overline{\mathbb{F}}_p)) = 0 \quad \text{for all } \ell \\ &\mathrm{Ext}(\mathbb{Z}/\ell^\infty, \pi_j\Gamma_0 B\mathcal{O}(\overline{\mathbb{F}}_p)) = \begin{cases} 0 & \ell \neq 2 \\ \pi_i\Gamma_0 B\mathcal{O}(\overline{\mathbb{F}}_p) & \ell = 2 \end{cases} \end{aligned}$$

Consequently we must have $\pi_i\Gamma_0 B\mathcal{O}(\overline{\mathbb{F}}_p) = \mathbb{Z}/2$ for $i \equiv 1,2 \pmod 8$.

For $j \equiv 0 \pmod 8$, (7.23) and (7.24) give us

$$(7.27)\quad \begin{aligned} 0 \to \Pi_{\ell\neq p}\mathrm{Ext}(\mathbb{Z}/\ell^\infty, \pi_j\Gamma_0 B\mathcal{O}(\overline{\mathbb{F}}_p)) &\to \Pi_{\ell\neq p}\hat{\mathbb{Z}}_{(\ell)} \\ &\to \Pi_{\ell\neq p}\mathrm{Hom}(\mathbb{Z}/\ell^\infty, \pi_{j-1}\Gamma_0 B\mathcal{O}(\overline{\mathbb{F}}_p)) \to 0 \end{aligned}$$

From (7.26) we have

$$\Pi_{\ell\neq p}\mathrm{Ext}(\mathbb{Z}/\ell^\infty, \pi_j\Gamma_0 B\mathcal{O}(\overline{\mathbb{F}}_p)) = \pi_j\Gamma_0 B\mathcal{O}(\overline{\mathbb{F}}_p)$$

is a $\mathbb{Z}/2$-vector space. Since $\Pi_{\ell\neq p}\hat{\mathbb{Z}}_{(\ell)}$ is torsion-free, (7.27) implies $\pi_j\Gamma_0 B\mathcal{O}(\overline{\mathbb{F}}_p) = 0$.

For $i \equiv 3,7 \pmod 8$, (7.23) and (7.24) give us

$$0 \to \Pi_{\ell\neq p}\mathrm{Ext}(\mathbb{Z}/\ell^\infty, \pi_i\Gamma_0 B\mathcal{O}(\overline{\mathbb{F}}_p)) \to 0 \to \Pi_{\ell\neq p}\mathrm{Hom}(\mathbb{Z}/\ell^\infty, \pi_{i-1}\Gamma_0 B\mathcal{O}(\overline{\mathbb{F}}_p)) \to 0$$

so that $\Pi_{\ell\neq p}\mathrm{Ext}(\mathbb{Z}/\ell^\infty, \pi_i\Gamma_0 B\mathcal{O}(\overline{\mathbb{F}}_p)) = 0$. Since $\pi_i\Gamma_0 B\mathcal{O}(\overline{\mathbb{F}}_p)$ is $\frac{1}{p}$-local, this implies $\pi_i\Gamma_0 B\mathcal{O}(\overline{\mathbb{F}}_p)$ is divisible. Since $\pi_i\Gamma_0 B\mathcal{O}(\overline{\mathbb{F}}_p)$ is also torsion and $\frac{1}{p}$-local, $\pi_i\Gamma_0 B\mathcal{O}(\overline{\mathbb{F}}_p)$ must be a direct sum of factors isomorphic to $\mathbb{Z}/\ell^\infty$ for various $\ell \neq p$. Now applying (7.23) and (7.24) in the case $j = i + 1 \equiv 0,4 \pmod 8$ we obtain (7.27). Since we already have shown $\pi_j\Gamma_0 B\mathcal{O}(\overline{\mathbb{F}}_p) = 0$, we obtain

$$\Pi_{\ell\neq p}\hat{\mathbb{Z}}_{(\ell)} \cong \Pi_{\ell\neq p}\mathrm{Hom}(\mathbb{Z}/\ell^\infty, \pi_i\Gamma_0 B\mathcal{O}(\overline{\mathbb{F}}_p))$$

Since we already know that $\pi_i \Gamma_0 B\mathcal{O}(\bar{\mathbb{F}}_p)$ is direct sum of various $\mathbb{Z}/\ell^\infty$, this implies $\pi_i \Gamma_0 B\mathcal{O}(\bar{\mathbb{F}}_p) = \oplus_{\ell \neq p} \mathbb{Z}/\ell^\infty = \bar{\mathbb{F}}_p^*$.

<u>Remark</u> 7.28. The reader may wonder why we went through this rather complicated proof involving Ext-computations, instead of deriving the result directly from Theorem 3.5. The problem is that although we know for instance that $\pi_i \Gamma_0 B\mathcal{O}(\bar{\mathbb{F}}_p) = \varinjlim \pi_i \Gamma_0 B\mathcal{O}(\mathbb{F}_{p^n})$ with the $\pi_i \Gamma_0 \mathcal{O}(\mathbb{F}_{p^n})$ completely calculated, we do not know the induced map $\pi_i \Gamma_0 B\mathcal{O}(\mathbb{F}_{p^m}) \to \pi_i \Gamma_0 B\mathcal{O}(\mathbb{F}_{p^n})$. For in passing from $B\mathcal{O}(\mathbb{F}_{p^n})$ to $\Gamma_0 B\mathcal{O}(\mathbb{F}_{p^n})$ we loose all control of the homotopy groups while retaining homological information. We use this homological information to obtain a geometric model $JO(p^n)$ for $\Gamma_0 B\mathcal{O}(\mathbb{F}_{p^n})$. It is from this geometric model that the homotopy groups $\pi_i \Gamma_0 B\mathcal{O}(\mathbb{F}_{p^n})$ are computed. The trouble is that is is unclear how to obtain (rigorously) a geometric model for the maps $\Gamma_0 B\mathcal{O}(\mathbb{F}_{p^m}) \to \Gamma_0 B\mathcal{O}(\mathbb{F}_{p^n})$.

Chapter IV

Calculations at the Prime 2

§1. Introduction

The purpose of this chapter is to give our principal calculations at the prime 2. All homology and cohomology groups are taken with $\mathbb{Z}/2$ coefficients and all spaces are localized at 2.

Specifically we compute the homology algebras $H_*B\overline{\mathcal{O}}(\mathbb{F}_q)$, establish an equivalence of infinite loop spaces

$$\Gamma_0 B\overline{\mathcal{O}}(\mathbb{F}_q) \to JO(q)$$

and compute the cohomology rings $H^*BO(n,\mathbb{F}_q)$ and $H^*BO(\overline{n},\mathbb{F}_q)$. Corresponding results are obtained for the other categories $\mathcal{S}p(\mathbb{F}_q)$, $\mathcal{GL}(\mathbb{F}_q)$, and $\mathcal{U}(\mathbb{F}_{q^2})$.

Throughout the chapter the characteristic of $\mathbb{F}_q$ is assumed odd.

§2. $H_*B\overline{\mathcal{O}}(\mathbb{F}_q)$ and $H_*\Gamma_0 B\overline{\mathcal{O}}(\mathbb{F}_q)$

Our goal in this section is to compute the mod 2 homology algebras $H_*B\overline{\mathcal{O}}(\mathbb{F}_q) = \oplus_{n\in\overline{\mathbb{N}}} H_*BO(n,\mathbb{F}_q)$ (in the notation of II 4.7) and $H_*B\overline{\mathcal{O}}(\mathbb{F}_q)$ as well as to establish the equivalences

$$\lambda: \Gamma_0 B\overline{\mathcal{O}}(\mathbb{F}_q) \to JO(q)$$

$$\lambda: \overset{+}{\Gamma_0} B\overline{\mathcal{O}}(\mathbb{F}_q) \to J\overline{O}(q)$$

of infinite loop spaces at 2.

According to VI 4.2

$$H_i BO(1,\mathbb{F}_q) = \mathbb{Z}/2 \text{ on generator } v_i \quad i \geq 0$$

$$H_i BO(\overline{1},\mathbb{F}_q) = \mathbb{Z}/2 \text{ on generator } y_i \quad i \geq 0$$

Note that in the notation of II 2.16(viii) we have $v_0 = [1]$, $y_0 = [\overline{1}]$.

Theorem 2.1. The mod-2 homology algebra $H_* B\overline{\mathcal{O}}(\mathbb{F}_q) = \oplus_{n \in \overline{\mathbb{N}}} H_* BO(n,\mathbb{F}_q)$ is a commutative algebra generated by v_i, y_i $i \geq 0$ subject only to the relations $v_i^2 = y_i^2$ $i \geq 0$.

Definition 2.2. Now let $\overline{v}_i$, $\overline{y}_i \in H_i B\overline{\mathcal{O}}(\mathbb{F}_q)$ denote the images of v_i, y_i under the maps

$$\overline{j}_1 : BO(1,\mathbb{F}_q) \to BO(2,\mathbb{F}_q) \to BO(\infty,\mathbb{F}_q) \to \Gamma_0 B\overline{\mathcal{O}}(\mathbb{F}_q)$$

$$\overline{j}_2 : BO(\overline{1},\mathbb{F}_q) \to BO(2,\mathbb{F}_q) \to BO(\infty,\mathbb{F}_q) \to \Gamma_0 B\overline{\mathcal{O}}(\mathbb{F}_q)$$

Since $O(1,\mathbb{F}_q) \cong O(\overline{1},\mathbb{F}_q) \cong \mathbb{Z}/2$, we may equally well describe $\overline{v}_i = j_{1*}(x_i)$, $\overline{y}_i = j_{2*}(x_i)$ where $j_1 : B\mathbb{Z}/2 \to \Gamma_0 B\overline{\mathcal{O}}(\mathbb{F}_q)$, $j_2 : B\mathbb{Z}/2 \to \Gamma_0 B\overline{\mathcal{O}}(\mathbb{F}_q)$ are the maps corresponding to $\overline{j}_1$, $\overline{j}_2$ and x_i is the generator. We now define a new map

$$j_3 = j_2 - j_1 : \quad B\mathbb{Z}/2 \to \Gamma_0 B\overline{\mathcal{O}}(\mathbb{F}_q)$$

and a corresponding family of elements $\overline{u}_i = j_{3*}(x_i) \in H_i \Gamma_0 B\overline{\mathcal{O}}(\mathbb{F}_q)$. Note that in the notation of II 2.16(viii) we have $\overline{v}_0 = \overline{y}_0 = [0]$.

Theorem 2.3. As an algebra $H_* \Gamma_0 B\overline{\mathcal{O}}(\mathbb{F}_q)$ has two equivalent descriptions

(i) $H_* \Gamma_0 B\overline{\mathcal{O}}(\mathbb{F}_q) = \mathbb{Z}/2[\overline{v}_1,\overline{v}_2,\ldots,\overline{y}_1,\overline{y}_2,\ldots]/\{\overline{v}_i^2 = \overline{y}_i^2 \mid i \geq 1\}$

(ii) $H_* \Gamma_0 B\overline{\mathcal{O}}(\mathbb{F}_q) = \mathbb{Z}/2[\overline{v}_1,\overline{v}_2,\ldots] \otimes E[\overline{u}_1,\overline{u}_2,\ldots]$

Theorem 2.4. Any H-space map

$$\lambda\colon \Gamma_0 B\overline{\mathcal{O}}(\mathbb{F}_q) \to JO(q)$$

which completes diagram III 2.14 is an equivalence at the prime 2; in particular there is an infinite loop equivalence λ at 2.

Theorem 2.5. Any infinite loop map

$$\lambda\colon \Gamma_0^+ B\overline{\mathcal{O}}(\mathbb{F}_q) \to J\overline{O}(q)$$

which completes diagram III 2.14 is an equivalence at the prime 2.

2.6. N.B. We should point out that the map λ is not uniquely determined by diagram III 2.14. In fact if λ is the infinite loop lift we chose, then λ_* is an isomorphism on π_1 by Theorem 2.4 so we can assume $\lambda_*(c_i) = \delta_i$, $i = 1,2,3$. (Here we use the notation of II 3.17 and I §1.) Now if we take $\gamma\colon \mathbb{R}P^\infty \to SO$ to be a map which is nontrivial on π_1 then

$$\tilde{\lambda} = \lambda + \tau\cdot\gamma\cdot(\Gamma_0 B\theta) \quad \text{(cf. II 3.17 re notation)}$$

also makes diagram III 2.14 commute and

$$\tilde{\lambda}_*(c_3) = \lambda_*(c_3) + \tau_*\cdot\gamma_*\cdot(\Gamma_0 B\theta)_*(c_3) = \delta_3 + \tau_*\cdot\gamma_*\cdot[(\Gamma_0 B\theta)\cdot\overline{j}_3]_*(\iota)$$

$$= \delta_3 + \tau_*\cdot\gamma_*(\iota) = \delta_3 + \delta_3 = 0$$

Hence $\tilde{\lambda}_*$ is not an isomorphism on π_1 and therefore cannot be an equivalence.

As outlined in Chapter III §6, we will prove these theorems by showing that

$$\lambda_*(\overline{v}_i) = \hat{v}_i, \quad \lambda_*(\overline{u}_i) = \hat{u}_i$$

for all $i > 0$ where $\{\hat{v}_i, \hat{u}_i\}$ are the generators of $H_*JO(q)$ defined in Prop. I 3.1. As a first step we analyze the relations between the elements $\overline{v}_i, \overline{y}_i$ and $\overline{u}_i$.

Definition 2.7. If X is an H-space, we shall denote by χ: $H_*X \to H_*X$ the algebra automorphism induced by the map -1: $X \to X$.

If Δ: $X \to X \times X$, μ: $X \times X \to X$ denote the diagonal and multiplication maps respectively, then we have for any $x \in \tilde{H}_*X$

$$(*) \qquad \mu_*(1 \otimes \chi)\Delta_*(x) = 0$$

which can be used to calculate the automorphism χ.

For instance if $X = \Gamma_0 B\overline{\mathcal{O}}(\mathbb{F}_q)$ then it follows from Def. 2.2 and Prop. VI 2.3 that

$$\Delta_*(\overline{v}_n) = \Sigma_{i=0}^{n}\overline{v}_i \otimes \overline{v}_{n-i}, \quad \Delta_*(\overline{y}_n) = \Sigma_{i=0}^{n}\overline{y}_i \otimes \overline{y}_{n-i}$$

Then (*) gives the inductive formulas

$$\chi(\overline{v}_0) = \overline{v}_0 = \overline{y}_0 = \chi(\overline{y}_0) = [0]$$

$$\chi(\overline{v}_n) = \Sigma_{i=0}^{n-1}\overline{v}_{n-i}\chi(\overline{v}_i), \quad \chi(\overline{y}_n) = \Sigma_{i=0}^{n-1}\overline{y}_{n-i}\chi(\overline{y}_i)$$

Proposition 2.8 (a). $\overline{u}_n = \Sigma_{i=0}^{n}\overline{y}_i\chi(\overline{v}_{n-i})$

(b) $\overline{y}_n = \Sigma_{i=0}^{n}\overline{v}_i\overline{u}_{n-i}$

Proof: We have by Def. 2.2

$$\bar{u}_n = j_{3*}(x_n) = (j_2 - j_1)_*(x_n) = \mu_*(1 \otimes \chi)(j_{2*} \otimes j_{1*})\Delta_*(x_n)$$

$$= \mu_*(1 \otimes \chi)(j_{2*} \otimes j_{1*})(\Sigma_{i=0}^n x_i \otimes x_{n-i})$$

$$= \Sigma_{i=0}^n \bar{y}_i \chi(\bar{v}_{n-i})$$

Similarly we have

$$\bar{y}_n = j_{2*}(x_n) = (j_1 + j_3)_*(x_n) = \mu_*(j_{1*} \otimes j_{3*})\Delta_*(x_n)$$

$$= \mu_*(j_{1*} \otimes j_{3*})(\Sigma_{i=0}^n x_i \otimes x_{n-i}) = \Sigma_{i=0}^n \bar{v}_i \bar{u}_{n-i}$$

Lemma 2.9. The following relations hold in $H_*\Gamma_0 B\bar{\mathcal{O}}(\mathbb{F}_q)$

(i) $\bar{v}_i^2 = \bar{y}_i^2$, $i \geq 0$

(ii) $\bar{u}_i^2 = 0$, $i \geq 1$

Moreover the relations (i) and (ii) imply each other.

Proof: (i) $\Longrightarrow$ (ii)

$$\bar{u}_n^2 = (\Sigma_{i=0}^n \bar{v}_i \chi(\bar{y}_{n-i}))^2 = \Sigma_{i=0}^n \bar{v}_i^2 \chi(\bar{y}_{n-i}^2)$$

$$= \Sigma_{i=0}^n \bar{v}_i^2 \chi(\bar{v}_{n-i}^{2\prime}) = (\Sigma_{i=0}^n \bar{v}_i \chi(v_{n-i}))^2 = 0$$

(ii) $\Longrightarrow$ (i)

$$\bar{y}_n^2 = (\Sigma_{i=0}^n \bar{v}_i \bar{u}_{n-i})^2 = \Sigma_{i=0}^n \bar{v}_i^2 \bar{u}_{n-i}^2 = \bar{v}_n^2 \bar{u}_0^2 = \bar{v}_n^2$$

Finally we show that relations (i) hold in $H_*\Gamma_0 B\bar{\mathcal{O}}(\mathbb{F}_q)$. We work in the infinite component version $H_*\Gamma B\bar{\mathcal{O}}(\mathbb{F}_q)$. By abuse of notation we will denote by v_i, y_i the images of the elements $v_i, y_i \in H_* B\bar{\mathcal{O}}(\mathbb{F}_q)$ under the natural map

$$H_*B\overline{\mathcal{O}}(\mathbb{F}_q) \to H_*\Gamma B\overline{\mathcal{O}}(\mathbb{F}_q)$$

We then have

$$\overline{v}_i = v_i*[-1] \quad \overline{y}_i = y_i*[-\overline{1}]$$

By Lemma VI 4.4 we have $v_i^2 = y_i^2$ so that

$$\overline{v}_i^2 = (v_i*[-1])^2 = v_i^2*[-2] = y_i^2*[-2] = (y_i*[-\overline{1}])^2 = \overline{y}_i^2$$

<u>Corollary</u> 2.10. $\chi(\overline{u}_n) = \overline{u}_n$ for all $n \geq 0$

<u>Proof</u>: Since the $\overline{u}_n$'s are images of generators of $H_*B\mathbb{Z}/2$ we have as in 2.7 the inductive formulas

$$(*) \qquad \chi(\overline{u}_n) = \Sigma_{i=0}^{n-1}\overline{u}_{n-i}\chi(\overline{u}_i)$$

Since $u_0 = [0]$ we have $\chi(\overline{u}_0) = \overline{u}_0$ to start the induction. Assuming $\chi(\overline{u}_i) = \overline{u}_i$ for $i < n$, we obtain from (*)

$$\chi(\overline{u}_n) = \begin{Bmatrix} \overline{u}_n & \text{if } n \text{ is odd} \\ \overline{u}_n + \overline{u}_k^2 & \text{if } n = 2k \end{Bmatrix} = \overline{u}_n$$

which completes the induction and proof.

<u>Proposition</u> 2.11 (a). As an algebra $H_*B\overline{\mathcal{O}}(\mathbb{F}_q) = \oplus_{n\in\mathbb{N}}H_*BO(n,\mathbb{F}_q)$ is generated by v_i, y_i $i \geq 0$.

(b) As an algebra $H_*\Gamma_0B\overline{\mathcal{O}}(\mathbb{F}_q)$ is generated by either of the following sets of generators

(i) $\overline{v}_i, \overline{y}_i \quad i \geq 1$

(ii) $\overline{v}_i, \overline{u}_i \quad i \geq 1$

Proof: Part (a) follows from Prop. VII 2.1 and Prop. VI 4.3(b). Hence any element of

$$H_*\Gamma_0 B\overline{\mathcal{O}}(\overline{\mathbb{F}}_q) \cong \lim_{\vec{n}} H_*BO(n,\overline{\mathbb{F}}_q)$$

can be expressed as a polynomial in $\{\overline{v}_i, \overline{y}_i \mid i \geq 1\}$. This proves b(i); and b(ii) follows from b(i) and Prop. 2.8.

Proposition 2.12. $\lambda_*(\overline{v}_i) = \hat{v}_i$

Proof: It is obvious that the Brauer character of

$$\mathbb{Z}/2 \cong O(1,\mathbb{F}_q) \to O(1,\overline{\mathbb{F}}_q)$$

is the character of the standard representation

$$\mathbb{Z}/2 \xrightarrow{\cong} O(1,\mathbb{R})$$

This implies that the following diagram commutes

$$\begin{array}{ccc} B\mathbb{Z}/2 & \xrightarrow{\cong} & BO(1,\mathbb{R}) \\ \downarrow{\scriptstyle j_1} & & \downarrow \\ \Gamma_0 B\overline{\mathcal{O}}(\overline{\mathbb{F}}_q) & \xrightarrow{\beta} & BO \\ & {\scriptstyle\lambda}\searrow \quad \nearrow{\scriptstyle r} & \\ & JO(q) & \end{array}$$

Chasing this diagram in homology we get

$$r_*\lambda_*(\overline{v}_i) = \beta_*(\overline{v}_i) = \overline{e}_i$$

where $\overline{e}_i$ are the standard generators of $H_*BO = \mathbb{Z}/2[\overline{e}_1, \overline{e}_2, \ldots]$. Since the generators $\hat{v}_i$ of $H_*JO(q)$ are defined in Prop. I.3.1 to be pre-images under r_* of the generators e_i we may as well take

$$\lambda_*(\overline{v}_i) = \hat{v}_i.$$

Proposition 2.13. If λ is an H-map which completes diagram III 2.14 then $\lambda_*(\overline{u}_i) = \hat{u}_i$.

Proof: It is obvious that the Brauer characters of both the representations

$$\mathbb{Z}/2 \cong O(1, \mathbb{F}_q) \to O(1, \overline{\mathbb{F}}_q)$$

$$\mathbb{Z}/2 \cong O(\overline{1}, \mathbb{F}_q) \to O(1, \overline{\mathbb{F}}_q)$$

are the characters of the standard representation

$$Z/2 \xrightarrow{\cong} O(1, \mathbb{R})$$

Hence the following diagram commutes

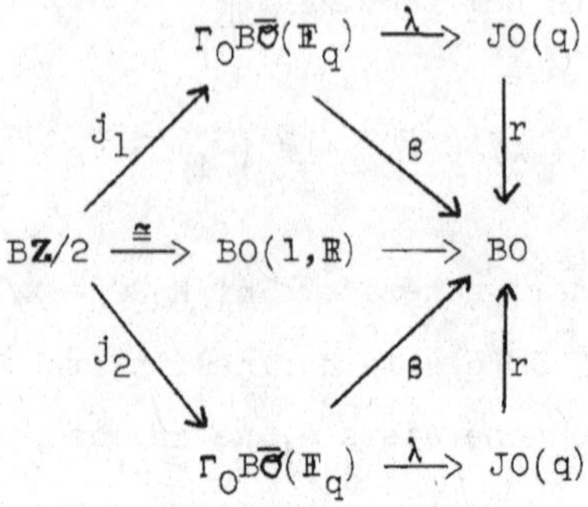

This implies that

$$r \cdot \lambda \cdot j_3 = \beta \cdot j_3 = \beta \cdot (j_2 - j_1) = 0 : B\mathbb{Z}/2 \to BO$$

This implies that there is a map $f: B\mathbb{Z}/2 \to SO$ such that $\tau \cdot f \simeq \lambda \cdot j_3$

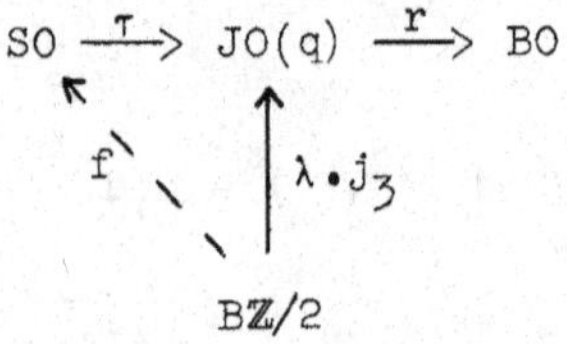

According to Lemma I 3.2 $f_*: H_*B\mathbb{Z}/2 \to H_*SO$ is either given by

(i) $f_*(x_i) = 0 \quad i \geq 1$

or by

(ii) $f_*(x_i) = u_i \quad i \geq 0$

where the u_i's are the standard generators of $H_*SO = E[u_1, u_2, \ldots]$.

We first eliminate the case (i). Assume that $f_*(x_i) = 0$ $i \geq 0$. We then would have

$$\lambda_*(\overline{u}_i) = \lambda_* \cdot j_{3*}(x_i) = \tau_* \cdot f_*(x_i) = 0$$

It then follows from Prop. 2.11, 2.12 and the fact that λ is an

H-map that

$$(2.14)\quad \mathrm{im}\lambda_* = \mathbb{Z}/2[\hat{v}_1,\hat{v}_2,\ldots] \subseteq H_*JO(q) = \mathbb{Z}/2[\hat{v}_1,\hat{v}_2,\ldots] \otimes E[\hat{u}_1,\hat{u}_2,\ldots]$$

According to Lemma I 7.4, we have a Bockstein operation

$$d_\nu(\hat{v}_2^2) = \hat{u}_1\hat{u}_2 \qquad \nu = \nu(\tfrac{1}{2}(q^2 - 1))$$

According to Lemma VI 4.5 we have that $d_\nu(v_2^2)$ is defined and hence so is $d_\nu(\overline{v}_2^2)$. We then have

$$\lambda_* d_\nu(\overline{v}_2^2) = d_\nu\lambda_*(\overline{v}_2^2) = d_\nu(\hat{v}_2^2) = \hat{u}_1\hat{u}_2$$

This contradicts 2.14. Hence case (i) cannot hold.

We must therefore have case (ii)

$$f_*(x_i) = u_i$$

Hence

$$\lambda_*(\overline{u}_i) = \lambda_*\cdot j_{3*}(x_i) = \tau_* f_*(x_i) = \tau_*(u_i) = \hat{u}_i.$$

2.15. Proof of Theorem 2.4. Consider the chain of maps

$$\mathbb{Z}/2[\overline{v}_1,\overline{v}_2,\ldots] \otimes E[\overline{u}_1,\overline{u}_2,\ldots] \xrightarrow{i_*} H_*(\Gamma_0 B\overline{\mathcal{O}}(\mathbb{F}_q))$$

$$\xrightarrow{\lambda_*} H_*JO(q) = \mathbb{Z}/2[\hat{v}_1,\hat{v}_2,\ldots] \otimes E[\hat{u}_1,\hat{u}_2,\ldots]$$

According to Prop. 2.11 i_* is an epimorphism. By Props. 2.12 and 2.13 $\lambda_*\cdot i_*$ is an isomorphism. It follows that i_* is 1-1 and hence an isomorphism. Consequently λ_* is an isomorphism and λ is an equivalence at 2.

2.16. Proof of Theorem 2.3. Part (ii) immediately follows from the proof of Theorem 2.4 above. Part (i) follows from part (ii), Prop. 2.8 and Lemma 2.9.

2.17. Proof of Theorem 2.1. We work with the infinite component version

$$\Gamma B\overline{\mathcal{O}}(\mathbb{F}_q) \simeq \Gamma_0 B\overline{\mathcal{O}}(\mathbb{F}_q) \times \mathbb{Z} \times \mathbb{Z}/2$$

It follows from Theorem 2.3 that

$$H_*\Gamma B\overline{\mathcal{O}}(\mathbb{F}_q) = \mathbb{Z}/2[[-1],[\gamma],[1],\overline{v}_1,\overline{v}_2,\ldots,\overline{y}_1,\overline{y}_2,\ldots]/R$$

where $\gamma = 1-\overline{1}$ and R denotes the relations generated by

$$\overline{v}_i^2 = \overline{y}_i^2 \quad i \geq 1, \ [-1][1] = [0], \quad [\gamma]^2 = [0]$$

Since $\overline{v}_i = v_i*[-1]$, $\overline{y}_i = y_i*[-\overline{1}]$ it follows that $H_*\Gamma B\overline{\mathcal{O}}(\mathbb{F}_q)$ may be equally well described as

$$H_*\Gamma B\overline{\mathcal{O}}(\mathbb{F}_q) = \mathbb{Z}/2[[-1],v_0,v_1,v_2,\ldots,y_0,y_1,y_2,\ldots]\Big/\begin{matrix} v_i^2 = y_i^2 \\ [-1]v_0 = [0]\end{matrix}$$

Now consider the chain of maps

$$\mathbb{Z}/2[v_0,v_1,v_2,\ldots,y_0,y_1,\ldots]/v_i^2 = y_i^2 \xrightarrow{i_*} H_*B\overline{\mathcal{O}}(\mathbb{F}_q) \xrightarrow{j_*} H_*\Gamma B\overline{\mathcal{O}}(\mathbb{F}_q).$$

Clearly j_*i_* is a monomorphism. By Prop. 2.11, i_* is an epimorphism.

This implies that i_* is an isomorphism.

2.18 Proof of Theorem 2.5. Since

$$\Gamma_0^+ B\bar{\mathcal{O}}(\mathbb{F}_q) = \Gamma_0 B\bar{\mathcal{O}}(\mathbb{F}_q) \amalg \Gamma_\gamma B\bar{\mathcal{O}}(\mathbb{F}_q) \cong \Gamma_0 B\mathcal{O}(\mathbb{F}_q) \times \mathbb{Z}/2$$

$$J\bar{O}(q) \cong JO(q) \times \mathbb{Z}/2$$

and in view of Theorem 2.4, it suffices to show that

$$\lambda\colon \Gamma_0^+ B\bar{\mathcal{O}}(\mathbb{F}_q) \to J\bar{O}(q)$$

induces an isomorphism on π_0.

Suppose it did not. Then λ would factor through the 0-components $JO(q) \cong \Gamma_0 B\bar{\mathcal{O}}(\mathbb{F}_q)$. But this would imply that $\lambda_*([\gamma]) = [0]$ which is impossible since

$$Q^n([0]) = 0 \quad n > 0$$

while

$$Q^n \lambda_*([\gamma]) = \lambda_*(\bar{u}_n) = \hat{u}_n \neq 0$$

by Chap. IX 2.2(c).

§3. The automorphism Φ

In this section we evaluate the automorphism $\Phi\colon \bar{\mathcal{O}}(\mathbb{F}_q) \to \bar{\mathcal{O}}(\mathbb{F}_q)$ of Def. II 4.4 on mod-2 homology. These results will be used in Chapter IX to compute homology operations in $H_*\Gamma B\bar{\mathcal{O}}(\mathbb{F}_q)$.

Proposition 3.1. The map $\Phi_*: H_*B\overline{\mathcal{O}}(\mathbb{F}_q) \to H_*B\overline{\mathcal{O}}(\mathbb{F}_q)$ is given by

$$\Phi_*(v_i) = y_i \quad \Phi_*(y_i) = v_i$$

Proof: This follows immediately from the fact that Φ maps $O(1,\mathbb{F}_q)$ isomorphically onto $O(\overline{1},\mathbb{F}_q)$ and $O(\overline{1},\mathbb{F}_q)$ isomorphically onto $O(1,\mathbb{F}_q)$. (cf. II 4.5).

Proposition 3.2. In $H_*\Gamma B\overline{\mathcal{O}}(\mathbb{F}_q)$ we have the following formulas

(i) $\Phi_*(\overline{v}_i) = \overline{y}_i = \Sigma_{i=0}^n \overline{v}_i\overline{u}_i$

(ii) $\Phi_*(\overline{y}_i) = \overline{v}_i$

(iii) $\Phi_*(\overline{u}_i) = \overline{u}_i$

(iv) $\Phi_*([\gamma]) = [\gamma]$ and $\Phi_*([-1]) = [-\overline{1}] = [-1]*[\gamma]$

(v) $(\Phi-1)_*(\overline{v}_i) = (\Phi-1)_*(\overline{y}_i) = \overline{u}_i$

(vi) $(\Phi-1)_*(\overline{u}_i) = 0$

(vii) $(\Phi-1)_*([\gamma]) = [0]$ and $(\Phi-1)_*([-1]) = [\gamma]$

Proof: By II 4.5 we have

$$\Phi_*([1]) = [\Phi(1)] = [\overline{1}] \quad \Phi_*([\overline{1}]) = [\Phi(\overline{1})] = [1]$$

$$\Phi_*([\gamma]) = [\Phi(\gamma)] = [\Phi(1) - \Phi(\overline{1})] = [\overline{1}-1] = [\gamma]$$

$$\Phi_*([-1]) = [-\Phi(1)] = [-\overline{1}] = [-1]*[\gamma]$$

which proves (iv). We have by Prop. 3.1

$$\Phi_*(\overline{v}_i) = \Phi_*(v_i*[-1]) = \Phi_*(v_i)\Phi_*([-1]) = y_i*[-\overline{1}] = \overline{y}_i$$

which together with Prop. 2.8 implies (i); (ii) is proved similarly. We also have

$$\begin{aligned}\Phi_*(\bar{u}_n) &= \Phi_*\chi(\bar{u}_n) = \Phi_*(\Sigma_{i=0}^n \bar{v}_i\chi(\bar{y}_{n-i})) = \Sigma_{i=0}^n \Phi_*(\bar{v}_i)\chi(\Phi_*(\bar{y}_{n-i})) \\ &= \Sigma_{i=0}^n \bar{y}_i\chi(\bar{v}_{n-i}) = \bar{u}_n\end{aligned}$$

which proves (iii).

To prove (v) we observe that

$$\begin{aligned}(\Phi-1)_*(\bar{v}_n) &= \mu_*(\Phi_* \otimes \chi)\Delta_*(\bar{v}_n) = \mu_*(\Phi_* \otimes \chi)(\Sigma_{i=0}^n \bar{v}_i \otimes \bar{v}_{n-i}) \\ &= \Sigma_{i=0}^n \bar{y}_i\chi(\bar{v}_{n-i}) = \bar{u}_n\end{aligned}$$

Similarly we prove $(\Phi-1)_*(\bar{y}_n) = \bar{u}_n$ as well as (vi) and (vii).

<u>Remark</u> 3.3. For n odd $\Phi\colon O(n,\mathbb{F}_q) \xrightarrow{\approx} O(\bar{n},\mathbb{F}_q)$. For n even, $O(n,\mathbb{F}_q)$ and $O(\bar{n},\mathbb{F}_q)$ are not isomorphic; however, $H_*BO(n,\mathbb{F}_q) \cong H_*BO(\bar{n},\mathbb{F}_q)$ as $\mathbb{Z}/2$-modules. In fact, it follows from Theorem 2.1 that for all $n > 0$

$$B_n = \left\{ v_{i_1}^{\epsilon_1} y_{i_1}^{\eta_1} \cdots v_{i_k}^{\epsilon_k} y_{i_k}^{\eta_k} \,\middle|\, \begin{array}{l} \epsilon_i \geq 0;\ \eta_i = 0,1;\ i_1 < i_2 < \cdots < i_k,\ \Sigma\epsilon_i + \Sigma\eta_i = n \\ \hfill \Sigma\eta_i = \text{even} \end{array} \right\}$$

is a basis for $H_*BO(n,\mathbb{F}_q)$ and that a similar basis $B_{\bar{n}}$ for $H_*BO(\bar{n},\mathbb{F}_q)$ is obtained by requiring $\Sigma\eta_i$ = odd. Since we may assume one of ϵ_1, η_1 is nonzero the correspondence $B_n \to B_{\bar{n}}$ given by

$$v_{i_1}^{\epsilon_1} y_{i_1}^{\eta_1} \cdots v_{i_k}^{\epsilon_k} y_{i_k}^{\eta_k} \to \begin{cases} y_{i_1}^{\epsilon_1+1} v_{i_2}^{\epsilon_2} y_{i_2}^{\eta_2} \cdots v_{i_k}^{\epsilon_k} y_{i_k}^{\eta_k} & \text{if } \eta_1 \neq 0 \\ v_{i_1}^{\epsilon_1-1} y_{i_1} v_{i_2}^{\epsilon_2} y_{i_2}^{\eta_2} \cdots v_{i_k}^{\epsilon_k} y_{i_k}^{\eta_k} & \text{if } \eta_1 = 0 \end{cases}$$

is obviously one-to-one. For n even we shall see in §4 that the cohomology algebras of $O(n,\mathbb{F}_q)$ and $O(\bar{n},\mathbb{F}_q)$ are very different.

§4. The cohomology rings $H^*BO(n,\mathbb{F}_q)$ and $H^*BO(\bar{n},\mathbb{F}_q)$

In this section we dualize the homology calculations of §3 and compute the mod-2 cohomology rings listed in the title. The point of departure for our calculations is Theorem 2.1 which states that the commutative homology algebra

$$H_*B\bar{\mathcal{O}}(\mathbb{F}_q) = H_*BO(0,\mathbb{F}_q) \oplus [\bigoplus_{n=1}^{\infty} (H_*BO(n,\mathbb{F}_q) \oplus H_*BO(\bar{n},\mathbb{F}_q))]$$

is generated by v_i, y_i $i \geq 0$ subject to the single relation

$$v_i^2 = y_i^2$$

The elements v_i, y_i are defined to be the nonzero elements of $H_iBO(1,\mathbb{F}_q)$ and $H_iBO(\bar{1},\mathbb{F}_q)$ respectively.

An immediate consequence of this theorem is that the direct sum homomorphisms

$$H_*(\coprod_{i=1}^{n} BO(1,\mathbb{F}_q)^{2i-1} \times BO(\bar{1},\mathbb{F}_q)^{2n-2i}) \to H_*BO(2n-1,\mathbb{F}_q)$$

$$H_*(\coprod_{i=0}^{n-1} BO(1,\mathbb{F}_q)^{2i} \times BO(\bar{1},\mathbb{F}_q)^{2n-2i-1}) \to H_*BO(\overline{2n-1},\mathbb{F}_q)$$

$$H_*(\coprod_{i=0}^{n} BO(1,\mathbb{F}_q)^{2i} \times BO(\bar{1},\mathbb{F}_q)^{2n-2i}) \to H_*BO(\overline{2n},\mathbb{F}_q)$$

$$H_*(\coprod_{i=1}^{n} (BO(1,\mathbb{F}_q)^{2i-1} \times BO(\bar{1},\mathbb{F}_q)^{2n-2i+1}) \to H_*BO(2n,\mathbb{F}_q)$$

are surjective, hence the corresponding maps in cohomology are injective. The cohomology rings of a disjoint union of spaces is the ring direct product of the cohomology rings ofthe components and the cohomology ring of a product is the tensor product of the cohomology rings of the factors. Since the cohomology rings

$H^*BO(1,\mathbb{F}_q)$ and $H^*BO(\overline{1},\mathbb{F}_q)$ are polynomial rings in one variable, it follows that $H^*BO(n,\mathbb{F}_q)$ and $H^*BO(\overline{n},\mathbb{F}_q)$ are subrings of a ring direct product of polynomial rings.

In order to fix notation we shall write

$$H^*BO(1,\mathbb{F}_q) = \mathbb{Z}/2[t], \quad H^*BO(\overline{1},\mathbb{F}_q) = \mathbb{Z}/2[\overline{t}]$$

and

$$H^*BO(1,\mathbb{F}_q)^n = \mathbb{Z}/2[T_n]$$

$$H^*BO(\overline{1},\mathbb{F}_q)^n = \mathbb{Z}/2[\overline{T}_n]$$

where $T_n = \{t_i\}_{i=1}^n$, $\overline{T}_n = \{\overline{t}_i\}_{i=1}^n$.

Let us denote by $\sigma_i(T_n)$ (resp. $\sigma_i(\overline{T}_n)$) the i-th elementary symmetric polynomials in the variables $\{t_i\}_{i=1}^n$ (resp. $\{\overline{t}_i\}_{i=1}^n$) if $1 \leq i \leq n$. By convention we set

$$\sigma_0(T_n) = 1 = \sigma_0(\overline{T}_n)$$

$$\sigma_i(T_n) = 0 = \sigma_i(\overline{T}_n) \quad \text{if} \quad i < 0 \quad \text{or} \quad i > n.$$

We further define elements $x_k(m,n)$, $\overline{x}_{2k-1}(m,n) \in \mathbb{Z}/2[T_m] \otimes \mathbb{Z}/2[\overline{T}_n]$ as follows:

$$x_{2k-1}(m,n) = \Sigma_{p+q=k}\sigma_{2p-1}(T_m) \otimes \sigma_{2q}(\overline{T}_n)$$

$$\overline{x}_{2k-1}(m,n) = \Sigma_{p+q=k}\sigma_{2p}(T_m) \otimes \sigma_{2q-1}(\overline{T}_n)$$

$$x_{2k}(m,n) = \Sigma_{p+q=k}\sigma_{2p}(T_m) \otimes \sigma_{2q}(\overline{T}_n)$$

The following two theorems are the main results of this section.

<u>Theorem</u> 4.1 (a). $H^*BO(2n-1,\mathbb{F}_q)$ is the subring of the ring direct product

$$\prod_{i=1}^{n}\mathbb{Z}/2[T_{2i-1}] \otimes \mathbb{Z}/2[\overline{T}_{2n-2i}]$$

generated by elements $\{x_k\}_{k=1}^{2n-1}$ and elements $\{\overline{x}_{2k-1}\}_{k=1}^{n-1}$ where $x_k = (x_k(2i-1,2n-2i))_{i=1}^{n}$ and $\overline{x}_{2k-1} = (\overline{x}_{2k-1}(2i-1,2n-2i))_{i=1}^{n}$.

(b) $H^*BO(\overline{2n-1},\mathbb{F}_q)$ is the subring of the ring direct product

$$\prod_{i=0}^{n-1}\mathbb{Z}/2[T_{2i}] \otimes \mathbb{Z}/2[\overline{T}_{2n-2i-1}]$$

generated by elements $\{x_k\}_{k=1}^{2n-2}$ and elements $\{\overline{x}_{2k-1}\}_{k=1}^{n}$ where $x_k = (x_k(2i,2n-2i-1))_{i=0}^{n-1}$ and $\overline{x}_{2k-1} = (\overline{x}_{2k-1}(2i,2n-2i-1))_{i=0}^{n-1}$

(c) $H^*BO(2n,\mathbb{F}_q)$ is the subring of the ring direct product

$$\prod_{i=0}^{n}\mathbb{Z}/2[T_{2i}] \otimes \mathbb{Z}/2[\overline{T}_{2n-2i}]$$

generated by elements $\{x_k\}_{k=1}^{2n}$ and elements $\{\overline{x}_{2k-1}\}_{k=1}^{n}$ where $x_k = (x_k(2i,2n-2i))_{i=0}^{n}$ and $\overline{x}_{2k-1} = (\overline{x}_{2k-1}(2i,2n-2i))_{i=0}^{n}$

(d) $H^*BO(\overline{2n},\mathbb{F}_q)$ is the subring of the ring direct product

$$\prod_{i=1}^{n}\mathbb{Z}/2[T_{2i-1}] \otimes \mathbb{Z}/2[\overline{T}_{2n-2i+1}]$$

generated by elements $\{x_k\}_{k=1}^{2n-1}$ and elements $\{\overline{x}_{2k-1}\}_{k=1}^{n}$ where $x_k = (x_k(2i-1,2n-2i+1))_{i=1}^{n}$ and $\overline{x}_{2k-1} = (\overline{x}_{2k-1}(2i-1,2n-2i+1))_{i=1}^{n}$.

Before stating the next theorem we define Stiefel-Whitney classes $w_i \in H^iBO(n,\mathbb{F}_q)$ and $w_i \in H^iBO(\overline{n},\mathbb{F}_q)$ as the images of the real Stiefel-Whitney classes under the maps induced in cohomology from Brauer lift

$$\beta\colon BO(n,\mathbb{F}_q) \to BO(\infty,\mathbb{F}_q) \to BO$$

$$\beta\colon BO(\overline{n},\mathbb{F}_q) \to BO(\infty,\mathbb{F}_q) \to BO$$

<u>Theorem</u> 4.2 (a) In $H^*BO(n,\mathbb{F}_q)$ or $H^*BO(\overline{n},\mathbb{F}_q)$ the Stiefel-Whitney classes w_i $i = 1,2,\ldots,n$ are given by the formula

$$w_i = x_i + \overline{x}_i \quad \text{if } i \text{ is odd}$$

and by the following recursion formulas if $i = 2k$ is even

$$R_{2k} = \Sigma_{p=1}^{k} x_{2p-1}\overline{x}_{2k-2p+1} + \Sigma_{q=1}^{k}(x_{2q} + w_{2q})x_{2k-2q} = 0$$

for $k = 1,2,\ldots,[n/2]$.

(b) $w_{2k} = 0$ for $k > [n/2]$ hence the formulas

$$R_{2k} = 0$$

reduce to formulas between the generators $\{x_k\}$, $\{x_{2k-1}\}$

(c) The relations (b) are the <u>only</u> relations in $H^*BO(n,\mathbb{F}_q)$ or $H^*BO(\overline{n},\mathbb{F}_q)$ with the single exception of $H^*BO(n,\mathbb{F}_q)$, n even, where there is an additional relation

$$x_n + w_n = 0.$$

In all cases the Poincaré series of $H^*BO(n,\mathbb{F}_q)$ and $H^*BO(\overline{n},\mathbb{F}_q)$ is

$$P(n,t) = \frac{\Pi_{i=1}^{n-1}(1 + t^i)}{\Pi_{i=1}^{n}(1 - t^i)}$$

<u>Corollary</u> 4.3. $H^*BO(\infty,\mathbb{F}_q)$ is a polynomial algebra on generators $\{x_k,\overline{x}_{2k-1}\}_{k=1}^{\infty}$.

Remark 4.4. The isomorphisms

$$\Phi\colon O(2n-1,\mathbb{F}_q) \to O(\overline{2n-1},\mathbb{F}_q)$$

$$\Phi\colon O(\overline{2n-1},\mathbb{F}_q) \to O(2n-1,\mathbb{F}_q)$$

$$\Phi\colon O(2n,\mathbb{F}_q) \to O(2n,\mathbb{F}_q)$$

$$\Phi\colon O(\overline{2n},\mathbb{F}_q) \to O(\overline{2n},\mathbb{F}_q)$$

(cf. II 4.5) induce the map on cohomology given by

$$\Phi^*(x_{2k-1}) = \overline{x}_{2k-1},\ \Phi^*(\overline{x}_{2k-1}) = x_{2k-1},\ \Phi^*(x_{2k}) = x_{2k}.$$

This is immediate since for $n = 1$, $\Phi^*(\overline{t}) = t$.

Thus $O(2n-1,\mathbb{F}_q)$ and $O(\overline{2n-1},\mathbb{F}_q)$ have isomorphic cohomology rings. For instance

$$H^*BO(3,\mathbb{F}_q) = \mathbf{Z}/2[x_1,\overline{x}_1,x_2,x_3]/\overline{x}_1x_3 + x_1\overline{x}_1x_2 = 0$$

$$H^*BO(\overline{3},\mathbb{F}_q) = \mathbf{Z}/2[x_1,\overline{x}_1,x_2,\overline{x}_3]/x_1\overline{x}_3 + x_1\overline{x}_1x_2 = 0.$$

Remark 4.5. As noted in Remark 3.3, although $O(2n,\mathbb{F}_q)$ and $O(\overline{2n},\mathbb{F}_q)$ are not isomorphic, they have isomorphic cohomology modules. However they have different ring structures. For example

$$H^*BO(2,\mathbb{F}_q) = \mathbf{Z}/2[x_1,\overline{x}_1,x_2]/x_1\overline{x}_1 = 0$$

$$H^*BO(\overline{2},\mathbb{F}_q) = \mathbf{Z}/2[x_1,\overline{x}_1]$$

$$H^*BO(4,\mathbb{F}_q) = \mathbf{Z}/2[x_1,\overline{x}_1,x_2,x_3,\overline{x}_3,x_4]\Big/ \begin{array}{l} x_1\overline{x}_3 + \overline{x}_1x_3 + x_1\overline{x}_1x_2 = 0 \\ x_3\overline{x}_3 + x_1\overline{x}_1x_4 = 0 \end{array}$$

$$H^*BO(\overline{4},\mathbb{F}_q) = \mathbf{Z}/2[x_1,\overline{x}_1,x_2,x_3\overline{x}_3]/x_3\overline{x}_3 + x_1\overline{x}_3x_2 + \overline{x}_1x_3x_2 + x_1\overline{x}_1x_2^2 = 0$$

Proof of Theorem 4.1: We shall confine ourselves to proving case (c) of the theorem. The proofs for the other cases are analogous.

According to Theorem 2.1, the homology $H_*BO(2n,\mathbb{F}_q)$ is obtained from the module

$$A = \oplus_{i=0}^{n}[\otimes_{p=1}^{2i} H_*BO(1,\mathbb{F}_q) \otimes \otimes_{q=1}^{2n-2i} H_*BO(\bar{1},\mathbb{F}_q)]$$

by dividing out by the actions of the following groups:

(1) For each $i = 0,1,2,\ldots,n$: the group $\Sigma_{2i} \times \Sigma_{2n-2i}$ acting on the direct summand $\otimes_{p=1}^{2i} H_*BO(1,\mathbb{F}_q) \otimes \otimes_{q=1}^{2n-2i} H_*BO(\bar{1},\mathbb{F}_q)$ by permuting the tensor factors, and acting trivially on the other summands.

(2) For each $i = 0,1,\ldots,n-1$: the group $\mathbb{Z}/2$ acting by interchanging an element of the form

$$v_{j_1} \otimes v_{j_2} \otimes \cdots \otimes v_{j_{2i}} \otimes y_j \otimes y_j \otimes y_{j_{2i+3}} \otimes \cdots \otimes y_{j_{2n}}$$

with the corresponding element of the form

$$v_{j_1} \otimes v_{j_2} \otimes \cdots \otimes v_{j_{2i}} \otimes v_j \otimes v_j \otimes y_{j_{2i+3}} \otimes \cdots \otimes y_{j_{2n}}$$

and fixing all the other standard basis elements of A.

By duality it follows that $H^*BO(2n,\mathbb{F}_q)$ is the submodule of

$$A^* = \oplus_{i=0}^{n}\mathbb{Z}/2[T_{2i}] \otimes \mathbb{Z}/2[\bar{T}_{2n-2i}]$$

invariant under the dual actions of the groups (1) and (2). But these dual actions are

(1)* For each $i = 0,1,\ldots,n$: the group $\Sigma_{2i} \times \Sigma_{2n-2i}$ acting by permuting the variables $T_{2i} \coprod \bar{T}_{2n-2i}$

(2)* For each $i = 0,1,\ldots,n-1$: the group $\mathbb{Z}/2$ acting by interchanging any monomial of the form

$$t_1^{j_1} t_2^{j_2} \cdots t_{2i}^{j_{2i}} \bar{t}_1^{j} \bar{t}_2^{j} \bar{t}_{2i+3}^{j_{2+3}} \cdots \bar{t}_{2n-2i}^{j_{2n}} \in \mathbb{Z}/2[T_{2i}] \otimes \mathbb{Z}/2[\bar{T}_{2n-2i}]$$

with the monomial

$$t_1^{j_1} t_2^{j_2} \cdots t_{2i}^{j_{2i}} t_{2i+1}^{j} t_{2i+2}^{j} \bar{t}_3^{j_{2i+3}} \cdots \bar{t}_{2n-2i-2}^{j_{2n}} \in \mathbb{Z}/2[T_{2i+2}] \otimes \mathbb{Z}/2[\bar{T}_{2n-2i-2}]$$

and fixing all other monomials.

It is clear that the elements $\{x_k\}_{k=1}^{2n}$ and $\{\bar{x}_{2k-1}\}_{k=1}^{n}$ are invariant under all the actions (1)* and (2)*. Hence these elements lie in $H^*BO(2n,\mathbb{F}_q)$. Since the inclusion

$$H^*BO(2n,\mathbb{F}_q) \to \Pi_{i=0}^{n} \mathbb{Z}/2[T_{2i}] \otimes \mathbb{Z}/2[\bar{T}_{2n-2i}]$$

is induced by the direct sum map

$$\coprod_{i=0}^{n} BO(1,\mathbb{F}_q)^{2i} \times BO(\bar{1},\mathbb{F}_q)^{2n-2i} \to BO(2n,\mathbb{F}_q)$$

the inclusion is an inclusion of rings. Hence the subalgebra $\mathcal{R}_{2n}$ of $\Pi_{i=1}^{n} \mathbb{Z}/2[T_{2i}] \otimes \mathbb{Z}/2[\bar{T}_{2n-2i}]$ generated by $\{x_k\}_{k=1}^{2n}$, $\{\bar{x}_{2k-1}\}_{k=1}^{n}$ is contained in $H^*BO(2n,\mathbb{F}_q)$.

It remains to show that $\mathcal{R}_{2n} = H^*BO(2n,\mathbb{F}_q)$. To see this, observe that Theorem 2.1 implies that the standard inclusions

$$BO(2n,\mathbb{F}_q) \to BO(2n+1,\mathbb{F}_q) \to \cdots \to BO(\infty,\mathbb{F}_q)$$

induce injections in homology and hence surjections in cohomology. Since these maps send $\mathcal{R}_{2n+2i}$ onto $\mathcal{R}_{2n}$, $i = 1,2,\ldots,\infty$, it suffices to check that $\mathcal{R}_\infty = \varprojlim \mathcal{R}_{2k}$ is all of $H^*BO(\infty,\mathbb{F}_q) = \varprojlim H^*BO(k,\mathbb{F}_q)$. But it is clear that in the inverse limit $\{x_k\}_{k=1}^\infty$, $\{\bar{x}_{2k-1}\}_{k=1}^\infty$ become algebraically independent. Hence $\mathcal{R}_\infty$ is a polynomial algebra on one generator in each even degree and two generators in each odd degree. But according to 2.4 and I 3.1, this is equal to $H^*BO(\infty,\mathbb{F}_q)$. Hence $H^*BO(2n,\mathbb{F}_q)$ is also equal to $\mathcal{R}_{2n}$.

<u>Proof of Theorem</u> 4.2: (a) Brauer lift is additive in the sense that the following diagram commutes up to homotopy

$$\begin{array}{ccccc} BO(1,\mathbb{F}_q)^n & \xrightarrow{\oplus} & BO(n,\mathbb{F}_q) & \xrightarrow{\beta} & BO \\ \wr & & & & \| \\ BO(1,\mathbb{R})^n & \xrightarrow{\oplus} & BO(n,\mathbb{R}) & \longrightarrow & BO \end{array}$$

(a similar diagram holds for the extraordinary orthogonal groups). Hence the coordinate of the Stiefel-Whitney class w_k in $\mathbb{Z}/2[T_i] \otimes \mathbb{Z}/2[\bar{T}_j]$ is

$$w_k(i,j) = \sigma_k(T_i \amalg \bar{T}_j)$$

the k-th elementary symmetric polynomial in all the variables $T_i \amalg \bar{T}_j$. It is well known that $\sigma_k(T_i \amalg \bar{T}_j) = \Sigma_{p+q=k}\sigma_p(T_i)\sigma_q(\bar{T}_j)$. Thus

$$\begin{aligned} w_{2k-1}(i,j) &= \Sigma_{p+q=2k-1}\sigma_p(T_i)\sigma_q(\bar{T}_j) \\ &= \Sigma_{p+q=k}\sigma_{2p-1}(T_i)\sigma_{2q}(\bar{T}_j) + \Sigma_{p+q=k}\sigma_{2p}(T_i)\sigma_{2q-1}(\bar{T}_j) \\ &= x_{2k-1}(i,j) + \bar{x}_{2k-1}(i,j) \end{aligned}$$

Hence for all k

$$w_{2k-1} = x_{2k-1} + \overline{x}_{2k-1}$$

The recursion relation for w_{2k} is more complicated. Since

$$w_{2k}(i,j) = \Sigma_{p+q=2k}\sigma_p(T_i)\sigma_q(\overline{T}_j)$$

and

$$x_{2k}(i,j) = \Sigma_{\substack{p+q=2k \\ p,q \text{ even}}}\sigma_p(T_i)\sigma \;\; \overline{T}_j)$$

we have

$$x_{2k}(i,j) + w_{2k}(i,j) = \Sigma_{\substack{p+q=2k \\ p,q \text{ odd}}}\sigma_p(T_i)\sigma_q(\overline{T}_j).$$

Consequently

$$\Sigma_{\substack{p+q=2k \\ p,q \text{ odd}}} x_p(i,j)\overline{x}_q(i,j) + \Sigma_{\substack{s+t=2k \\ s,t \text{ even}}}(x_s(i,j) + w_s(i,j))x_t(i,j)$$

$$= \Sigma_{\substack{p+q=2k \\ p,q \text{ odd}}}\Sigma_{\substack{a+b=p \\ a \text{ odd} \\ b \text{ even}}}\sigma_a(T_i)\sigma_b(\overline{T}_j)\Sigma_{\substack{c+d=q \\ c \text{ even} \\ d \text{ odd}}}\sigma_c(T_i)\sigma_d(\overline{T}_j)$$

$$+ \Sigma_{\substack{s+t=2k \\ s,t \text{ even}}}\Sigma_{\substack{e+f=s \\ e,f \text{ odd}}}\sigma_e(T_i)\sigma_f(\overline{T}_j)\Sigma_{\substack{g+h=t \\ g,h \text{ even}}}\sigma_g(T_i)\sigma_h(\overline{T}_j)$$

$$= \Sigma_{\substack{a+b+c+d=2k \\ a,d \text{ odd} \\ b,c \text{ even}}}\sigma_a(T_i)\sigma_b(\overline{T}_j)\sigma_c(T_i)\sigma_d(\overline{T}_j)$$

$$+ \Sigma_{\substack{e+f+g+h=2k \\ e,f \text{ odd} \\ g,h \text{ even}}} \sigma_e(T_i)\sigma_f(\overline{T}_j)\sigma_g(T_i)\sigma_h(\overline{T}_j)$$

$$= 0.$$

It follows that for all k

$$R_{2k} = \Sigma_{p=1}^{k} x_{2p-1}\overline{x}_{2k-2p+1} + \Sigma_{q=1}^{k}(x_{2q} + w_{2q})x_{2k-2q} = 0.$$

It is obvious that $w_{2k} = 0$ for $k > [n/2]$, hence the equations $R_{2k} = 0$, $[n/2] < k \leq 2[n/2]$ give relations between the generators $\{x_i\}$ and $\{\overline{x}_{2i-1}\}$. The relation $x_{2n} + w_{2n} = 0$ which holds in $H^*BO(2n,\mathbb{F}_q)$ is also easily verified.

It remains to show that these are the only relations. Let $n^\epsilon = n, \overline{n}$ and let $\mathcal{R}(O(n^\epsilon,\mathbb{F}_q))$ denote the algebra on the generators and relations specified in the theorem. Then we have just shown that $\mathcal{R}(O(n^\epsilon,\mathbb{F}_q))$ maps onto $H^*BO(n^\epsilon,\mathbb{F}_q)$ so that in terms of Poincaré series

(1) $$P(H^*(BO(n^\epsilon,\mathbb{F}_q),t) \leq P(\mathcal{R}(O(n^\epsilon,\mathbb{F}_q)),t)$$

The proof will be complete once we show these series are equal.

It is clear that the classes $\{w_k\}_{k=1}^{n}$ are algebraically independent in $H^*BO(n^\epsilon,\mathbb{F}_q)$. A fortiori they are algebraically independent in $\mathcal{R}(O(n^\epsilon,\mathbb{F}_q))$ and determine a subalgebra $\mathcal{B}$ whose Poincaré series is

$$P(\mathcal{B},t) = \Pi_{i=1}^{n}(1-t^i)^{-1}$$

Let W be the ideal of $\mathcal{R}(O(n^\epsilon,\mathbb{F}_q))$ generated by $\{w_k\}_{k=1}^{n}$. Then $\mathcal{R}(O(n^\epsilon,\mathbb{F}_q))/W$ is generated by $\{x_i\}_{i=1}^{n-1}$ (note that $x_{2i-1} = \overline{x}_{2i-1} \bmod W$). The relations $R_{2k} = 0$ reduce mod W to the relations

$$x_k^2 + x_{2k} = 0$$

But these generators and relations are the same as those for $H^*(SO(n,\mathbb{K}))$ (cf. [39, Ch. IV 4.5]). Hence in terms of Poincaré series

$$P(\mathcal{R}(O(n^\epsilon,\mathbb{F}_q))/W,t) = \Pi_{i=1}^{n-1}(1 + t^i).$$

Consequently

$$(2) \quad P(\mathcal{R}(O(n^\epsilon,\mathbb{F}_q),t) \le P(\mathcal{B},t)P(\mathcal{R}(O(n^\epsilon,\mathbb{F}_q))/W,t) = \frac{\Pi_{i=1}^{n-1}(1 + t^i)}{\Pi_{i=1}^{n}(1 - t^i)}$$

This gives an upper bound on $P(H^*BO(n^\epsilon,\mathbb{F}_q),t)$. To get a lower bound we consider the homology algebra

$$\mathcal{A} = H_*BO(0,\mathbb{F}_q) \oplus \oplus_{n=1}^{\infty}(H_*BO(n,\mathbb{F}_q) \oplus H_*BO(\bar{n},\mathbb{F}_q))$$

which according to Theorem 2.1 can be expressed as

$$\mathcal{A} = \mathbb{Z}/2[v_0,v_1,\ldots] \otimes E[v_0 + y_0, v_1 + y_1,\ldots]$$

Let us agree to bigrade $\mathcal{A}$ as follows: we assign to an element in $H_m(BO(n^\epsilon,\mathbb{F}_q))$ the bidegree (n,m). Then the Poincaré series of $\mathcal{A}$ is

$$P(\mathcal{A},s,t) = \frac{\Pi_{i=0}^{\infty}(1 + t^i s)}{\Pi_{i=0}^{n}(1 - t^i s)}$$

where the variable s is assigned the bidegree $(1,0)$ and the variable t is assigned the bidegree $(0,1)$. A rather involved calculation then shows that

$$[\Pi_{i=0}^{\infty}(1 - t^i s)^{-1}] = \Sigma_{K=0}^{\infty}(\Pi_{j=1}^{n}(1 - t^j))^{-1}s^n$$

and that

$$\Pi_{i=0}^{\infty}(1 + t^i s) = \Sigma_{n=0}^{\infty} t^{(1/2)n(n-1)}(\Pi_{j=1}^{n}(1 - t^j))^{-1}s^n$$

Multiplying together we get

$$P(\mathfrak{A}, s, t) = \Sigma_{n=0}^{\infty}\left[\frac{2\Pi_{i=1}^{n-1}(1 + t^i)}{\Pi_{i=1}^{n}(1 - t^i)}\right]s^n$$

In view of (1) and (2) it follows that

$$P(H^*BO(n^\epsilon, \mathbb{F}_q), t) = P(\mathcal{R}(O(n^\epsilon, \mathbb{F}_q)), t) = \frac{\Pi_{i=1}^{n-1}(1 + t^i)}{\Pi_{i=1}^{n}(1 - t^i)} .$$

Remark 4.6 (Added May, 1978). By picking different generators for $H^*BO(n, \mathbb{F}_q)$ and $H^*BO(\overline{n}, \mathbb{F}_q)$, we can write the relations of Theorem 4.2 in a slightly simpler form:

Instead of picking the generators $\{x_i, \overline{x}_{2j-1}\}$ as in Theorem 4.1, we can pick as generators the elements $\{w_i, z_i \mid 1 \leq i \leq n\}$ where w_i are the Stiefel-Whitney classes of 4.2 and where z_i are defined as follows: In the case of $H^*BO(n, \mathbb{F}_q)$, we define $z_i = x_i$, $1 \leq i \leq n$. In the case of $H^*BO(\overline{n}, \mathbb{F}_q)$, we define

$$z_i = \Phi_*(x_i) = \begin{cases} x_i & \text{if } i \text{ is even} \\ \overline{x}_i & \text{if } i \text{ is odd} \end{cases}$$

for $1 \leq i \leq n$. With respect to this choice of generators, Theorems 4.1 and 4.2 may be restated as follows:

Theorem 4.7. The cohomology rings $H^*BO(n, \mathbb{F}_q)$ and $H^*BO(\overline{n}, \mathbb{F}_q)$ are generated by the elements $\{w_i, z_i \mid 1 \leq i \leq n\}$ subject only to the relations

(1) $z_i^2 = \Sigma_{j=0}^{2i} z_j w_{2i-j} \quad 1 \leq i \leq n.$

(2) In the cases $H^*BO(n,\mathbb{F}_q)$ n even or odd, $H^*BO(\overline{n},\mathbb{F}_q)$ n odd, there is an additional relation

$$w_n = z_n$$

(3) In the case $H^*BO(\overline{n},\mathbb{F}_q)$ n even, there is an additional relation

$$z_n = 0.$$

Proof. It is clear from the relations of Theorem 4.2

$$w_{2k-1} = x_{2k-1} + \overline{x}_{2k-1}$$

$$R_{2k} = \Sigma_{p=1}^{k} x_{2p-1}\overline{x}_{2k-2p+1} + \Sigma_{q=1}^{k}(x_{2q} + w_{2q})x_{2k-2q} = 0$$

that the generators $\{x_i, \overline{x}_{2j-1}\}$ can be expressed in terms of the elements $\{w_i, z_i\}$ and vice-versa. Hence $\{w_i, z_i \mid 1 \leq i \leq n\}$ generates $H^*BO(n,\mathbb{F}_q)$ and $H^*BO(\overline{n},\mathbb{F}_q)$.

It is clear that the relation $R_{2k} = 0$ may be rewritten as

$$\Sigma_{p=1}^{k}(z_{2p-1} + w_{2p-1})z_{2k-2p+1} + \Sigma_{q=1}^{k}(x_{2q} + w_{2q})x_{2k-2q} = 0$$

$$\Sigma_{j=0}^{2k}(z_j + w_j)z_{2k-j} = 0$$

$$\Sigma_{j=0}^{2k} z_j z_{2k-j} + \Sigma_{j=0}^{2k} w_j z_{2k-j} = 0$$

$$z_k^2 + \Sigma_{j=0}^{2k} w_j z_{2k-j} = 0$$

which gives the relations (1). The relations (2) and (3) are either explicitly or implicitly stated in Theorem 4.2. Finally it is equally clear that the relations (1), (2), (3) imply all the relations of Theorem 4.2.

This alternate statement of the structure of $H^*BO(n,\mathbb{F}_q)$ and $H^*BO(\overline{n},\mathbb{F}_q)$ is essentially due to de Concine [42]. However he mistakenly omits relation (3) and claims relation (2) holds in all cases.

Comparing the two versions of the result we see that Theorem 4.7 uses more generators and relations then Theorems 4.1 and 4.2 with the advantage of simplifying the relations somewhat. The version of Theorem 4.1 and 4.2 is more economical in its use of generators and relations and enables us to see immediately that $H^*JO(q) \cong H^*BO(\infty,\mathbb{F}_q)$ is a polynomial algebra.

§5. $H_*B\mathcal{S}p(\mathbb{F}_q)$ and $H_*\Gamma_0 B\mathcal{S}p(\mathbb{F}_q)$

In this section we compute the mod-2 homology algebras $H_*B\mathcal{S}p(\mathbb{F}_q)$ and $H_*\Gamma_0 B\mathcal{S}p(\mathbb{F}_q)$ and establish the equivalence

$$\Gamma_0 B\mathcal{S}p(\mathbb{F}_q) \to JSp(q)$$

of infinite loop spaces at 2.

(5.1) According to Proposition VI 5.4, $H_i BSp(2,\mathbb{F}_q) = 0,0,\mathbf{Z}/2,\mathbf{Z}/2$ for $i \equiv 1,2,3,4 \bmod 4$. Let $\sigma_j \in H_{4j}BSp(2,\mathbb{F}_q)$ $j \geq 0$ and $\tau_k \in H_{4k-1}BSp(2,\mathbb{F}_q)$ denote generators.

Theorem 5.2. $H_*B\mathcal{S}p(\mathbb{F}_q) = \bigoplus_{n=0}^{\infty} H_*BSp(2n,\mathbb{F}_q)$ is a commutative algebra generated by σ_j $j \geq 0$, τ_k $k > 0$ subject to $\tau_k^2 = 0$, i.e.

$$H_*B\mathcal{Sp}(\mathbb{F}_q) = \mathbb{Z}/2[\sigma_0,\sigma_1,\sigma_2,\ldots] \otimes E[\tau_1,\tau_2,\ldots].$$

Let $\bar{\sigma}_j, \bar{\tau}_k$ $j,k > 0$ denote the images of σ_j, τ_k under the homomorphism $H_*BSp(2,\mathbb{F}_q) \to H_*\Gamma_0 B\mathcal{Sp}(\mathbb{F}_q)$.

Theorem 5.3. $H_*\Gamma_0 B\mathcal{Sp}(\mathbb{F}_q) = \mathbb{Z}/2[\bar{\sigma}_1,\bar{\sigma}_2,\ldots] \otimes E[\bar{\tau}_1,\bar{\tau}_2,\ldots]$.

Theorem 5.4. Any H map

$$\lambda: \Gamma_0 B\mathcal{Sp}(\mathbb{F}_q) \to JSp(q)$$

which makes diagram III 2.15 commute is an equivalence at 2; in particular any infinite loop map λ is an equivalence at 2.

The first step in the proofs of these theorems is to analyze the Brauer character. The Sylow 2-subgroup of $Sp(2,\mathbb{F}_q)$ is the generalized quaternion group Q_t (cf. Lemma VI 5.1). Let

$$E: Q_t \to Sp(2,\bar{\mathbb{F}}_q)$$

be the natural inclusion. Then the Brauer character of E is given by

$$\chi_E(x) = \beta(\theta_1) + \beta(\theta_2)$$

$$\chi_E(y) = \beta(\varphi_1) + \beta(\varphi_2)$$

where θ_i (resp. φ_i) are 2^t-th (resp. 4-th) roots of unity in $\bar{\mathbb{F}}_q$ and $\beta: \bar{\mathbb{F}}_q^* \to S^1 \subseteq \mathbb{C}^*$ is an imbedding. Since $Q_t \subseteq Sp(2,\mathbb{F}_q) = SL(2,\mathbb{F}_q)$, $\det x = \det y = 1$. Thus $\theta_1 = \theta_2^{-1}$, $\varphi_1 = \varphi_2^{-1}$ and so for a suitable choice of β

$$\chi_E(x) = 2\cos(\pi i/2^{t-1})$$

$$\chi_E(y) = 0$$

Let $\rho\colon Q_t \to S^3 = Sp(1) \subseteq U(2,\mathbb{C})$ be the representation of Remark VI 5.3. Then we have

Lemma 5.5. When restricted to Q_t, $\chi_E = x_\rho$

Proof: This is immediate since the inclusion $S^3 \to U(2,\mathbb{C})$ is given by

$$e^{i\theta} \to \begin{pmatrix} e^{i\theta} & 0 \\ 0 & e^{-i\theta} \end{pmatrix} \qquad j \to \begin{pmatrix} 0 & -1 \\ 1 & 0 \end{pmatrix}$$

Corollary 5.6. $\lambda_*(\bar{\sigma}_i) = \hat{g}_i$

Proof: According to Lemma 5.5 we have the commutative diagram

$$\begin{array}{ccccccccc} & & & \nearrow & BSp(2,\mathbb{F}_q) & \to & BSp(\infty,\mathbb{F}_q) & \to & \Gamma_0 B\mathcal{S}p(\mathbb{F}_q) & \overset{\lambda}{\searrow} & \\ B\mathbb{Z}/2 & \to & BQ_t & & & & & & \downarrow{\scriptstyle\beta} & & JSp(q) \\ & & & \underset{\rho}{\searrow} & BS^3 = BSp(1) & & \longrightarrow & & BSp & \overset{r}{\swarrow} & \end{array}$$

Chasing the generator $x_{4k} \in H_{4k}B\mathbb{Z}/2$ along the diagram we obtain

$$\begin{array}{ccccccccc} & & & \nearrow & \sigma_k & \longrightarrow & \bar{\sigma}_k & \overset{\lambda_*}{\searrow} & \\ x_{4k} & \to & x_{4k} & & & & \downarrow{\scriptstyle\beta_*} & & ? \\ & & & \searrow & g_k & \longrightarrow & g_k & \overset{r_*}{\swarrow} & \end{array}$$

so that

$$r_* \lambda_* (\bar{\sigma}_k) = \beta_* (\bar{\sigma}_k) = g_k$$

Since in Prop. I 4.2 we defined $\hat{g}_k$ merely to be a preimage of g_k under r_* we may as well take

$$\lambda_* (\bar{\sigma}_k) = \hat{g}_k .$$

Lemma 5.7. $\lambda_* (\bar{\tau}_i) = \hat{h}_i$ + decomposables.

Proof: According to Lemma I 8.2 we have Bockstein relations

$$d_\nu (\hat{g}_i) = \hat{h}_i + \text{decomposables} \quad i = 1,2$$

where $\nu = \nu(q^2 - 1)$ is the largest integer such that 2^ν divides q^2-1. By Prop. VI 5.7

$$d_\nu (\bar{\sigma}_i) = \bar{\tau}_i \quad \text{for all } i$$

Hence

$$\lambda_* (\bar{\tau}_i) = \lambda_* d_\nu (\bar{\sigma}_i) = d_\nu \lambda_* (\bar{\sigma}_i) = d_\nu (\hat{g}_i) = \hat{h}_i + \text{decomposables } i = 1,2.$$

Now let $\bar{M}_*$ be the $\mathbf{Z}/2$-submodule of $H_* BSp(2, \mathbb{F}_q)$ generated by $\{\tau_i | i \geq 2\}$ and let $\bar{N}_*$ be the $\mathbf{Z}/2$-submodule generated by τ_2. Let M_* be the $\mathbf{Z}/2$-submodule of $H_* JSp(q)$ generated by $\{\hat{h}_i | i \geq 2\}$ and let $\bar{N}_*$ be the $\mathbf{Z}/2$-submodule generated by $\hat{h}_2$.

According to Prop. I 4.2, (M_*, N_*) is an $\mathcal{A}_*$-module pair where $\mathcal{A}_*$ denotes the dual Steenrod algebra. Similarly by Prop. VI 5.7, $(\bar{M}_*, \bar{N}_*)$ is an $\mathcal{A}_*$-module pair. Also the formulas of Prop. I 4.2 and Prop. VI 5.7 show that $g_*: (\bar{M}_*, \bar{N}_*) \to (M_*, N_*)$ given by $g_*(\tau_i) = \hat{h}_i$ is

$\mathcal{A}_*$ isomorphism. By Lemma VI 5.8, $CL(\overline{N}_*) = \overline{M}_*$. Since Steenrod operations send decomposable elements to decomposable elements and λ is an H-map, the composite map

$$\lambda': BSp(2,\mathbb{F}_q) \to BSp(\infty,\mathbb{F}_q) \to \Gamma_0 B\mathcal{S}p(\mathbb{F}_q) \xrightarrow{\lambda} JSp(q)$$

induces an $\mathcal{A}_*$-map

$$\lambda'_*: (\overline{M}_*,\overline{N}_*) \to (M_*,N_*) \text{ mod decomposables}$$

By the above we have that $\lambda'_*|\overline{N}_* = g_*|\overline{N}_*$. Hence by Lemma III 6.7 $\lambda'_* = g_*$ so that

$$\lambda_*(\overline{\tau}_i) = \lambda'_*(\tau_i) = \hat{h}_i + \text{decomposables for all } i$$

(Note that we could not take $\overline{M}_*$ to be the $\mathbb{Z}/2$ module of $H_*\Gamma_0 B\mathcal{S}p(\mathbb{F}_q)$ generated by $\{\overline{\tau}_i | i \geq 2\}$ since we did not yet know that $\overline{\tau}_1 \neq 0$.)

<u>Proof of Theorem 5.3 and 5.4.</u> By Theorem VII 3.1, $H_*BSp(\infty,\mathbb{F}_q)$ is generated by the images of σ_i,τ_i under $H_*BSp(2,\mathbb{F}_q) \to H_*BSp(\infty,\mathbb{F}_q)$. Since $H_*\Gamma_0 B\mathcal{S}p(\mathbb{F}_q) \cong H_*BSp(\infty,\mathbb{F}_q)$ as algebras, it follows that $\overline{\sigma}_i,\overline{\tau}_i$ generate $H_*\Gamma_0 B\mathcal{S}p(\mathbb{F}_q)$. Since $\tau_k^2 = 0$ by Prop. VI 5.6, it follows that $\overline{\tau}_k^2 = 0$.

Hence in the chain of maps

$$\mathbb{Z}/2[\overline{\sigma}_1,\overline{\sigma}_2,\ldots] \otimes E[\overline{\tau}_1,\overline{\tau}_2,\ldots] \xrightarrow{i_*} H_*\Gamma_0 B\mathcal{S}p(\mathbb{F}_q)$$

$$\xrightarrow{\lambda_*} H_*JSp(q) = \mathbb{Z}/2[\hat{g}_1,\hat{g}_2,\ldots] \otimes E[\hat{h}_1,\hat{h}_2,\ldots]$$

i_* is an epimorphism, while by Cor. 5.6 and Lemma 5.7, $\lambda_* \cdot i_*$ is an isomorphism. Consequently both i_* and λ_* are isomorphisms. This proves both Theorem 5.3 and 5.4.

Proof of Theorem 5.2. By Theorem 5.3 we obtain that

$$H_*\Gamma_0 B\mathcal{Sp}(\mathbb{F}_q) = \mathbb{Z}/2[[2],\bar\sigma_1,\bar\sigma_2,\ldots] \otimes E[\bar\tau_1,\bar\tau_2,\ldots]$$

Now consider the chain of maps

$$\mathbb{Z}/2[\sigma_0,\sigma_1,\sigma_2,\ldots] \otimes E[\tau_1,\tau_2,\ldots] \xrightarrow{i_*} H_*B\mathcal{Sp}(\mathbb{F}_q) \xrightarrow{j_*} H_*\Gamma B\mathcal{Sp}(\mathbb{F}_q)$$

Obviously $j_*\cdot i_*$ is a monomorphism. By Theorem VII 3.1, i_* is an epimorphism. Hence i_* is an isomorphism.

§6. The cohomology rings $H^*BSp(2m,\mathbb{F}_q)$

First consider the case m = 1; this ring is computed explicitly in Proposition VI 5.4:

$$H^*BSp(2,\mathbb{F}_q) = \mathbb{Z}/2[P] \otimes E[\tilde{x}]$$

where $\deg P = 4$, $\deg \tilde{x} = 3$. For the general case we observe that (according to Ch. VII 3.1) the direct sum group homomorphism

$$i\colon Sp(2,\mathbb{F}_q) \times \cdots \times Sp(2,\mathbb{F}_q) \to Sp(2m,\mathbb{F}_q)$$

induces a monomorphism in mod-2 cohomology. Thus

$$i^*\colon H^*BSp(2m;\mathbb{F}_q) \to \bigotimes_1^m H^*BSp(2,\mathbb{F}_q) = \mathbb{Z}/2[P_1,P_2,\ldots,P_m] \otimes E[x_1,\ldots,x_m]$$

where $\deg P_j = 4$, $\deg x_j = 3$. Let

$$g_j = \Sigma_{i_1<\cdots<i_j} P_{i_1}\cdots P_{i_j} \qquad 1 \le j \le m$$

$$h_j = \Sigma_{\substack{i_1<\cdots<i_j \\ 1\le k\le j}} P_{i_1}\cdots \hat{P}_{i_k}\cdots P_{i_j} x_{i_k} \qquad 1 \le j \le m$$

then deg $g_j = 4j$, deg $h_j = 4j - 1$.

Theorem 6.1. $H^*BSp(2m,\mathbb{F}_q) = \mathbb{Z}/2[g_1,\ldots,g_m] \otimes E[h_1,\ldots,h_m]$.

Proof: By 5.2

$$\bigoplus_{m=0}^{\infty} H_*BSp(2m,\mathbb{F}_q) = \mathbb{Z}/2[\sigma_0,\sigma_1,\ldots] \otimes E[\tau_1,\tau_2,\ldots].$$

Thus it is easily seen that the elements $g_j, h_j \in \mathrm{Im}\ i^*$ for $1 \le j \le m$. Except for the relations $h_j^2 = 0$ these elements are algebraically independent [35, Lemma 9] and thus generate the subalgebra $\mathcal{R}_m = \mathbb{Z}/2[g_1,\ldots,g_m] \otimes E[h_1,h_2,\ldots,h_m]$. It remains to show $\mathcal{R}_m = H^*BSp(2m,\mathbb{F}_q)$. By 5.3, $H_*BSp(\infty,\mathbb{F}_q) = \mathbb{Z}/2[\bar{\sigma}_1,\bar{\sigma}_2,\ldots] \otimes E[\bar{\tau}_1,\bar{\tau}_2,\ldots]$ and so $\mathcal{R}_\infty = H^*BSp(\infty,\mathbb{F}_q)$ since they have the same Poincare series. This now implies $\mathcal{R}_m = H^*BSp(2m,\mathbb{F}_q)$ by the argument of 4.1. Q E D

§7. $H_*B\mathcal{GL}(\mathbb{F}_q)$ and $H_*\Gamma_0B\mathcal{GL}(\mathbb{F}_q)$; $H_*B\mathcal{U}(\mathbb{F}_{q^2})$ and $H_*\Gamma_0B\mathcal{U}(\mathbb{F}_{q^2})$

We continue our calculations by computing the mod-2 homology algebras listed in the title and establishing the equivalences

$$\Gamma_0B\mathcal{GL}(\mathbb{F}_q) \to JU(q)$$

$$\Gamma_0B\mathcal{U}(\mathbb{F}_{q^2}) \to JU(-q)$$

of infinite loop spaces at 2. Because the general linear and unitary cases are so similar we shall treat them in tandem and prove only one case, the other being entirely analogous.

(7.1) We begin by observing that $GL(1,\mathbb{F}_q) = \mathbb{F}_q^*$ and $U(1,\mathbb{F}_{q^2})$ is the subgroup of $\mathbb{F}_{q^2}^*$ consisting of elements x such that $x\bar{x} = 1$; thus these groups are cyclic. Their orders are $q - 1$ and $q + 1$ respectively (by II§6). Let $\alpha_i \in H_{2i}BGL(1,\mathbb{F}_q)$ $i \geq 0$, $\beta_i \in H_{2i-1}BGL(1,\mathbb{F}_q)$ $i > 0$ denote the nonzero elements. Similarly let $\xi_i \in H_{2i}BU(1,\mathbb{F}_{q^2})$ $i \geq 0$ $\eta_i \in H_{2i-1}BU(1,\mathbb{F}_{q^2})$ $i > 0$ denote the nonzero elements.

<u>Theorem</u> 7.2 (1) $H_*B\mathcal{GL}(\mathbb{F}_q) = \mathbb{Z}/2[\alpha_0,\alpha_1,\ldots] \otimes E[\beta_1,\beta_2,\ldots]$

(2) $H_*B\mathcal{U}(\mathbb{F}_q) = \mathbb{Z}/2[\xi_0,\xi_1,\ldots] \otimes E[\eta_1,\eta_2,\ldots]$

Let $\bar{\alpha}_i, \bar{\beta}_i$ denote the images of α_i, β_i under $H_*BGL(1,\mathbb{F}_q) \to H_*\Gamma_0B\mathcal{GL}(\mathbb{F}_q)$ and let $\bar{\xi}_i, \bar{\eta}_i$ denote the images of ξ_i, η_i under $H_*BU(1,\mathbb{F}_{q^2}) \to H_*\Gamma_0B\mathcal{U}(\mathbb{F}_{q^2})$.

<u>Theorem</u> 7.3 (1) $H_*\Gamma_0B\mathcal{GL}(\mathbb{F}_q) = \mathbb{Z}/2[\bar{\alpha}_1,\bar{\alpha}_2,\ldots] \otimes E[\bar{\beta}_1,\bar{\beta}_2,\ldots]$

(2) $H_*\Gamma_0B\mathcal{U}(\mathbb{F}_q) = \mathbb{Z}/2[\bar{\xi}_1,\bar{\xi}_2,\ldots] \otimes E[\bar{\eta}_1,\bar{\eta}_2,\ldots]$

<u>Theorem</u> 7.4. The maps

(1) $\lambda\colon \Gamma_0B\mathcal{GL}(\mathbb{F}_q) \to JU(q)$

(2) $\lambda\colon \Gamma_0B\,\mathcal{U}(\mathbb{F}_q) \to JU(-q)$

are equivalences of infinite loop spaces at 2.

As a preliminary step in the proof of these theorems we next prove

<u>Lemma</u> 7.5 (1) $\lambda_*(\bar{\alpha}_n) = \hat{a}_n$ + decomposables

(2) $\lambda_*(\bar{\xi}_n) = \hat{a}_n$ + decomposables

<u>Proof</u>: (2) From the definition of Brauer lifting we obtain the commutative diagram

$$\begin{array}{ccccc} BU(1,\mathbb{F}_{q^2}) & \longrightarrow & \Gamma_0 B\mathcal{U}(\mathbb{F}_{q^2}) & & \\ \Vert & & & \searrow^{\lambda} & \\ B\mathbb{Z}/_{q+1} & & \downarrow{\scriptstyle \beta} & & JU(-q) \\ \downarrow{\scriptstyle i} & & & \swarrow_{r} & \\ BS^1 = CP^\infty & \longrightarrow & BU & & \end{array}$$

Now $i: B\mathbb{Z}/_{q+1} \to BS^1$ is induced by the inclusion $\mathbb{Z}/_{q+1} \to S^1$. It is known that the induced map $H_*(BU(1,\mathbb{F}_{q^2})) \to H_*(BS^1)$ is monomorphic in even degrees. Hence by applying $H_{2n}(\cdot)$ to the above diagram we obtain

$$r_*\lambda_*(\bar{\xi}_n) = \beta_*(\bar{\xi}_n) = a_n.$$

Hence we get $\lambda_*(\bar{\xi}_n) = \hat{a}_n$ + decomposables by Proposition I 4.5. The proof of (1) is similar.

<u>Lemma 7.6</u> (1) $\lambda_*(\bar{\beta}_k) = \hat{b}_k$ + decomposables

(2) $\lambda_*(\bar{\eta}_k) = \hat{b}_k$ + decomposables

<u>Proof</u>: (2) According to Lemma I 8.2 we have the Bockstein relation

$$d_\nu(\hat{a}_1) = \hat{b}_1$$

where $\nu = \nu(q + 1)$ is the largest integer such that 2^ν divides $q + 1$. By Prop. VI 2.4

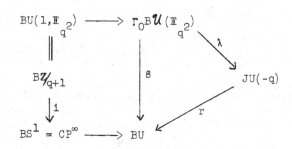

$$d_\nu(\bar{\xi}_1) = \hat{\eta}_1$$

Hence

$$\lambda_*(\hat{\eta}_1) = \lambda_* d_\nu(\bar{\xi}_1) = d_\nu \lambda_*(\bar{\xi}_1) = d_\nu(\hat{a}_1) = \hat{b}_1$$

According to Lemma I 4.6 we have a homology operation

$$Q^2(\hat{b}_1) = \hat{b}_2$$

According to Theorem IX 4.4 we have

$$Q^2(\bar{\eta}_1) = \bar{\eta}_2 + \bar{\xi}_1\bar{\eta}_1$$

Since λ is an infinite loop map

$$\lambda_*(\bar{\eta}_2) = \lambda_*[Q^2(\bar{\eta}_1) + \bar{\xi}_1\bar{\eta}_1] = Q^2\lambda_*(\bar{\eta}_1) + \hat{a}_1\hat{b}_1$$
$$= Q^2(\hat{b}_1) + \hat{a}_1\hat{b}_1 = \hat{b}_2 + \hat{a}_1\hat{b}_1.$$

Now let $\bar{M}_*$ be the $\mathbb{Z}/2$-submodule of $H_*BU(1,\mathbb{F}_{q^2})$ generated by $\{\eta_i | i \geq 2\}$ and let $\bar{N}_*$ be the $\mathbb{Z}/2$-submodule generated by η_2. Let M_* be the $\mathbb{Z}/2$-submodule of $H_*JU(-q)$ generated by $\{\hat{b}_i | i \geq 2\}$ and let N_* be the $\mathbb{Z}/2$-submodule generated by $\hat{b}_2$.

According to Prop. I 4.6, (M_*, N_*) is an $\mathcal{A}_*$-module pair where $\mathcal{A}_*$ denotes the dual Steenrod algebra. Similarly by Prop. VI 2.4 $(\bar{M}_*, \bar{N}_*)$ is an $\mathcal{A}_*$-module pair. Also the formulas of Prop. I 4.6 and Prop. VI 2.4 show that $g_*: (\bar{M}_*, \bar{N}_*) \to (M_*, N_*)$ given by $g_*(\bar{\eta}_i) = \hat{b}_i$ is an $\mathcal{A}_*$ isomorphism. By Lemma VI 2.6, $CL(\bar{N}_*) = \bar{M}_*$. Since Steenrod operations send decomposable elements to decomposable elements and λ is an H-map, the composite map

$$\lambda': BU(1,\mathbb{F}_{q^2}) \to BU(\infty,\mathbb{F}_{q^2}) \to \Gamma_0 B\mathcal{U}(\mathbb{F}_{q^2}) \xrightarrow{\lambda} JU(-q)$$

induces an $\mathcal{a}_*$-map

$$\lambda_*': (\overline{M}_*,\overline{N}_*) \to (M_*,N_*) \text{ mod decomposables}$$

By the above we have that $\lambda_*'|\overline{N}_* = g_*|\overline{N}_*$. Hence by Lemma III 6.7 $\lambda_*' = g_*$ so that

$$\lambda_*(\overline{\eta}_i) = \lambda_*'(\eta_i) = \hat{b}_i + \text{decomposables for all } i.$$

This proves (2). The proof of (1) is similar.

Proofs of Theorems 7.3 and 7.4. The proofs are similar to those of Theorems 5.3 and 5.4 except that we use Theorems VII 4.1, 4.2 in place of Theorem VII 3.1.

Proof of Theorem 7.2. This is derived from Theorem 7.3 in the same way that Theorem 5.2 is derived from Theorem 5.3. Again we use Theorems VII 4.1, 4.2 in place of Theorem VII 3.1.

§8. The cohomology rings $H^*BGL(n,\mathbb{F}_q)$ and $H^*BU(n,\mathbb{F}_{q^2})$

The groups $GL(1,\mathbb{F}_q)$ and $U(1,\mathbb{F}_{q^2})$ are cyclic of orders $q-1$ and $q+1$ respectively (see 7.1). Thus by Proposition VI 2.1 we have

$$H^*BGL(1,\mathbb{F}_q) = \begin{cases} \mathbb{Z}/2[x] & q \equiv 3 \pmod 4 \\ \mathbb{Z}/2[y] \otimes E[x] & q \equiv 1 \pmod 4 \end{cases}$$

(1)

$$H^*BU(1,\mathbb{F}_{q^2}) = \begin{cases} \mathbb{Z}/2[y] \otimes E[x] & q \equiv 3 \pmod 4 \\ \mathbb{Z}/2[x] & q \equiv 1 \pmod 4 \end{cases}$$

where deg x = 1, deg y = 2.

For a general n we will use the fact that the direct sum group homomorphisms

$$(2)\qquad \begin{array}{l} i\colon GL(1,\mathbb{F}_q) \times \cdots \times GL(1,\mathbb{F}_q) \to GL(n,\mathbb{F}_q) \\ i\colon U(1,\mathbb{F}_{q^2}) \times \cdots \times U(1,\mathbb{F}_{q^2}) \to U(n,\mathbb{F}_{q^2}) \end{array}$$

induce monomorphisms in mod-2 cohomology. This is implied by Theorem 7.2.

In the case of $G = GL(n,\mathbb{F}_q)$ $q \equiv 1 \pmod 4$ or $G = U(n,\mathbb{F}_{q^2})$ $q \equiv 3 \pmod 4$, we have an inclusion of algebras

$$i^*\colon H^*BG \to \mathbb{Z}/2[y_1,y_2,\ldots,y_n] \otimes E[x_1,x_2,\ldots,x_n]$$

where deg $x_j = 1$, deg $y_j = 2$. Let

$$a_{2k} = \Sigma_{i_1<\cdots<i_k}\, y_{i_1}\cdots y_{i_k} \qquad 1 \le k \le n$$

$$b_{2k-1} = \Sigma_{\substack{i_1<\cdots<i_k \\ 1\le j\le k}}\, y_{i_1}\cdots \hat{y}_{i_j}\cdots y_{i_k}\, x_{i_j} \qquad 1 \le k \le n$$

<u>Theorem</u> 8.1. For $q \equiv 1 \pmod 4$

$$H^*BGL(n,\mathbb{F}_q) = \mathbb{Z}/2[a_2,a_4,\ldots,a_{2n}] \otimes E[b_1,b_3,\ldots,b_{2n-1}]$$

and for $q \equiv 3 \pmod 4$

$$H^*BU(n,\mathbb{F}_q) = \mathbb{Z}/2[a_2,a_4,\ldots,a_{2n}] \otimes E[b_1,b_3,\ldots,b_{2n-1}]$$

Proof: We give the proof for $U(n,\mathbb{F}_{q^2})$; the proof for $GL(n,\mathbb{F}_q)$ is identical. By 7.2

$$\bigoplus_{n=1}^{\infty} H_*BU(n,\mathbb{F}_q) = \mathbb{Z}/2[\xi_0,\xi_1,\ldots,] \otimes E[\eta_1,\eta_2,\ldots]$$

where $\deg \xi_i = 2i$, $\deg \eta_i = 2i-1$. Thus it follows by duality that $a_{2k}, b_{2k-1} \in \operatorname{Im} i^*$. Except for the relations $b_{2k-1}^2 = 0$ these elements are algebraically independent [35, Lemma 9], thus they generate the subalgebra $\mathcal{R}_n = \mathbb{Z}/2[a_2,\ldots,a_{2n}] \otimes E[b_1,\ldots,b_{2n-1}]$. We must show $\mathcal{R}_n$ is all of $H^*BU(n,\mathbb{F}_{q^2})$. By 7.3 we have

$$H_*BU(\infty,\mathbb{F}_{q^2}) = \mathbb{Z}/2[\bar{\xi}_1,\bar{\xi}_2,\ldots] \otimes E[\bar{\eta}_1,\bar{\eta}_2,\ldots]$$

thus $\mathcal{R}_\infty = H_*BU(\infty,\mathbb{F}_{q^2})$ since they have the same Poincaré series. This now implies $\mathcal{R}_n = H_*BU(n,\mathbb{F}_{q^2})$ by the argument of 4.1. Q.E.D.

In the case of $G = GL(n,\mathbb{F}_q)$ $q \equiv 3 \pmod 4$ and $G = U(n,\mathbb{F}_{q^2})$ $q \equiv 1 \pmod 4$, (2) gives an inclusion of algebras

$$i^*: H^*BG \to \mathbb{Z}/2[x_1,x_2,\ldots,x_n]$$

where $\deg x_i = 1$. Let

$$a_{4k} = \Sigma_{i_1<\cdots<i_{2k}} x_{i_1}^2 \cdots x_{i_{2k}}^2 = \sigma_{2k}(x_1,\ldots,x_n)^2 \qquad 1 \le k \le [n/2]$$

$$b_{2k-1} = \sum_{j=1}^{n} x_i^{2k-1} \qquad 1 \le k \le n$$

To find relations among these elements we observe that the Newton polynomial expressing b_{2k-1} in terms of the elementary symmetric polynomials $\sigma_i = \sigma_i(x_1,\ldots,x_n)$ has the form

$$b_{2k-1} = \sigma_{2k-1} + \text{decomposable terms in the } \sigma_i\text{'s.}$$

Thus $\{b_1, b_3, \ldots, b_{2[\frac{n+1}{2}]-1}, \sigma_2, \sigma_4, \ldots, \sigma_{2[\frac{n}{2}]}\}$ are algebraically independent generators of the symmetric subalgebra of $\mathbb{Z}/2[x_1, \ldots, x_n]$; hence for $[\frac{n+1}{2}] < k \leq n$

$$b_{2k-1} = p_{2k-1}(b_1, \ldots, b_{2[\frac{n+1}{2}]-1}, \sigma_2, \ldots, \sigma_{2[\frac{n}{2}]})$$

for some polynomial p_{2k-1}. Upon squaring this polynomial we have

3) $$b_{2k-1}^2 = p_{2k-1}(b_1^2, b_3^2, \ldots, b^2_{2[\frac{n+1}{2}]-1}, a_4, a_8, \ldots, a_{4[\frac{n}{2}]})$$

<u>Theorem</u> 8.2. For $q \equiv 3 \pmod 4$

$$H^*BGL(n, \mathbb{F}_q) = \mathbb{Z}/2[a_4, a_8, \ldots, a_{4[\frac{n}{2}]}, b_1, b_3, \ldots, b_{2n-1}]/R$$

and for $q \equiv 1 \pmod 4$

$$H^*BU(n, \mathbb{F}_{q^2}) = \mathbb{Z}/2[a_4, a_8, \ldots, a_{4[\frac{n}{2}]}, b_1, b_3, \ldots, b_{2n-1}]/R$$

where R is the relations ideal generated by 3) for $[\frac{n+1}{2}] < k \leq n$.

<u>Corollary</u> 8.3. For $n = \infty$ the algebras of Theorem 8.3 are polynomial algebras on generators $\{a_{4k}, b_{2k-1}\}_{k=1}^{\infty}$.

In the case of $H^*BGL(n, \mathbb{F}_q)$ $q \equiv 3 \pmod 4$, Quillen [35] chose different generators and relations. His presentation has the advantage of providing explicit formulas for the relations. However, it has the serious disadvantage of obscuring the fact that $H^*BGL(\infty, \mathbb{F}_q)$ $q \equiv 3 \pmod 4$ is a polynomial algebra.

<u>Proof of Theorem</u> 8.2: We give the proof for $GL(n, \mathbb{F}_q)$; the proof for

$U(n, \mathbb{F}_{q^2})$ is identical. Let $\mathcal{R}_n$ be the subalgebra of $\mathbb{Z}/2[x_1, \ldots, x_n]$ generated by $\{a_4, \ldots, a_{4[\frac{n}{2}]}, b_1, \ldots, b_{2n-1}\}$. According to 7.2

$$\bigoplus_{k=1}^{\infty} H_* BGL(k, \mathbb{F}_q) = \mathbb{Z}/2[\alpha_0, \alpha_1, \ldots,] \otimes E[\beta_1, \beta_2, \ldots]$$

where $\deg \alpha_i = 2i$, $\deg \beta_i = 2i-1$. Dualizing this result we see that $\mathcal{R}_n$ is a subalgebra of $H^*BGL(n, \mathbb{F}_q)$. First we show $\mathcal{R}_n = H^*BGL(n, \mathbb{F}_q)$. Since $\{a_4, \ldots, a_{4[\frac{n}{2}]}, b_1, \ldots, b_{2[\frac{n+1}{2}]-1}\}$ are algebraically independent $\mathcal{R}_\infty$ is a polynomial algebra on $\{a_{4k}, b_{2k-1}\}_{k=1}^{\infty}$. Thus its Poincaré series is

$$P.S.(\mathcal{R}_\infty, t) = \prod_{k=1}^{\infty} \frac{1}{1-t^{4k}} \frac{1}{1-t^{2k-1}}.$$

According to 7.3, $H_* BGL(\infty, \mathbb{F}_q) = \mathbb{Z}/2[\bar{\alpha}_1, \bar{\alpha}_2, \ldots] \otimes E[\bar{\eta}_1, \bar{\eta}_2, \ldots]$ thus

$$P.S.(H^*BGL(\infty, \mathbb{F}_q), t) = \prod_{k=1}^{\infty} \frac{1 + t^{2k-1}}{1 - t^{2k}} = \prod_{k=1}^{\infty} \frac{1}{1-t^{4k}} \frac{1}{1-t^{2k-1}}$$

Thus $\mathcal{R}_\infty = H^*BGL(\infty, \mathbb{F}_q)$ which implies $\mathcal{R}_n = H^*BGL(n, \mathbb{F}_q)$ by the argument of 4.1.

It remains to show that there are no relations in $H^*BGL(n, \mathbb{F}_q)$ other than 3) for $[\frac{n+1}{2}] < k \leq n$. Let $\mathcal{A}$ denote the algebra on generators and relations specified in the theorem. We have shown that $\mathcal{A}$ maps onto $H^*BGL(n, \mathbb{F}_q)$ and so the proof will be complete if we can show they have the same Poincaré series. According to 7.2 the Poincaré series for $H_* BGL(n, \mathbb{F}_q)$ is the same for $q \equiv 1$ or $3 \pmod 4$ thus

$$P.S.(H^*BGL(n, \mathbb{F}_q), t) = \prod_{k=1}^{n} \frac{1 + t^{2k-1}}{1 - t^{2k}} .$$

Let $\mathcal{B} \subset \mathcal{A}$ be the subalgebra generated by

$\{a_4, \ldots, a_{4[\frac{n}{2}]}, b_1, \ldots, b_{2[\frac{n+1}{2}]-1}\}$. These elements are algebraically independent in $H^*BGL(n,\mathbb{F}_q)$ and hence also in $\mathcal{B}$; thus

$$P.S.(\mathcal{B},t) = \prod_{k=1}^{[\frac{n}{2}]} \frac{1}{1-t^{4k}} \prod_{k=1}^{[\frac{n+1}{2}]} \frac{1}{1-t^{2k-2}}.$$

Let W be the ideal of $\mathcal{O}\!\iota$ generated by the generators of $\mathcal{B}$; then $\mathcal{O}\!\iota/W = E[b_{2[\frac{n+1}{2}]+1}, \ldots, b_{2n-1}]$, thus

$$P.S.(\mathcal{O}\!\iota/W,t) = \prod_{k=[\frac{n+1}{2}]+1}^{n} (1 + t^{2k-1})$$

Hence

$$P.S.(\mathcal{O}\!\iota,t) \leq P.S.(\mathcal{B},t)\,P.S.(\mathcal{O}\!\iota/W,t)$$

$$= \left(\prod_{k=1}^{[\frac{n}{2}]} \frac{1}{1-t^{4k}}\right)\left(\prod_{k=1}^{[\frac{n+1}{2}]} \frac{1}{1-t^{2k-1}}\right)\left(\prod_{k=[\frac{n+1}{2}]+1}^{n} (1 + t^{2k-1})\right)$$

$$= \prod_{k=1}^{n} \frac{1+t^{2k-1}}{1-t^{2k}} = P.S.(H^*BGL(n,\mathbb{F}_q),t).$$

However since $\mathcal{O}\!\iota$ maps onto $H^*BGL(n,\mathbb{F}_q)$ equality must hold.

Chapter V

Calculations at Odd Primes

§1. Introduction

The purpose of this chapter is to give our principal calculations at odd primes ℓ. All homology and cohomology groups are taken with $\mathbb{Z}/\ell$ coefficients and all spaces are localized at ℓ.

Specifically we compute the homology algebras $H_*B\bar{\mathcal{O}}(\mathbb{F}_q)$, establish an equivalence of infinite loop spaces

$$\Gamma_0 B\bar{\mathcal{O}}(\mathbb{F}_q) \to JO(q)$$

and compute the cohomology rings $H^*BO(n,\mathbb{F}_q)$, $H^*BO(\bar{n},\mathbb{F}_q)$. Corresponding results are obtained for the other categories $\mathcal{S}p(\mathbb{F}_q)$, $\mathcal{OGL}(\mathbb{F}_q)$, $\mathcal{GL}(\mathbb{F}_q)$, and $\mathcal{U}(\mathbb{F}_{q^2})$.

Throughout the chapter, ℓ is assumed different from the characteristic of $\mathbb{F}_q$.

§2. $H_*B\bar{\mathcal{O}}(\mathbb{F}_q)$ and $H_*\Gamma_0 B\bar{\mathcal{O}}(\mathbb{F}_q)$

In this section we compute the mod-ℓ homology algebras $H_*B\bar{\mathcal{O}}(\mathbb{F}_q)$ and $H_*\Gamma_0 B\bar{\mathcal{O}}(\mathbb{F}_q)$ and establish that

$$\lambda\colon \Gamma_0 B\bar{\mathcal{O}}(\mathbb{F}_q) \to JO(q)$$

is an equivalence of infinite loop spaces at ℓ. (In case q is even, we will understand $\bar{\mathcal{O}}(\mathbb{F}_q)$ to stand for $\bar{\mathcal{O}}^{ev}(\mathbb{F}_q)$.) Throughout this section we will use the notation of II 4.7.

Throughout this section, as in Def. VIII 4.1, d will denote the minimal positive integer d such that $q^{2d} \equiv 1 \pmod{\ell}$. If q is even we will denote $\varepsilon = 1$. If q is odd we will denote $\varepsilon = \pm 1$ according as $q \equiv \pm 1 \pmod{4}$. We will denote by c the element in

$\tilde{\mathbb{N}}$ (cf. II 4.7) given by $c = 2d$ if $q^d \equiv \epsilon^d \pmod{\ell}$ and by $c = \overline{2d}$ if $q \equiv -\epsilon^d \pmod{\ell}$. According to Lemma VIII 4.2, c is the smallest element of $\tilde{\mathbb{N}}$ such that ℓ divides the order of $O(c,\mathbb{F}_q)$.

According to Theorem VIII 4.4

$$H_n BO(c,\mathbb{F}_q) = \begin{cases} \text{cyclic on generator } \gamma_i & \text{if } n = 4di \\ \text{cyclic on generator } \delta_i & \text{if } n = 4di-1 \\ 0 & \text{otherwise.} \end{cases}$$

Note that at this point we do not yet know that $\gamma_i \neq 0$ and $\delta_i \neq 0$. As in II 2.16(viii) we will denote by $[n]$ the generator of $H_0 BO(n,\mathbb{F}_q)$.

Theorem 2.1(a). If q is odd $H_* B\overline{\mathcal{O}}(\mathbb{F}_q) = \oplus_{n\in\tilde{\mathbb{N}}} H_* BO(n,\mathbb{F}_q)$ is a commutative algebra generated by γ_i, δ_i $i > 0$, $[1]$ and $[\overline{1}]$ subject to the single relation $[1]^2 = [\overline{1}]^2$, i.e.

$$H_* B\overline{\mathcal{O}}(\mathbb{F}_q) = \mathbb{Z}/\ell[[1],[\overline{1}],\gamma_1,\gamma_2,\ldots] \otimes E[\delta_1,\delta_2,\ldots]/[1]^2 = [\overline{1}]^2$$

(b) If q is even $H_* B\overline{\mathcal{O}}(\mathbb{F}_q) = \oplus_{n\in\tilde{\mathbb{N}}^{ev}} H_* BO(n,\mathbb{F}_q)$ is a commutative algebra generated by γ_i, δ_i $i > 0$, $[2]$ and $[\overline{2}]$ subject to the single relation $[2]^2 = [\overline{2}]^2$, i.e.

$$H_* B\overline{\mathcal{O}}^{ev}(\mathbb{F}_q) = \mathbb{Z}/\ell[[2],[\overline{2}],\gamma_1,\gamma_2,\ldots] \otimes E[\delta_1,\delta_2,\ldots]/[2]^2 = [\overline{2}]^2$$

(In particular we must have $\gamma_i \neq 0$, $\delta_i \neq 0$ all $i > 0$).

Let $\overline{\gamma}_i, \overline{\delta}_i \in H_* \Gamma_0 B\overline{\mathcal{O}}(\mathbb{F}_q)$ denote the images of γ_i, δ_i under the map

$$BO(c,\mathbb{F}_q) \to \Gamma_0 B\overline{\mathcal{O}}(\mathbb{F}_q)$$

Theorem 2.2. $H_* \Gamma_0 B\overline{\mathcal{O}}(\mathbb{F}_q) = \mathbb{Z}/\ell[\overline{\gamma}_1,\overline{\gamma}_2,\ldots] \otimes E[\overline{\delta}_1,\overline{\delta}_2,\ldots]$.

<u>Theorem</u> 2.3. Any H-map

$$\lambda\colon \Gamma_0 B\overline{\mathcal{O}}(\mathbb{F}_q) \to JO(q)$$

which completes diagram III 2.14 (or III 2.21) is an equivalence at ℓ. In particular there is an infinite loop equivalence λ at ℓ.

As outlined in Chapter III §6, we will prove these theorems by showing that

$$\lambda_*(\overline{\gamma}_i) = s_i + \text{decomposables}, \quad \lambda_*(\overline{\delta}_i) = t_i + \text{decomposables}$$

for $i > 0$, where $\{s_i, t_i\}$ are the generators of $H_* JO(q)$ defined in Prop. I 5.2.

Our first step will be to analyze the Brauer character of $O(c,\mathbb{F}_q)$. According to Theorem VIII 4.4, the Sylow ℓ-subgroup of $O(c,\mathbb{F}_q)$ is a cyclic group $\mathbb{Z}/\ell^\nu$ where $\nu = \nu_\ell(q^{2d}-1)$. Let E denote the natural inclusion

$$E\colon O(c,\mathbb{F}_q) \to O(2d,\overline{\mathbb{F}}_q)$$

According to Chapter III §2, the Brauer character χ_E is the character of a unique virtual real representation of $O(c,\mathbb{F}_q)$. Let P denote the real representation $P = \rho + \rho^q + \cdots + \rho^{q^{d-1}}$ where $\rho\colon \mathbb{Z}/\ell^\nu \to S^1 \xrightarrow{\cong} SO(2,\mathbb{R})$ is the standard representation. Then we have

<u>Proposition</u> 2.4. When restricted to $\mathbb{Z}/\ell^\nu$, $\chi_E = \chi_P$.

<u>Proof</u>. Let A denote a generator of $\mathbb{Z}/\ell^\nu$. Let $\mu \in \mathbb{F}_{q^{2d}}$ denote a primitive ℓ^ν root of unity.

There are two cases to consider: (1) $q^d \equiv -1 \pmod{\ell}$ and

(2) $q^d \equiv 1 \pmod{\ell}$. In the first case we have the chain of inclusions

$$\mathbb{Z}/\ell^\nu \to U(1,\mathbb{F}_{q^{2d}}) \xrightarrow{i} O(\hat{c},\mathbb{F}_{q^d}) \xrightarrow{j} O(c,\mathbb{F}_q)$$

(cf. Theorem VIII 4.4(a)) which is induced by regarding $\mathbb{F}_{q^{2d}}$ as a vector space over $\mathbb{F}_q$ and regarding A as multiplication by μ. Obviously the minimal polynomial of A (over $\mathbb{F}_q$) is the minimal polynomial $p(x)$ of μ over $\mathbb{F}_q$. Since μ generates $\mathbb{F}_{q^{2d}}$ over $\mathbb{F}_q$, $p(x)$ has degree 2d and is therefore also the characteristic polynomial of A. Hence the eigenvalues of A are

$$\{\mu,\mu^q,\ldots,\mu^{q^{d-1}},\mu^{q^d},\mu^{q^{d+1}},\ldots,\mu^{q^{2d-1}}\}$$

$$= \{\mu,\mu^q,\ldots,\mu^{q^{d-1}},\mu^{-1},\mu^{-q},\ldots,\mu^{-q^{d-1}}\}$$

The eigenvalues of $\beta(A)$ are

$$\{\beta(\mu),\beta(\mu)^q,\ldots,\beta(\mu)^{q^{d-1}},\beta(\mu)^{-1},\beta(\mu)^{-q},\ldots,\beta(\mu)^{-q^{d-1}}\}$$

Thus the representation χ_E is equivalent to $\rho + \rho^q + \rho^{q^2} + \cdots + \rho^{q^{d-1}}$.

In case (2) we have the chain of inclusions

$$\mathbb{Z}/\ell^\nu \to GL(1,\mathbb{F}_{q^d}) \xrightarrow{h} O(\hat{c},\mathbb{F}_{q^d}) \xrightarrow{j} O(c,\mathbb{F}_q)$$

(cf. Theorem VIII 4.4(b)) which is induced by regarding $\mathbb{F}_{q^d} \oplus \mathbb{F}_{q^d}$ as a vector space over $\mathbb{F}_q$ with A acting as the diagonal matrix (over $\mathbb{F}_{q^d}$)

$$\begin{pmatrix} \mu & 0 \\ 0 & \mu^{-1} \end{pmatrix}$$

It is clear that the minimal polynomial of A over $\mathbb{F}_q$ is $p(x)$ where $p(x)$ is the minimal polynomial of μ over $\mathbb{F}_q$. Since μ generates $\mathbb{F}_{q^d}$ over $\mathbb{F}_q$, $p(x)$ has degree d. Since the characteristic polynomial of A has the same roots as the minimal polynomial up to multiplicity and has degree $2d$, it follows that the characteristic polynomial of A is $p(x)^2$. Thus A has eigenvalues

$$\{\mu, \mu^q, \ldots, \mu^{q^{d-1}}, \mu^{-1}, \mu^{-q}, \ldots, \mu^{-q^{d-1}}\}.$$

By the same reasoning as above, χ_E is again equivalent to the representation $\rho + \rho^q + \rho^{q^2} + \cdots + \rho^{q^{d-1}}$.

Lemma 2.5. $\lambda_*(\bar{\gamma}_k) = s_k$.

Proof. We have the commutative diagram

$$\begin{array}{ccccccc}
B\mathbb{Z}/\ell^\nu & \longrightarrow & BO(c,\mathbb{F}_q) & \longrightarrow & \Gamma_0 B\bar{O}(\mathbb{F}_q) & \xrightarrow{\lambda} & JO(q) \\
\downarrow{\scriptstyle (1\times q\times\cdots\times q^{d-1})} & & & & & \searrow^{\beta} & \downarrow{\scriptstyle r} \\
B\mathbb{Z}/\ell^\nu \times \cdots \times B\mathbb{Z}/\ell^\nu & & \xrightarrow{\rho\times\cdots\times\rho} & & BO \times \cdots \times BO & \xrightarrow{\oplus} & BO
\end{array}$$

Now let $x_i \in H_i B\mathbb{Z}/\ell^\nu$ denote a generator. Then applying $H_*(\cdot)$ to the preceding diagram and chasing the element x_{4kd} using Lemma III 6.1 to evaluate ρ_*, we obtain

$$\begin{aligned} r_*\lambda_*(\bar{\gamma}_k) &= \Sigma_{i=0}^{d-1} q^{2dki} \rho_*(x_{4kd}) + \text{decomposables} \\ &= \Sigma_{i=0}^{d-1} p_{dk} + \text{decomposables} \end{aligned}$$

$$= d\, p_{dk} + \text{decomposables}$$

where $H_*BO = \mathbb{Z}/\ell[p_1, p_2, \ldots]$. (In the calculation $q^{2dki} \equiv 1$ in $\mathbb{Z}/\ell$ since $q^{2d} \equiv 1 \pmod{\ell}$.). Since $d \not\equiv 0 \pmod{\ell}$ and since the generators $s_i \in H_*JO(q)$ of Chapter I §5 were defined to be preimages under r_* of algebra generators of H_*BO, we may as well take $s_k = \lambda_*(\overline{\gamma}_k)$.

<u>Lemma</u> 2.6. $\lambda_*(\overline{\delta}_k) = t_k$ + decomposables.

<u>Proof</u>. According to Lemma I 7.5 we have the Bockstein relations

$$d_\nu(s_i) = t_i + \text{decomposables} \quad i = 1,2,\ldots,\ell$$

where $\nu = \nu_\ell(q^{2d}-1)$. Since the $\overline{\gamma}_i, \overline{\delta}_i$ are images of generators of $H_*B\mathbb{Z}/\ell^\nu$, according to Prop. VI 2.4 we have

$$d_\nu(\overline{\gamma}_i) = \overline{\delta}_i \text{ for all } i.$$

Hence

$$\lambda_*(\overline{\delta}_i) = \lambda_* d_\nu(\overline{\gamma}_i) = d_\nu \lambda_*(\overline{\gamma}_i) = d_\nu(s_i) = t_i + \text{decomposables}$$

for $1 \leq i \leq \ell$.

Now let M_* be the $\mathbb{Z}/\ell$-submodule of $H_*JO(q)$ generated by $\{t_i \mid i \geq 1\}$ and let N_* be the $\mathbb{Z}/\ell$-submodule generated by $\{t_i \mid 1 \leq i \leq \ell\}$. Let $\overline{M}_*$ be the $\mathbb{Z}/\ell$-submodule of $H_*B\mathbb{Z}/\ell^\nu$ generated by $\{x_{4id-1} \mid i \geq 1\}$ and let $\overline{N}_*$ be the $\mathbb{Z}/\ell$-submodule generated by $\{x_{4id-1} \mid 1 \leq i \leq \ell\}$.

By Prop. I 5.2, (M_*, N_*) is an $\mathcal{A}_*$-module pair where $\mathcal{A}_*$ denotes the dual mod-ℓ Steenrod algebra. By Prop. VI 2.4 and Prop. VIII 2.6 (which implies that 2d divides ℓ-1), $(\overline{M}_*, \overline{N}_*)$ is also an $\mathcal{A}_*$-module pair. Moreover comparing the formulas of Prop. I 5.2 and Prop. VI 2.4, we see that $g_*: (\overline{M}_*, \overline{N}_*) \to (M_*, N_*)$ given by $g_*(x_{4id-1}) = t_i$ is an $\mathcal{A}_*$-isomorphism. Since Steenrod operations send decomposable elements to decomposable elements and λ is an H-map, the composite map

$$\lambda': B\mathbb{Z}/\ell^{\nu} \to BO(c, \mathbb{F}_q) \to BO(\infty, \mathbb{F}_q) \to \Gamma_0 B\mathcal{O}(\mathbb{F}_q) \xrightarrow{\lambda} JO(q)$$

induces an $\mathcal{A}_*$-map

$$\lambda_*': (\overline{M}_*, \overline{N}_*) \to (M_*, N_*) \text{ mod decomposables}$$

By the above $\lambda_*' | \overline{N}_* = g_* | \overline{N}_*$. Lemma III 6.7 will imply that $\lambda_*' = g_*$ and

$$\lambda_*(\overline{\delta}_i) = \lambda_*'(x_{4id-1}) = t_i + \text{decomposables for all } i$$

once we prove the following result

<u>Lemma</u> 2.7. $CL(\overline{N}_*) = \overline{M}_*$

<u>Proof</u>. It is immediate from Prop. VI 2.4 that in the notation of VI 2.5

$$\overline{M}_* = \oplus_{i=1}^{s} \overline{M}_*^{2id}$$

where $s = \frac{1}{2d}(\ell-1)$, and that this is a splitting of $\mathcal{A}_*$-modules. It then follows from Prop. VI 2.5 that $CL(\tilde{N}_*) = \overline{M}_*$ where $\tilde{N}_*$ is the $\mathbb{Z}/\ell$-module generated by $\{x_{4id-1} | 1 \le i \le s\}$. Since $\tilde{N}_* \subseteq \overline{N}_*$, a fortiori $CL(\overline{N}_*) = \overline{M}_*$.

<u>Proof of Theorems</u> 2.2 <u>and</u> 2.3. By Theorem VIII 4.5, $H_*BO(\infty, \mathbb{F}_q)$ is generated by the images of γ_i, δ_i under $H_*BO(c, \mathbb{F}_q) \to H_*BO(\infty, \mathbb{F}_q)$. Since

$H_*\Gamma_0 B\bar{\mathcal{O}}(\mathbb{F}_q) \cong H_* BO(\infty, \mathbb{F}_q)$ as algebras, it follows that $\bar{\gamma}_i, \bar{\delta}_i$ generate $H_*\Gamma_0 B\bar{\mathcal{O}}(\mathbb{F}_q)$. Hence in the chain of maps

$$\mathbb{Z}/\ell[\bar{\gamma}_1, \bar{\gamma}_2, \ldots] \otimes E[\bar{\delta}_1, \bar{\delta}_2, \ldots] \xrightarrow{i_*} H_*\Gamma_0 B\bar{\mathcal{O}}(\mathbb{F}_q)$$

$$\xrightarrow{\lambda_*} H_* JO(q) = \mathbb{Z}/\ell[s_1, s_2, \ldots] \otimes E[t_1, t_2, \ldots]$$

i_* is an epimorphism, while by Lemma 2.5 and 2.6, $\lambda_* i_*$ is an isomorphism. Hence both i_* and λ_* must be isomorphisms. This proves Theorems 2.2 and 2.3.

Proof of Theorem 2.1(a) Theorem 2.2 implies that

$$H_*\Gamma B\bar{\mathcal{O}}(\mathbb{F}_q) = \mathbb{Z}/\ell[[-1], [-\bar{1}], [1], [\bar{1}], \gamma_1, \gamma_2, \ldots] \otimes E[\delta_1, \delta_2, \ldots]/R$$

where R denotes the relations generated by

$$[-1]*[1] = [-\bar{1}]*[\bar{1}] = [0] = 1 \qquad [\bar{1}]^2 = [1]^2$$

Here we use the relation $\bar{\gamma}_i = \gamma_i*[-c]$, $\delta_i*[-c]$ in

$$H_*\Gamma B\bar{\mathcal{O}}(\mathbb{F}_q) = H_* B\bar{\mathcal{O}}(\mathbb{F}_q)[\tilde{\bar{N}}]^{-1}$$

cf. II 2.17 and II 4.7. Let $\mathfrak{A}$ denote the abstract algebra

$$\mathfrak{A} = \mathbb{Z}/\ell[[1], [\bar{1}], \gamma_1, \gamma_2, \ldots] \otimes E[\delta_1, \delta_2, \ldots]/[1]^2 = [\bar{1}]^2$$

Let $\pi\colon \mathfrak{A} \to H_* B\bar{\mathcal{O}}(\mathbb{F}_q)$ denote the projection map. By Theorem VIII 4.5 π is surjective. But by the above the composite

$$\mathfrak{a} \xrightarrow{\pi} H_*B\bar{\mathcal{O}}(\mathbb{F}_q) \to H_*\Gamma B\bar{\mathcal{O}}(\mathbb{F}_q)$$

is injective. Hence π is an isomorphism.

The proof of case (b) is identical.

§3. $H_*B\mathcal{Sp}(\mathbb{F}_q)$ and $H_*\Gamma_0 B\mathcal{Sp}(\mathbb{F}_q)$

In this section we compute the mod-ℓ homology algebras $H_*B\mathcal{Sp}(\mathbb{F}_q)$ and $H_*\Gamma_0 B\mathcal{Sp}(\mathbb{F}_q)$ and establish that

$$\lambda\colon \Gamma_0 B\mathcal{Sp}(\mathbb{F}_q) \to JSp(q)$$

is an equivalence of infinite loop spaces at ℓ. In many cases our proofs will be so analogous to those of §2 that little or no comment need be made.

Throughout this section d will denote the minimal positive integer such that $q^{2d} \equiv 1 \pmod{\ell}$. Alternately d can be described as the minimal positive integer d such that ℓ divides the order of $Sp(2d,\mathbb{F}_q)$ (cf. VIII 5.2).

According to Theorem VIII 5.4

$$H_*BSp(2d,\mathbb{F}_q) = \begin{cases} \text{cyclic on generator } \sigma_i & \text{if } n = 4di \\ \text{cyclic on generator } \tau_i & \text{if } n = 4di - 1 \\ 0 & \text{otherwise} \end{cases}$$

Again at this point we do not yet know that $\sigma_i \neq 0$, $\tau_i \neq 0$. As in II 2.16(viii) we will denote by $[2n]$ the generator of $H_0BSp(2n,\mathbb{F}_q)$.

<u>Theorem 3.1.</u> $H_*B\mathcal{Sp}(\mathbb{F}_q) = \oplus_{n=0}^{\infty} H_*BSp(2n,\mathbb{F}_q)$ is a free commutative algebra generated by σ_i, τ_i $i > 0$ and $[2]$, i.e.

$$H_*B\mathcal{Sp}(\mathbb{F}_q) = \mathbb{Z}/\ell[[2],\sigma_1,\sigma_2,\ldots] \otimes E[\tau_1,\tau_2,\ldots]$$

(In particular $\sigma_i \neq 0$, $\tau_i \neq 0$ for all $i > 0$.)

Let $\bar{\sigma}_i, \bar{\tau}_i \in H_*\Gamma_0 B\mathcal{Sp}(\mathbb{F}_q)$ denote the images of σ_i, τ_i under the map

$$BSp(2d,\mathbb{F}_q) \to BSp(\infty,\mathbb{F}_q) \to \Gamma_0 B\mathcal{Sp}(\mathbb{F}_q)$$

<u>Theorem</u> 3.2. $H_*\Gamma_0 B\mathcal{Sp}(\mathbb{F}_q) = \mathbb{Z}/\ell[\bar{\sigma}_1,\bar{\sigma}_2,\ldots] \otimes E[\bar{\tau}_1,\bar{\tau}_2,\ldots]$

<u>Theorem</u> 3.3. Any H-map

$$\lambda\colon \Gamma_0 B\mathcal{Sp}(\mathbb{F}_q) \to JSp(q)$$

which completes diagram III 2.15 is an equivalence at ℓ. In particular there is an infinite loop equivalence λ at ℓ.

By Prop. I 6.1, $JSp(q)$ is equivalent away from 2 to $JO(q)$. Hence by Prop. I 5.2

$$H_*JSp(q) = \mathbb{Z}/\ell[s_1,s_2,\ldots] \otimes E[t_1,t_2,\ldots]$$

Now our proof of Theorems 3.1-3.3 proceeds as in §2 by showing that

$$\lambda_*(\bar{\sigma}_i) = s_i + \text{decomposables} \qquad \lambda_*(\bar{\tau}_i) = t_i + \text{decomposables}$$

Again we begin by analyzing the Brauer character of $Sp(2d,\mathbb{F}_q)$. According to VIII 5.3, the Sylow ℓ-subgroup of $Sp(2d,\mathbb{F}_q)$ is a cyclic subgroup $\mathbb{Z}/\ell^\nu$ where $\nu = \nu_\ell(q^{2d} - 1)$. Let E denote the natural inclusion

$$E\colon Sp(2d,\mathbb{F}_q) \to Sp(2d,\overline{\mathbb{F}}_q)$$

According to Chapter III §2, the Brauer character χ_E is the character of a unique virtual quaternionic representation of $Sp(2d,\mathbb{F}_q)$. Let P denote the quaternionic representation $P = \rho + \rho^2 + \cdots + \rho^{q^{d-1}}$ where $\rho\colon \mathbb{Z}/\ell^\nu \to S^1 \to S^3 = Sp(1)$ is the standard representation. Then we have

Proposition 3.4. When restricted to $\mathbb{Z}/\ell^\nu$, $\chi_E = \chi_P$.

Proof. Identical to the proof of Prop. 2.4.

Proposition 3.5. $\lambda_*(\overline{\sigma}_k) = s_k$.

Proof. Analogous to the proof of Prop. 2.5

Proposition 3.6. $\lambda_*(\overline{\tau}_k) = t_k$ + decomposables.

Proof. Analogous to the proof of Prop. 2.6.

The proofs of Theorems 3.1-3.3 now proceed in exactly the same way as those of Theorems 2.1-2.3.

§4. $H_*B\mathcal{O}\mathcal{B}\mathcal{L}(\mathbb{F}_q)$ q even.

In this section we exploit the results of §3 to compute the mod-ℓ homology algebra

$$H_*B\mathcal{O}\mathcal{B}\mathcal{L}(\mathbb{F}_q) = \oplus_{n=0}^{\infty} H_*BEO(n,\mathbb{F}_q) \oplus \oplus_{n=1}^{\infty} H_*BSp(2n,\mathbb{F}_q) \text{ for } q$$

even (cf. Def. II 7.5).

As in §3, d will denote the minimal positive integer d such that $q^{2d} \equiv 1 \pmod{\ell}$.

<u>Lemma</u> 4.1. The natural inclusions (cf. Chap II 7.2 and 7.3)

(a) $Sp(2m,\mathbb{F}_q) \to EO(2m+1,\mathbb{F}_q)$

(b) $Sp(2m,\mathbb{F}_q) \to EO(2m+2,\mathbb{F}_q)$

induce isomorphisms in mod-ℓ homology and cohomology.

<u>Proof</u>. According to Remark II 7.3 the inclusion (a) is actually an isomorphism of groups and consequently must induce an isomorphism in mod-ℓ homology and cohomology.

Now according to II 7.2 and II 6.3

$$|Sp(2m,\mathbb{F}_q)| = N_m \qquad |EO(2m+2,\mathbb{F}_q)| = q^{2m-1}N_m$$

where $N_m = \Pi_{i=1}^{m}(q^{2i}-1)q^{2i-1}$. Consequently $Sp(2m,\mathbb{F}_q)$ contains a Sylow ℓ-subgroup of $EO(2m+2,\mathbb{F}_q)$, and inclusion (b) induces an epimorphism in mod-ℓ homology.

Now consider the commutative diagram of inclusions

$$\begin{array}{ccc} Sp(2m,\mathbb{F}_q) & \to & Sp(2m+2,\mathbb{F}_q) \\ \downarrow & & \downarrow \\ EO(2m+2,\mathbb{F}_q) & \to & EO(2m+3,\mathbb{F}_q) \end{array}$$

The right hand arrow induces an isomorphism in mod-ℓ homology. By Theorem 3.1 the top arrow induces a monomorphism in mod-ℓ homology. Consequently the left hand arrow must be injective in mod-ℓ homology. It follows that (b) induces an isomorphism in mod-ℓ homology and hence also in mod-ℓ cohomology.

Combining Theorem 3.1 and Lemma 4.1 we get a complete description of the homology algebra $H_*B\mathcal{OBL}(\mathbb{F}_q)$. In what follows we denote by [2n] the generator of $H_0BSp(2n,\mathbb{F}_q)$ and by $[\bar{n}]$ the generator of

$H_0BEO(n,\mathbb{F}_q)$. The elements $\sigma_i, \tau_i \in H_*BSp(2d,\mathbb{F}_q)$ will denote the generators of Theorem 3.1.

Theorem 4.2. $H_*B\mathcal{OBL}(\mathbb{F}_q) = \oplus_{n=0}^{\infty} H_*BEO(n,\mathbb{F}_q) \oplus \oplus_{n=1}^{\infty} H_*BSp(2n,\mathbb{F}_q)$ is a commutative algebra generated by σ_i, τ_i $i > 0$, $[2]$ and $[\tilde{1}]$ subject to the single relation $[2]*[\tilde{1}] = [\tilde{1}]^3$, i.e.:

$$H_*B\mathcal{OBL}(\mathbb{F}_q) = \mathbb{Z}/\ell[[\tilde{1}],[2],\sigma_1,\sigma_2,\ldots] \otimes E\{\tau_1,\tau_2,\ldots\}/[2]*[\tilde{1}] = [\tilde{1}]^3$$

Proof: This is immediate from Theorem 3.1 and, Lemma 4.1. The relation $[2]*[\tilde{1}] = [\tilde{1}]^3 = [\tilde{3}]$ comes from the isomorphism

$$(\mathbb{F}_q^2,A) \oplus (\mathbb{F}_q,E) \cong (\mathbb{F}_q,E) \oplus (\mathbb{F}_q,E) \oplus (\mathbb{F}_q,E) \cong (\mathbb{F}_q^3,E)$$

cf. Chap. II 7.2.

§5. $H_*B\mathcal{GL}(\mathbb{F}_q)$ and $H_*\Gamma_0B\mathcal{GL}(\mathbb{F}_q)$; $H_*B\mathcal{U}(\mathbb{F}_{q^2})$ and $H_*\Gamma_0B\mathcal{U}(\mathbb{F}_{q^2})$

In this section we compute the mod-ℓ homology algebras listed in the title and establish the equivalences

$$\lambda\colon \Gamma_0B\mathcal{GL}(\mathbb{F}_q) \to JU(q)$$

$$\lambda\colon \Gamma_0B\mathcal{U}(\mathbb{F}_{q^2}) \to JU(-q)$$

of infinite loop spaces at ℓ. Since the two cases are so similar, we will treat them in tandem and prove results only for the unitary case (which is slightly more complicated), the proof for the general linear case being entirely analogous. The methods used in this section are slight variations of those used in §2 and §3.

Throughout this section $u(+)$, $u(-)$ will denote the minimal positive integers such that $q^{u(+)} \equiv 1 \pmod{\ell}$, $(-q)^{u(-)} \equiv 1 \pmod{\ell}$

respectively. Alternately u(+), u(-) may be described as the minimal positive integers such that ℓ divides the order of $GL(r,\mathbb{F}_q)$, $U(u,\mathbb{F}_{q^2})$ respectively. (cf. Lemmas VIII 6.2 and 7.2).

According to Theorem VIII 6.4 and 7.4

$$H_n BGL(u(+),\mathbb{F}_q) = \begin{cases} \text{cyclic on generator } \alpha_i \text{ if } n = 2iu(+) \\ \text{cyclic on generator } \beta_i \text{ if } n = 2iu(+) - 1 \\ 0 \quad \text{otherwise.} \end{cases}$$

$$H_n BU(u(-),\mathbb{F}_{q^2}) = \begin{cases} \text{cyclic on generator } \xi_i \text{ if } n = 2iu(-) \\ \text{cyclic on generator } \eta_i \text{ if } n = 2iu(-) - 1 \\ 0 \quad \text{otherwise.} \end{cases}$$

Again we do not yet know that $\alpha_i \neq 0$, $\beta_i \neq 0$, $\xi_i \neq 0$, $\eta_i \neq 0$. As in II 2.16(viii) we will denote by [n] the generator of $H_0 BGL(n,\mathbb{F}_q)$ or $H_0 BU(n,\mathbb{F}_{q^2})$.

<u>Theorem</u> 5.1 (a) $H_* B\mathcal{GL}(\mathbb{F}_q) = \oplus_{n=0}^{\infty} H_* BGL(n,\mathbb{F}_q)$ is a free commutative algebra generated by α_i, β_i $i > 0$ and [1], i.e.

$$H_* B\mathcal{GL}(\mathbb{F}_q) = \mathbb{Z}/\ell[[1],\alpha_1,\alpha_2,\ldots] \otimes E[\beta_1,\beta_2,\ldots]$$

(b) $H_* B\mathcal{U}(\mathbb{F}_{q^2}) = \oplus_{n=0}^{\infty} H_* BU(n,\mathbb{F}_{q^2})$ is a free commutative algebra generated by ξ_i, η_i $i > 0$ and [1], i.e.

$$H_* B\mathcal{U}(\mathbb{F}_{q^2}) = \mathbb{Z}/\ell[[1],\xi_1,\xi_2,\ldots] \otimes E[\eta_1,\eta_2,\ldots]$$

(In particular $\alpha_i \neq 0$, $\beta_i \neq 0$, $\xi_i \neq 0$, $\eta_i \neq 0$ for all $i > 0$)

Let $\bar{\alpha}_i, \bar{\beta}_i \in H_* \Gamma_0 B\mathcal{GL}(\mathbb{F}_q)$, resp. $\bar{\xi}_i, \bar{\eta}_i \in H_* \Gamma_0 B\mathcal{U}(\mathbb{F}_{q^2})$ denote the images of α_i, β_i resp. ξ_i, η_i under the maps

$$BGL(u(+),\mathbb{F}_q) \to BGL(\infty,\mathbb{F}_q) \to \Gamma_0 B\mathcal{GL}(\mathbb{F}_q)$$

$$BU(u(-),\mathbb{F}_{q^2}) \to BU(\infty,\mathbb{F}_{q^2}) \to \Gamma_0 B\mathcal{U}(\mathbb{F}_{q^2})$$

<u>Theorem</u> 5.2 (a) $H_*\Gamma_0 B\mathcal{GL}(\mathbb{F}_q) = \mathbb{Z}/\ell[\bar{\alpha}_1,\bar{\alpha}_2,\ldots] \otimes E[\bar{\beta}_1,\bar{\beta}_2,\ldots]$

(b) $H_*\Gamma_0 B\mathcal{U}(\mathbb{F}_{q^2}) = \mathbb{Z}/\ell[\bar{\xi}_1,\bar{\xi}_2,\ldots] \otimes E[\bar{\eta}_1,\bar{\eta}_2,\ldots]$

<u>Theorem</u> 5.3. Any H-map

$$\lambda\colon \Gamma_0 B\mathcal{GL}(\mathbb{F}_q) \to JU(q)$$

resp.

$$\lambda\colon \Gamma_0 B\mathcal{U}(\mathbb{F}_{q^2}) \to JU(-q)$$

which completes diagram III 2.13 (resp. III 2.19) is an equivalence at ℓ. In particular there are infinite loop equivalences λ at ℓ.

By Prop. I 6.3

$$H_*JU(\pm q) = \mathbb{Z}/\ell[\hat{a}_1,\hat{a}_2,\ldots] \otimes E[\hat{b}_1,\hat{b}_2,\ldots]$$

Now our proof of Theorems 5.1-5.3 proceeds as in §2 by showing that

$$\lambda_*(\bar{\alpha}_i) = \hat{a}_i + \text{decomposables}, \quad \lambda_*(\bar{\beta}_i) = \hat{b}_i + \text{decomposables}$$

$$\lambda_*(\bar{\xi}_i) = \hat{a}_i + \text{decomposables}, \quad \lambda_*(\bar{\eta}_i) = \hat{b}_i + \text{decomposables}$$

Again we begin by analyzing the Brauer character of $GL(u(+),\mathbb{F}_q)$, respectively $U(u(-),\mathbb{F}_{q^2})$. According to VIII 6.4 and 7.4, the Sylow ℓ-subgroup of both $GL(u(+),\mathbb{F}_q)$ and $U(u(-),\mathbb{F}_{q^2})$ is a cyclic subgroup

$\mathbb{Z}/\ell^\nu$ where

$$\nu = \nu_\ell(q^{u(+)}-1) = \nu_\ell((-q)^{u(-)}-1) \text{ (cf. Lemma VIII 2.7)}$$

Let E,F denote the natural inclusions

$$E\colon GL(u(+),\mathbb{F}_q) \to GL(u(+),\overline{\mathbb{F}}_q)$$

$$F\colon U(u(-),\mathbb{F}_{q^2}) \to GL(u(-),\overline{\mathbb{F}}_q)$$

According to Chapter III §2, the Brauer characters χ_E, χ_F are the characters of unique complex representations of $GL(u(+),\mathbb{F}_q)$, resp. $U(u(-),\mathbb{F}_{q^2})$. Let

$$P = \rho + \rho^q + \cdots + \rho^{q^{u-1}}$$

$$Q = \rho + \rho^{q^2} + \cdots + \rho^{q^{2(u-1)}}$$

where $\rho\colon \mathbb{Z}/\ell^\nu \to S^1 = U(1)$ is the standard representation and $u = u(\pm)$ depending on the case. Then we have

Proposition 5.4. When restricted to $\mathbb{Z}/\ell^\nu$:

(a) $\chi_E = \chi_P$

(b) $\chi_F = \begin{cases} \chi_P & \text{if } u(-) \text{ is even} \\ \chi_Q & \text{if } u(-) \text{ is odd} \end{cases}$

Proof: We prove (b). If u(-) is odd the inclusion

$$\mathbb{Z}/\ell^\nu \to U(u(-),\mathbb{F}_{q^2})$$

is given by considering $\mathbb{Z}/\ell^{\nu}$ as a subgroup of $U(1,\mathbb{F}_{q^{2u(-)}})$ and then viewing $\mathbb{F}_{q^{2u(-)}}$ as a vector space over $\mathbb{F}_{q^2}$ (cf. VIII 7.4(a)). Arguing as in the proof of Prop. 2.4 we have

$$\chi_F(z) = z + z^{q^2} + \cdots + z^{q^{2(u(-)-1)}} \qquad z \in \mathbb{Z}/\ell^{\nu} \subseteq S^1$$

Thus $\chi_F = \chi_Q$ on $\mathbb{Z}/\ell^{\nu}$.

If $u(-)$ is even the inclusion

$$\mathbb{Z}/\ell^{\nu} \to U(u(-),\mathbb{F}_{q^2})$$

is given by considering $\mathbb{Z}/\ell^{\nu}$ as a subgroup of $Sp(u(-),\mathbb{F}_q)$ as in Prop. 3.4 and then extending scalars to $\mathbb{F}_{q^2}$ (cf. VIII 7.4(b)). It now immediately follows from Prop. 3.4 that $\chi_E = \chi_P$ on $\mathbb{Z}/\ell^{\nu}$.

The proof for case (a) is similar to the proof of case (b) $u(-)$ odd.

<u>Lemma</u> 5.5 (a) $\lambda_*(\bar{\alpha}_i) = \hat{a}_i$.

(b) $\lambda_*(\bar{\xi}_i) = \hat{a}_i$.

<u>Proof</u>. Analogous to that of Lemma 2.5

<u>Lemma</u> 5.6 (a) $\lambda_*(\bar{\beta}_i) = \hat{b}_i$ + decomposables

(b) $\lambda_*(\bar{\eta}_i) = \hat{b}_i$ + decomposables.

<u>Proof</u>. Analogous to that of Lemma 2.6.

The proofs of Theorems 5.1-5.3 now proceed in exactly the same way as those of Theorems 2.1-2.3.

§6. The cohomology rings $H^*BO(n,\mathbb{F}_q)$, $H^*BO(\overline{n},\mathbb{F}_q)$, $H^*BSp(2n,\mathbb{F}_q)$, $H^*BGL(n,\mathbb{F}_q)$, $H^*BU(n,\mathbb{F}_{q^2})$.

In this section we dualize the results on mod ℓ homology that we obtained in §2-5, to obtain a complete description of the mod-ℓ cohomology rings listed in the title.

Let a be the smallest index for which ℓ divides the order of $G(a)$: for the groups in question this is $O(c,\mathbb{F}_q)$, $Sp(2d,\mathbb{F}_q)$, $GL(u(+),\mathbb{F}_q)$ $U(u(-),\mathbb{F}_{q^2})$ respectively. As shown in Chapter VIII the Sylow ℓ-subgroup of $G(a)$ is a cyclic group $\mathbb{Z}/\ell^\nu$. Hence

$$H_*B\mathbb{Z}/\ell^\nu \to H\,BG(a)$$

is surjective. Moreover

$$H_nBG(a) = \begin{cases} \mathbb{Z}/\ell \text{ on generator } \alpha_i \text{ if } n = 2ai \\ \mathbb{Z}/\ell \text{ on generator } \beta_i \text{ if } n = 2ai - 1 \\ 0 \quad \text{otherwise} \end{cases}$$

with the generators α_i, β_i coming from $H_*B\mathbb{Z}/\ell^\nu$. Dualizing these results we obtain that

$$H^*BG(a) \to H^*B\mathbb{Z}/\ell^\nu$$

is injective. By Chapter VI 2.1 we have $H^*B\mathbb{Z}/\ell^\nu = \mathbb{Z}/\ell[y] \otimes E[x]$ where $\deg x = 1$, $\deg y = 2$. Hence $H^*BG(a) = \mathbb{Z}/\ell[w] \otimes E[v]$ where $w \to y^{ai} \in H^{2ai}B\mathbb{Z}/\ell^\nu$ and $v \to xy^{ai-1} \in H^{2ai-1}B\mathbb{Z}/\ell^\nu$.

Let n be a fixed positive integer (or fixed element in $\tilde{\mathbb{N}}$ in the case of the orthogonal groups). Let m denote the maximal positive integer such that $ma \leq n$. Then according to Chapter VIII 4.1, 5.1, 6.1 and 7.1 the direct sum homomorphism

$$G(a) \times G(a) \times \cdots \times G(a) \xrightarrow{\oplus} G(ma) \to G(n)$$

induces epimorphisms in homology and hence monomorphisms in cohomology. Thus $H^*BG(n)$ may be regarded as a subring of

$$H^*(BG(a))^m = \mathbb{Z}/\ell[w_1,w_2,\ldots,w_m] \otimes E[v_1,v_2,\ldots,v_m]$$

<u>Theorem</u> 6.1. $H^*BG(n) = \mathbb{Z}/\ell[s_1,s_2,\ldots,s_m] \otimes E[e_1,e_2,\ldots,e_m]$ where

$$s_j = \Sigma_{i_1<i_2<\cdots<i_j} w_{i_1}w_{i_2}\cdots w_{i_j} \in H^{2aj}BG(n) \subseteq H^{2aj}(BG(a)^m)$$

$$e_j = \Sigma_{\substack{i_1<i_2<\cdots<i_j \\ 1\leq k\leq j}} w_{i_1}\cdots\hat{w}_{i_k}\cdots w_{i_j}v_{i_k} \in H^{2aj-1}BG(n) \subseteq H^{2aj-1}(BG(a)^m)$$

The mod-ℓ cohomology ring $H^*BEO(n,\mathbb{F}_q)$ can be computed from the cohomology ring $H^*BSp(2n,\mathbb{F}_q)$ using Lemma 4.1.

<u>Proof</u>. It follows from Theorems 2.1, 3.1, 4.2 that $H_*BG(n)$ is obtained from $H_*(BG(a)^m) = H_*BG(a) \otimes \cdots \otimes H_*BG(a)$ by dividing out by the action of the symmetric group $\mathcal{S}_m$ acting on $H_*(BG(a)^m)$ by permuting the factors.

Dualizing we obtain that $H^*BG(n)$ is the subring of invariants of $H^*(BG(a)^m)$ under the action of $\mathcal{S}_m$ acting by permuting factors. Since the elements s_i,e_i $1 \leq i \leq m$ are clearly invariant under the action of $\mathcal{S}_m$, it follows that $s_i,e_i \in H^*BG(n)$. According to [35, Lemma 9], the elements s_i,e_i $1 \leq i \leq m$ are algebraically independent. Hence

$$\mathcal{R}_m = \mathbb{Z}/\ell[s_1,s_2,\ldots,s_m] \otimes E[e_1,e_2,\ldots,e_m] \subseteq H^*BG(n)$$

for all $1 \leq n \leq \infty$. By Theorems 2.2, 3.2, 4.2 we have

$$H_*BG(\infty) \cong H_*\Gamma_0 B\mathcal{A} = \mathbb{Z}/\ell[\bar{\alpha}_1, \bar{\alpha}_2, \ldots] \otimes E[\bar{\beta}_1, \bar{\beta}_2, \ldots]$$

where deg $\bar{\alpha}_i$ = deg s_i and deg $\bar{\beta}_i$ = deg e_i. Comparing Poincaré series we see that $H^*BG(\infty) = \mathcal{R}_\infty$. Arguing as in Theorem IV 4.1 we obtain that $H^*BG(n) = \mathcal{R}_m$.

Chapter VI

The Homology of Certain Finite Groups

§1. Introduction

In this chapter we determine generators and relations for the homology of the classical groups G(k), k small. These results provide the necessary input for our main calculations in Chapters IV and V. The calculations proceed by first analyzing Sylow subgroups. Except in §2, all homology and cohomology groups are taken with coefficients in $\mathbb{Z}/2$.

In more detail: In §2 we recall standard facts about the mod-ℓ cohomology of cyclic groups and prove two lemmas about the action of the Steenrod algebra. The dihedral groups are treated in §3 and are used in §4 to study the orthogonal groups. §5 is devoted to $SL(2,\mathbb{F}_q) = Sp(2,\mathbb{F}_q)$, a relation in $H_*BSp(4,\mathbb{F}_q)$ and a lemma about Steenrod operations. Finally in §6 we consider $GL(2,\mathbb{F}_q)$ and the closely related group $U(2,\mathbb{F}_q)$.

§2. Cyclic groups

In this section we recall some elementary results about the (co-) homology of cyclic groups for use throughout the paper. Let $C = C_r$ be a cyclic group of order r. Let ℓ be a prime dividing r. The following result is well known [12].

Proposition 2.1. The cohomology of C is given by

(a) $$H^*(BC;\ \mathbb{Z}/\ell) = \begin{cases} \mathbb{Z}/\ell[y] \otimes E[x] & \text{if } \ell > 2 \text{ or } \ell = 2 \text{ and } r \equiv 0 \pmod 4 \\ \mathbb{Z}/\ell[y,x]/(y-x^2) \simeq \mathbb{Z}/\ell[x] & \text{if } \ell = 2 \text{ and } r \equiv 2 \pmod 4 \end{cases}$$

where deg x = 1 and deg y = 2.

(b) $H^*(BC;\ \mathbb{Z}) = C[y]$

where deg y = 2.

(c) The multiplication map $\mu: C \times C \to C$ induces on mod-ℓ cohomology the map given by

$$\mu^*(y^n) = \Sigma_{i=0}^{n}(i,n-i)y^i \otimes y^{n-i}$$

$$\mu^*(xy^n) = \Sigma_{i=0}^{n}(i,n-i)[xy^i \otimes y^{n-i} + y^i \otimes xy^{n-i}]$$

Bockstein operations and Steenrod operations are very useful in studying homomorphisms involving the (co-) homology of cyclic groups. In the following result the statement about the Bockstein operations follows from Prop. 2.1(b) (cf. also III 6.3) while the statement about the Steenrod operations may be found in [39].

Proposition 2.2 (a) The differentials in the mod-ℓ Bockstein cohomology spectral sequence of C are given by

$$d_k^*(y^m) = 0 \text{ for all } k \text{ and all } m$$
$$d_k^*(xy^m) = 0 \text{ for all } m \text{ and all } k < \nu$$
$$d_\nu^*(xy^m) = y^{m+1} \text{ for all } m$$

where $\nu = \nu_\ell(r)$ is the largest positive integer for which ℓ^ν divides r.

(b) If $\ell > 2$, then the Steenrod operations in $H^*(BC; \mathbb{Z}/\ell)$ are given by

$$P^i(y^k) = (i,k-i)y^{k+i(\ell-1)}$$

$$P^i(xy^k) = (i,k-i)xy^{k+i(\ell-1)}$$

for all $i,k \geq 0$.

(c) If $\ell = 2$ and $r \equiv 2 \pmod 4$ then the Steenrod operations in $H^*(BC; \mathbb{Z}/2)$ are given by

$$Sq^i(x^k) = (i,k-i)x^{k+i}$$

for all $i,k \geq 0$.

(d) If $\ell = 2$ and $r \equiv 0 \pmod 4$ then the Steenrod operations in $H^*(BC; \mathbb{Z}/2)$ are given by

$$\begin{aligned} Sq^{2i+1}(y^k) &= Sq^{2i+1}(xy^k) = 0 \\ Sq^{2i}(y^k) &= (i,k-i)y^{k+i} \\ Sq^{2i}(xy^k) &= (i,k-i)xy^{k+i} \end{aligned}$$

for all $i,k \geq 0$.

Dualizing we obtain corresponding results in homology (cf. also [12]).

<u>Proposition</u> 2.3. The homology of C is given by

(a) $H_n(BC; \mathbb{Z}/\ell) = \mathbb{Z}/\ell$ on generator x_n for all $n \geq 0$ where x_n is given by the dual pairing $\langle x_n, y^{k-1}x\rangle = 1$ if $n = 2k-1$ and by $\langle x_n, y^k\rangle = 1$ if $n = 2k$.

(b) If $\ell = 2$ and $r = 2 \pmod 4$ then the coproduct

$$\Delta_*: H_*BC \to H_*BC \otimes H_*BC$$

induced by the diagonal $\Delta: C \to C \times C$ is given by

$$\Delta_*(x_n) = \Sigma_{i=0}^n x_i \otimes x_{n-i}$$

(c) If $\ell > 2$ or $\ell = 2$ and $r \equiv 0 \pmod 4$ then the coproduct in H_*BC is given by

$$\Delta_*(x_{2n-1}) = \Sigma_{i=0}^{2n-1} x_i \otimes x_{2n-1-i}$$

$$\Delta_*(x_{2n}) = \Sigma_{i=0}^{n} x_{2i} \otimes x_{2n-2i}$$

(d) The integral homology of C is given by

$$\tilde{H}_i(BC;\ \mathbb{Z}) = \begin{cases} 0 & \text{if } i \text{ is even} \\ C & \text{if } i \text{ is odd} \end{cases}$$

(e) The multiplication map μ: C x C → C induces on mod-ℓ homology the map given by

$$\mu_*(x_n \otimes x_m) = (n,m)x_{n+m}$$

Proposition 2.4 (a) The differentials in the mod-ℓ Bockstein homology spectral sequence of C are given by

$$\begin{aligned} d_k(x_{2m-1}) &= 0 \quad \text{for all } k \text{ and all } m \geq 1 \\ d_k(x_{2m}) &= 0 \quad \text{for all } m \geq 0 \text{ and all } k < \nu \\ d_\nu(x_{2m}) &= x_{2m-1} \quad \text{for all } m \geq 1 \end{aligned}$$

where $\nu = \nu_\ell(r)$ is the largest positive integer for which ℓ^ν divides r.

(b) If $\ell > 2$, then the dual Steenrod operations in $H_*(BC;\ \mathbb{Z}/\ell)$ are given by

$$P^i_*(x_{2k}) = (i,k-\ell i)x_{2(k-(\ell-1)i)}$$

$$P^i_*(x_{2k-1}) = (i,k-1-\ell i)x_{2(k-(\ell-1)i)-1}$$

for all $i,k \geq 0$.

(c) If $\ell = 2$ and $r \equiv 2 \pmod 4$ then the dual Steenrod operations in $H_*(BC;\ \mathbb{Z}/2)$ are given by

$$Sq^i_*(x_k) = (i,k-2i)x_{k-i}$$

for all $i,k \geq 0$.

(d) If $\ell = 2$ and $r \equiv 0 \pmod 4$ then the dual Steenrod operations in $H_*(BC;\ \mathbb{Z}/2)$ are given by

$$Sq_*^{2i+1}(x_k) = 0$$

$$Sq_*^{2i}(x_{2k}) = (i,k-2i)x_{2(k-i)}$$

$$Sq_*^{2i}(x_{2k-1}) = (i,k-1-2i)x_{2(k-i)-1}$$

for all $i,k \geq 0$.

The following result is due to J. P. May and the second author. We use the notation of Def. III 6.6.

Proposition 2.5. Let $\ell > 2$. Let α be one of the integers $1,2,\ldots,\ell-2,\ell-1$. Let $\overline{M}_*^\alpha$ denote the $\mathbb{Z}/\ell$-submodule of $H_*(BC;\ \mathbb{Z}/\ell)$ generated by $\{x_{2i-1} \mid i \equiv \alpha \pmod{\ell-1}\}$ and let $\overline{N}_*^\alpha$ denote the $\mathbb{Z}/\ell$-submodule generated by $x_{2\alpha-1}$, if $\alpha \neq 1$ and the $\mathbb{Z}/\ell$-submodule generated by x_1 and x_ℓ if $\alpha = 1$. Then $\overline{M}_*^\alpha$ and $\overline{N}_*^\alpha$ are $\mathcal{A}_*$-submodules of $H_*(BC;\ \mathbb{Z}/\ell)$ and $CL(\overline{N}_*^\alpha) = \overline{M}_*^\alpha$.

Proof. The fact that $\overline{M}_*^\alpha$, $\overline{N}_*^\alpha$ are $\mathcal{A}_*$-submodules of $H_*(BC;\ \mathbb{Z}/\ell)$ is immediate from Proposition 2.4(b). As an induction assumption suppose $x_{2j-1} \in CL(\overline{N}_*^\alpha)$ whenever $j < i$ and $j \equiv \alpha \pmod{\ell-1}$.

If $i \not\equiv 0 \pmod{\ell}$ then

$$P_*^1(x_{2i-1}) = (1,i-1-\ell)x_{2(i-\ell+1)-1} = ix_{2(i-\ell+1)} \neq 0$$

Hence x_{2i-1} is Steenrod related to $CL(\overline{N}_*^\alpha)$ and hence $x_{2i-1} \in CL(\overline{N}_*^\alpha)$.

Now suppose $i = u\ell^k$ where $u \not\equiv 0 \pmod{\ell}$. If $k \geq 2$, then

$$P_*^\ell(x_{2(i+\ell-1)-1}) = (\ell,i-\ell^2+\ell-2)x_{2(i-(\ell-1)^2)-1} \neq 0$$

so $x_{2(i+\ell-1)-1}$ is Steenrod related to $CL(\overline{N}_*^{\alpha})$ and hence $x_{2(i+\ell-1)-1} \in CL(\overline{N}_*^{\alpha})$. Since

$$P_*^1(x_{2(i+\ell-1)-1}) = (1,i-2)x_{2i-1} = -x_{2i-1}$$

it follows that $x_{2i-1} \in CL(\overline{N}_*^{\alpha})$.

If $i = u\ell$ where $u \not\equiv 0 \pmod{\ell}$ and $u \geq \ell$ then

$$P_*^{\ell}(x_{2i-1}) = (\ell, i-\ell^2-1)x_{2(i-\ell(\ell-1))-1} \neq 0$$

so x_{2i-1} is Steenrod related to $CL(\overline{N}_*^{\alpha})$ and hence $x_{2i-1} \in CL(\overline{N}_*^{\alpha})$.

If $i = u\ell$ where $1 < u \leq \ell - 1$, then

$$P_*^{\ell}(x_{2(i+(\ell-1)^2)-1}) = (\ell, i-2\ell)x_{2(i-\ell+1)-1} = (u-1)x_{2(i-\ell+1)-1} \neq 0$$

so $x_{2(i+(\ell-1)^2)-1}$ is Steenrod related to $CL(\overline{N}_*^{\alpha})$ and hence $x_{2(i+(\ell-1)^2)-1} \in CL(\overline{N}_*^{\alpha})$. Since

$$P_*^{\ell-1}(x_{2(i+(\ell-1)^2)-1}) = (\ell-1, i-\ell)x_{2i-1} \neq 0$$

it follows that $x_{2i-1} \in CL(\overline{N}_*^{\alpha})$.

The only remaining case is $i = \ell$. But $x_{2\ell-1} \in \overline{N}_*^1$ by hypothesis. This completes the induction and proof.

<u>Lemma</u> 2.6. Let $\ell = 2$. Let $\overline{M}_*$ denote the $\mathbb{Z}/2$-submodule of $H_*(BC; \mathbb{Z}/2)$ generated by $\{x_{2i-1} | i \geq 2\}$ and let $\overline{N}_*$ denote the $\mathbb{Z}/2$-submodule generated by x_3. Then $\overline{M}_*$ and $\overline{N}_*$ are $\mathcal{A}_*$-submodules of $H_*(BC; \mathbb{Z}/2)$ and $CL(\overline{N}_*) = \overline{M}_*$.

Proof. We observe that

$$Sq_*^{2n+1}(x_{2i-1}) = 0$$

$$Sq_*^{2n}(x_{2i-1}) = (n,i-1-2n)x_{2(i-n)-1}$$

This is immediate from Prop. 2.4(c) if $r \equiv 0 \pmod 4$. If $r \equiv 2 \pmod 4$, then according to Prop. 2.4(b)

$$Sq_*^{2n+1}(x_{2i-1}) = (2n+1,2i-4n-3)x_{2(i-n)-2} = 0$$

$$\begin{aligned} Sq_*^{2n}(x_{2i-1}) &= (2n,2i-4n-1)x_{2(i-n)-1} \\ &= [(2n,2i-4n-2) + (2n-1,2i-4n-1)]x_{2(i-n)-1} \\ &= (2n,2i-4n-2)x_{2(i-n)-1} \\ &= (n,i-2n-1)x_{2(i-n)-1} \end{aligned}$$

It follows that $\overline{M}_*$ and $\overline{N}_*$ are $\mathcal{A}_*$-submodules of $H_*(BC; \mathbb{Z}/2)$.

Now as an induction assumption, suppose that $x_{2j-1} \in CL(\overline{N}_*)$ for $1 < j < i$. If i is odd then

$$Sq_*^2(x_{2i-1}) = (1,i-1-2)x_{2i-3} = x_{2i-3}$$

so that x_{2i-1} is Steenrod related to $CL(\overline{N}_*)$ and hence $x_{2i-1} \in CL(\overline{N}_*)$.

If $i \equiv 0 \pmod 4$ say $i = 4k$ then

$$Sq_*^4(x_{2i+1}) = (2,i-4)x_{2i-3} = (2,4k-4)x_{2i-3} = (1,2k-2)x_{2i-3} = x_{2i-3}$$

so that x_{2i+1} is Steenrod related to $CL(\overline{N}_*)$ and hence $x_{2i+1} \in CL(\overline{N}_*)$. Consequently

$$Sq_*^2(x_{2i+1}) = (1,i-2)x_{2i-1} = x_{2i-1} \in CL(\overline{N}_*).$$

If $i \equiv 2 \pmod 4$ then $i = 4k + 2 \quad k \geq 0$ and

$$Sq^4_*(x_{2i-1}) = (2,i-5)x_{2i-5} = (2,4k-3)x_{2i-5} = x_{2i-5}$$

provided $k \geq 1$. Hence x_{2i-1} is Steenrod related to $CL(\overline{N}_*)$ and so $x_{2i-1} \in CL(\overline{N}_*)$.

The only remaining case is $i = 2$. But then $x_{2i-1} = x_3 \in \overline{N}_*$ by hypothesis. This concludes the induction and proof.

We conclude this section by discussing inclusions of cyclic groups and passing to the consequent direct limits. These results were used in Chapter III §7.

<u>Proposition</u> 2.7. Let A be a (necessarily cyclic) subgroup of C. Let $j: A \to C$ be the inclusion. Let us denote by $\{x_A, y_A\}$ (respectively x_i^A) the generators of $H^*(BA; \mathbb{Z}/\ell)$ (respectively $H_*(BA; \mathbb{Z}/\ell)$). Then

(a) If the index $[C,A]$ is relatively prime to ℓ then

$$j^*: H^*(BC; \mathbb{Z}/\ell) \to H^*(BA; \mathbb{Z}/\ell)$$

$$j_*: H_*(BA; \mathbb{Z}/\ell) \to H_*(BC; \mathbb{Z}/\ell)$$

are isomorphisms, so that

$$j^*(x) = x_A \quad j^*(y) = y_A$$

$$j_*(x_i^A) = x_i \text{ for all } i$$

(b) If ℓ divides $[C,A]$, then

$$j^*(x) = 0 \quad j^*(y) = y_A$$

$$j_*(x_i^A) = \begin{cases} x_i & \text{if } i \text{ is even} \\ 0 & \text{if } i \text{ is odd} \end{cases}$$

<u>Proof</u>: According to Lemma III 6.1, the composite inclusion

$$A \xrightarrow{j} C \longrightarrow S^1$$

induces isomorphisms in even dimensional homology groups. Consequently we must have $j_*(x_{2i}^A) = x_{2i}$ for all $i \geq 0$. Let us denote by η (respectively ν) the largest power ℓ^η (respectively ℓ^ν) which divides the order of A (respectively C). According to Prop. 2.4(a) we have

(a) in case (a) $\nu = \eta$ so that

$$j_*(x_{2i-1}^A) = j_* d_\eta(x_{2i}^A) = d_\nu j_*(x_{2i}^A) = d_\nu(x_{2i}) = x_{2i-1}$$

(b) in case (b) $\eta < \nu$ so that

$$j_*(x_{2i-1}^A) = j_* d_\eta(x_{2i}^A) = d_\eta j_*(x_{2i}^A) = d_\eta(x_{2i}) = 0$$

This proves the result in homology. The cohomological result follows by dualizing.

<u>Remark</u> 2.8. We shall denote by $\mathbb{Z}/\ell^\infty$ the direct limit $\lim_{n\to\infty} \mathbb{Z}/\ell^n$. Alternately $\mathbb{Z}/\ell^\infty$ may be described as

$$\mathbb{Z}/\ell^\infty = \mathbb{Z}[1/\ell]/\mathbb{Z} \quad \text{or} \quad \mathbb{Z}/\ell^\infty = \mathbb{Q}/\mathbb{Z}_{(\ell)}$$

The following result is now an immediate consequence of Prop. 2.7 and Lemma III 6.1.

Proposition 2.9 (a). $H^*(B\mathbb{Z}/\ell^\infty;\mathbb{Z}/\ell) = \mathbb{Z}/\ell\,[y]$ where degree $y = 2$

(b) $$H_i(B\mathbb{Z}/\ell^\infty;\ \mathbb{Z}/\ell) = \begin{cases} \mathbb{Z}/\ell \text{ on generator } x_i & \text{if } i \text{ is even} \\ 0 & \text{if } i \text{ is odd} \end{cases}$$

(c) If $i\colon \mathbb{Z}/\ell^\infty \to S^1$ is an embedding, then the induced maps in mod-ℓ homology and cohomology are isomorphisms.

§3. Dihedral groups D_r

The purpose of this section is to record the computation of the mod-2 cohomology of dihedral groups D_r (of order 2^{r+1}). These results were obtained by Quillen [34]. We present them here for completeness and because his sketch contains several crucial misprints.

Let $D = D_r$ denote the dihedral group of order 2^{r+1}, $r \geq 2$

$$D = \{s,b\colon s^{2^r} = 1,\ b^2 = 1,\ bsb = s^{-1}\}$$

and let $\sigma\colon D \to O(2,\mathbb{R})$ be the standard representation of D on the plane where D acts via reflection and rotation, i.e.

$$\sigma(s) = \begin{pmatrix} \cos\theta & -\sin\theta \\ \sin\theta & \cos\theta \end{pmatrix} \qquad \theta = 2\pi/2^r$$

$$\sigma(b) = \begin{pmatrix} 0 & 1 \\ 1 & 0 \end{pmatrix}$$

Now define $x,y \in H^1BD$ by $\langle x,s\rangle = 1 = \langle y,b\rangle$; $\langle x,b\rangle = 0 = \langle y,s\rangle$ and let $w \in H^2BD$ denote the second Stiefel-Whitney class of σ (i.e. $w = \sigma^*(w_2)$ where $w_2 \in H^2BO(2,\mathbb{R})$ is universal Stiefel-Whitney class).

Proposition 3.1. $H^*BD = \mathbb{Z}/2[x,y,w]/(x^2 + xy)$.

Proof: Let A be the normal subgroup of D generated by s^2 so that

$A \approx \mathbb{Z}/2^{r-1}$. Consider the Serre-Hochschild spectral sequence of the extension

$$A \to D \to D/A$$

Since A is cyclic, D/A acts trivially on H^*BA and we have

$$E_2 = H^*B(D/A) \otimes H^*BA \Longrightarrow H^*BD.$$

Let $H^*BA = \mathbb{Z}/2[u] \otimes E[v]$ where $\dim u = 2$, $\dim v = 1$. Since $D/A \cong \mathbb{Z}/2 \times \mathbb{Z}/2$ generated by s and b, we have $H^*B(D/A) = \mathbb{Z}/2[x,y]$. We shall show that $d_2 v = x^2 + xy$, $d_3 u = 0$. Thus $E_3 = E_\infty$ and the result follows. To determine $d_2 v$ one either computes directly with the bar resolution or more indirectly by using the induced extensions obtained by the three non-trivial homomorphisms $\mathbb{Z}/2 \to D/A$ and the naturality of transgression. To show $d_3 u = 0$ it suffices to show that ω pulls back to a non-zero class in H^*BA. In fact more is true: let $A' \xrightarrow{i} D$ be the group of order 2 generated by $s^{2^{r-1}}$ then $H^*BA' = \mathbb{Z}/2[z]$. Now $\sigma(s^{2^{r-1}}) = -I_2$ and so

$$\text{(3.2)} \qquad i^*\sigma^*(\omega_2) = z^2.$$

This completes the proof.

Recall that a 2-primary abelian group is said to be elementary if each nontrivial element has order 2.

<u>Proposition</u> 3.3. The elementary abelian 2-subgroups of D detect H^*BD. In fact, $E_1 = \{b, s^{2^{r-1}}\}$ and $E_2 = \{sb, s^{2^{r-1}}\}$ detect.

<u>Proof:</u> $H^*BE_i = \mathbb{Z}_2[\alpha_i, \beta_i]$ for $i = 1,2$, where $\langle \alpha_i, s^{2^{r-1}} \rangle = \langle \beta_1, b \rangle =$

$\langle \beta_2, sb\rangle = 1$ and $\langle \alpha_1, b\rangle = \langle \beta_1, s^{2^{r-1}}\rangle = \langle \alpha_2, sb\rangle = 0$. Let $j_i\colon E_i \to D$ denote inclusion and let j^* denote $j_1^* \times j_2^*$. Then clearly

$$j^*(x) = (0, \beta_2)$$

$$j^*(y) = (\beta_1, \beta_2)$$

Further 3.2 implies

$$j^*(\omega) = (\alpha_1^2, \alpha_2^2).$$

Thus
$$j^*(x^m\omega^n) = (0, \beta_2^m\alpha_2^{2n})$$

$$j^*(y^k\omega^\ell) = (\beta_1^k\alpha_1^{2\ell}, \beta_2^k\alpha_2^{2\ell})$$

and so j^* is injective as required.

§4. $O(2,\mathbb{F}_q)$.

In this section we compute the mod-2 homology $H_*BO(2,\mathbb{F}_q)$ and derive relations in $H_*BO(2,\mathbb{F}_q)$. As a first step we show that

Lemma 4.1. The group $SO(2,\mathbb{F}_q)$ is cyclic of order $q-\epsilon$ where $\epsilon = \pm 1$ according as $q \equiv \pm 1 \pmod 4$.

Proof: The group $SO(2,\mathbb{F}_q)$ is the kernel of the determinant map det: $O(2,\mathbb{F}_q) \to \mathbb{Z}/2$. Since according to II 3.1

$$|O(2,\mathbb{F}_q)| = 2(q-\epsilon)$$

it follows that the order of $SO(2,\mathbb{F}_q)$ is $q-\epsilon$.

Now suppose

$$\begin{pmatrix} x & z \\ y & w \end{pmatrix} \in SO(2,\mathbb{F}_q)$$

Then the equations

$$xz + yz = 0, \quad xw-yz = 1, \quad x^2 + y^2 = 1$$

imply that $w = x$ and $z = -y$. In other words

$$SO(2,\mathbb{F}_q) = \{\begin{pmatrix} x & -y \\ y & x \end{pmatrix} | x^2 + y^2 = 1\}$$

<u>Case 1</u>. $q \equiv 1 \pmod 4$. (and hence $\varepsilon = 1$).

Then $\mathbb{F}_q$ contains a square root of -1; call it δ. Then $x^2 + y^2 = (x + y\delta)(x - y\delta)$ and

$$\begin{pmatrix} x & -y \\ y & x \end{pmatrix} \to x + y\delta$$

is easily seen to define a monomorphism $SO(2,\mathbb{F}_q) \to \mathbb{F}_q^* = \mathbb{Z}/q-1$. Comparing orders we see that $SO(2,\mathbb{F}_q) \cong \mathbb{F}_q^* = \mathbb{Z}/q-1$.

<u>Case 2</u>. $q \equiv -1 \pmod 4$ (and hence $\varepsilon = -1$)

In this case $\mathbb{F}_q$ does not contain a square root of -1. We then consider the field extension $\mathbb{F}_{q^2} = \mathbb{F}_q[\delta]$ where $\delta^2 = -1$: Again we see that $x^2 + y^2 = (x + y\delta)(x - y\delta) = N(x + y\delta)$ where $N\colon \mathbb{F}_{q^2}^* \to \mathbb{F}_q^*$ denotes the norm homomorphism. Also

$$\begin{pmatrix} x & -y \\ y & x \end{pmatrix} \to x + y\delta$$

defines a monomorphism $SO(2,\mathbb{F}_q) \to \ker(N\colon \mathbb{F}_{q^2}^* \to \mathbb{F}_q^*) \cong \mathbb{Z}/q + 1$. Comparing orders we see that $SO(2,\mathbb{F}_q) \cong \mathbb{Z}/q + 1$. This completes the proof.

Definition 4.2. Since $O(1,\mathbb{F}_q) \cong O(\overline{1},\mathbb{F}_q) \cong \mathbb{Z}/2$

$$H_i BO(1,\mathbb{F}_q) \cong H_i BO(\overline{1},\mathbb{F}_q) \cong \mathbb{Z}/2$$

for all $i \geq 0$. Define $v_i \in H_i BO(1,\mathbb{F}_q)$, $y_i \in H_i BO(\overline{1},\mathbb{F}_q)$ to be the generators.

Proposition 4.3(a). The Sylow 2-subgroup of $O(2,\mathbb{F}_q)$ is the dihedral group D_ν where $\nu = \nu(q-\epsilon)$ is the largest integer for which 2^ν divides $q-\epsilon$ (where again $\epsilon = \pm 1$ according as $q \equiv \pm 1 \pmod 4$.)

(b) The elements $\{v_i v_j, y_i y_j\} i,j \geq 0$ span $H_* BO(2,\mathbb{F}_q)$.

Proof: Let s be a generator of the Sylow 2-subgroup of $SO(2,\mathbb{F}_q) = \mathbb{Z}/q-\epsilon$. Then it follows that

$$D_\nu = \{s,b \mid b = \begin{pmatrix} 0 & 1 \\ 1 & 0 \end{pmatrix},\ s^{2^\nu} = 1 = b^2,\ bsb = s^{-1}\}$$

is a Sylow 2-subgroup of $O(2,\mathbb{F}_q)$. This proves (a).

Now according to Prop. 3.3 the elementary abelian 2-subgroups of D_ν detect $H_* BD_\nu$. It follows that the elementary abelian 2-subgroups of $O(2,\mathbb{F}_q)$ detect $H_* BO(2,\mathbb{F}_q)$. But according to Prop. VII 2.5 any elementary abelian 2-subgroup of $O(2,\mathbb{F}_q)$ is conjugate to a subgroup either of $T_{2,0} = O(1,\mathbb{F}_q) \times O(1,\mathbb{F}_q)$ or of $T_{0,2} = O(\overline{1},\mathbb{F}_q) \times O(\overline{1},\mathbb{F}_q)$. Hence the subgroups $T_{2,0}$ and $T_{0,2}$ detect $H_* BO(2,\mathbb{F}_q)$ and

$$H_* BT_{2,0} \oplus H_* BT_{0,2} \to H_* BO(2,\mathbb{F}_q)$$

is surjective. This implies part (b).

We now establish an important relation in $H_* BO(2,\mathbb{F}_q)$.

<u>Lemma</u> 4.4. For all $i \geq 0$: $v_i^2 = y_i^2$

<u>Proof</u>: We have a commutative diagram

$$\begin{array}{ccccc}
 & \overset{\Delta}{\nearrow} & \mathbb{Z}/2 \times \mathbb{Z}/2 = O(1,\mathbb{F}_q) \times O(1,\mathbb{F}_q) & \searrow & \\
\mathbb{Z}/2 & \longrightarrow & \mathbb{Z}/q-\varepsilon = SO(2,\mathbb{F}_q) & \longrightarrow & O(2,\mathbb{F}_q) \\
 & \underset{\Delta}{\searrow} & \mathbb{Z}/2 \times \mathbb{Z}/2 = O(\bar{1},\mathbb{F}_q) \times O(\bar{1},\mathbb{F}_q) & \nearrow &
\end{array}$$

Applying $H_*B(\cdot)$ to this diagram and pursuing the generator $x_{2i} \in H_{2i}B\mathbb{Z}/2$ across the top we get v_i^2; pursuing along the bottom we get y_i^2. Consequently $v_i^2 = y_i^2$.

We conclude with some results on Bockstein operations in $H_*BO_{\pm}(1,\mathbb{F}_q)$ and $H_*BO(2,\mathbb{F}_q)$.

<u>Lemma</u> 4.5(a) $d_1(v_{2i}) = v_{2i-1}$, $d_1(v_{2i-1}) = 0$ $\quad i \geq 1$

(b) $d_1(y_{2i}) = y_{2i-1}$, $d_1(y_{2i-1}) = 0$ $\quad i \geq 1$

(c) $d_k(v_i^2) = 0$ $\quad 1 \leq k < \nu$

where $\nu = \nu(q-\varepsilon)$ is the largest integer for which 2^ν divides $q-\varepsilon$ (or equivalently $\nu = \nu(\frac{1}{2}(q^2-1))$ is the largest integer for which 2^ν divides $\frac{1}{2}(q^2-1)$.).

<u>Proof</u>: Since $O(1,\mathbb{F}_q) \cong O(\bar{1},\mathbb{F}_q) \cong \mathbb{Z}/2$, parts (a) and (b) follow immediately from Prop. 2.4(a).

To prove part (c) we observe that as in Lemma 4.4 there is a commutative diagram

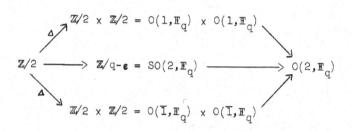

$$\begin{array}{ccccc} & & \mathbb{Z}/2 \times \mathbb{Z}/2 = O(1,\mathbb{F}_q) \times O(1,\mathbb{F}_q) & & \\ & \overset{\Delta}{\nearrow} & & \searrow & \\ \mathbb{Z}/2 & \xrightarrow{i} & \mathbb{Z}/q-\epsilon = SO(2,\mathbb{F}_q) & \xrightarrow{j} & O(2,\mathbb{F}_q) \end{array}$$

Applying $H_*B(\cdot)$ to the diagram we get

$$j_* i_*(x_{2n}) = v_n^2$$

But recording to Prop. 2.4(a) we have

$$d_k i_*(x_{2n}) = 0 \text{ for } 1 \leq k < \nu$$

Hence

$$d_k(v_n^2) = d_k j_* i_*(x_{2n}) = j_* d_k i_*(x_{2n}) = 0 \text{ for } 1 \leq k < \nu.$$

§5. $SL(2,\mathbb{F}_q)$ and $Sp(4,\mathbb{F}_q)$

In this section we compute the mod-2 cohomology ring of $SL(2,\mathbb{F}_q)$ (= $Sp(2,\mathbb{F}_q)$ by II 6.3) q odd and establish a certain relation in the mod-2 homology of $Sp(4,\mathbb{F}_q)$. The structure of $H^*BSL(2,\mathbb{F}_q)$ was first derived by Quillen [35]. We include these results for the sake of completeness and because our proof of the key fact that $H^2BSL(2,\mathbb{F}_q) = 0$ is more elementary.

As in the preceding section we make the convention that $\epsilon = \pm 1$ according as $q \equiv \pm 1 \pmod 4$.

Lemma 5.1. The Sylow 2-subgroup of $SL(2,\mathbb{F}_q)$ is isomorphic to the generalized quaternion group

$$Q_t = \{x,y \mid x^{2^{t-1}} = y^2,\ x^{2^t} = 1, yxy^{-1} = x^{-1}\}$$

where $t = \nu(q-\varepsilon)$ is the largest integer such that 2^t divides $q-\varepsilon$.

Proof: Consider the subgroup of $SL(2,\mathbb{F}_q)$

$$SO(2,\mathbb{F}_q) = \{\begin{pmatrix} u & -v \\ v & u \end{pmatrix} \mid u^2 + v^2 = 1\}$$

According to Lemma 4.1, $SO(2,\mathbb{F}_q) \cong \mathbb{Z}/(q-\varepsilon)$. Let x denote a generator of the Sylow 2-subgroup of $SO(2,\mathbb{F}_q)$. Then $x^{2^t} = 1$.

Now according to Lemma II 3.11 we can find $a,b \in \mathbb{F}_q$ such that $a^2 + b^2 = -1$. Then the element

$$y = \begin{pmatrix} -a & b \\ b & a \end{pmatrix}$$

lies in $SL(2,\mathbb{F}_q)$ and

$$y^2 = x^{2^{t-1}} = \begin{pmatrix} -1 & 0 \\ 0 & -1 \end{pmatrix}$$

Moreover it is easily checked that

$$y\begin{pmatrix} u & -v \\ v & u \end{pmatrix}y^{-1} = \begin{pmatrix} u & v \\ -v & u \end{pmatrix} = \begin{pmatrix} u & v \\ -v & u \end{pmatrix}^{-1}$$

for all u,v. Hence $yxy^{-1} = x^{-1}$.

Consequently we have an imbedding $Q_t \to SL(2,\mathbb{F}_q)$. Since $|Q_t| = 2^{t+1}$ and $|SL(2,\mathbb{F}_q)| = (q^2-1)q = (q-\varepsilon)(q+\varepsilon)q$, it follows that Q_t is a Sylow 2-subgroup of $SL(2,\mathbb{F}_q)$.

Next we recall some basic results about the (co-) homology of generalized quarternion group (cf. [12]).

<u>Proposition</u> 5.2 (a) The mod-2 cohomology of Q_t is given by

$$H^n BQ_t = \begin{cases} \mathbb{Z}/2 \oplus \mathbb{Z}/2 & \text{if} \quad n \equiv 1 \text{ or } 2 \pmod 4 \\ \mathbb{Z}/2 & \text{if} \quad n \equiv 0 \text{ or } 3 \pmod 4 \end{cases}$$

There is a periodicity isomorphism $H^i \xrightarrow{\cong} H^{i+4}$ given by multiplica tion by the nonzero class $P \in H^4 BQ_t$

(b) The integral cohomology of Q_t is given by

$$\tilde{H}^n(BQ_t;\ \mathbb{Z}) = \begin{cases} \mathbb{Z}/2 \oplus \mathbb{Z}/2 & \text{if} \quad n \equiv 2 \pmod 4 \\ \mathbb{Z}/2^{t+1} & \text{if} \quad n \equiv 0 \pmod 4 \\ \text{otherwise.} & \end{cases}$$

<u>Remark</u> 5.3. There is a faithful symplectic representation

$$\rho\colon\ Q_t \to S^3 = Sp(1)$$

given by

$$\rho(x) = \exp(2\pi i/2^t) \qquad \rho(y) = j$$

The class $P \in H^4 BQ_t$ is the image of the generator of $H^4 BS^3$ under the induced map ρ^*. This follows from the commutative diagram

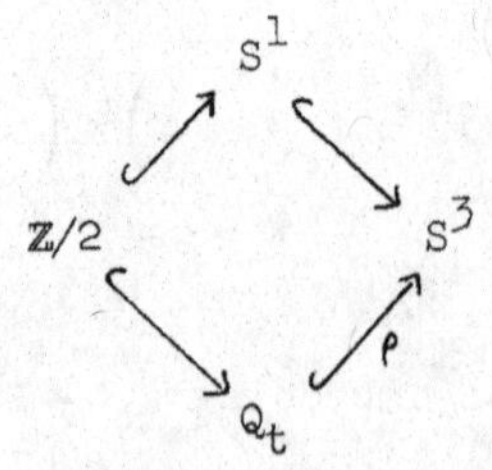

and Lemma III 6.1.

We now turn to the calculation of $H^* BSL(2, \mathbb{F}_q)$. Let $\tilde{x} \in H^3 BQ_t$

denote the generator. We shall consider $H^*BSL(2,\mathbb{F}_q)$ as a subalgebra under the injection $i^*\colon H^*BSL(2,\mathbb{F}_q) \to H^*BQ_t$ induced by the inclusion $i\colon Q_t \to SL(2,\mathbb{F}_q)$ of Sylow 2-subgroups.

<u>Proposition</u> 5.4. $H^*BSL(2,\mathbb{F}_q) = \mathbb{Z}_2[P] \otimes E[\tilde{x}]$.

<u>Proof:</u> Swan [40, Theorem 1] has shown that any group G with 2-Sylow subgroup Q_t is 4-periodic (i.e. $H^4(BG;\ \mathbb{Z})$ has an element of order 2^{t+1} and $H^jBG \approx H^{j+4}BG$). Since H^4BQ_t is generated by P it follows that the periodicity isomorphism is given by $x \to x\cdot P$ where by abuse of notation $P \in H^4BG$ denotes a class which maps to P. Thus for $G = SL(2,\mathbb{F}_q)$ we are reduced to showing that $\tilde{x} \in \operatorname{Im} i^*$ and $H^jBSL(2,\mathbb{F}_q) = 0$ for $j = 1,2$. Since $SL(2,\mathbb{F}_q)$ is generated by the transvections

$$\begin{pmatrix} 1 & a \\ 0 & 1 \end{pmatrix} \quad \begin{pmatrix} 1 & 0 \\ a & 1 \end{pmatrix} \qquad a \in \mathbb{F}_q$$

which are of order q odd, it follows that $H^1BSL(2,\mathbb{F}_q) = 0$. Since the integral cohomology groups of $SL(2,\mathbb{F}_q)$ are finite, the Bockstein spectral sequence converges to zero in positive dimensions. Thus P must be a boundary since $H^5BSL(2,\mathbb{F}_q) = 0$ by periodicity and so $H^3BSL(2,\mathbb{F}_q) \neq 0$ and $\tilde{x} \in \operatorname{Im} i^*$. It follows that $\tilde{x}$ is not a boundary. This together with $H^1BSL(2,\mathbb{F}_q) = 0$ allows us to conclude $H^2BSL(2,\mathbb{F}_q) = 0$.

<u>Remark</u> 5.5. It follows from Prop. 5.4 that

$$H_iBSL(2,\mathbb{F}_q) = \begin{cases} \mathbb{Z}/2 & \text{if } i \equiv 0 \text{ or } 3 \pmod 4 \\ 0 & \text{otherwise.} \end{cases}$$

Let $\sigma_j \in H_{4j}BSL(2,\mathbb{F}_q)$ $j \geq 0$ and $\tau_k \in H_{4k-1}BSL(2,\mathbb{F}_q)$ $k \geq 1$ denote the generators. The coproduct

$$\Delta_*: H_*BSL(2,\mathbb{F}_q) \to H_*BSL(2,\mathbb{F}_q) \otimes H_*BSL(2,\mathbb{F}_q)$$

induced by the diagonal Δ: $SL(2,\mathbb{F}_q) \to SL(2,\mathbb{F}_q) \times SL(2,\mathbb{F}_q)$ is given by

$$\Delta_*(\sigma_n) = \Sigma_{i=0}^{n}\sigma_i \otimes \sigma_{n-i}$$

$$\Delta_*(\tau_n) = \Sigma_{i=0}^{n-1}(\sigma_i \otimes \tau_{n-i} + \tau_{i+1} \otimes \sigma_{n-i-1})$$

Moreover it follows from the diagram of Remark 5.3 and Prop. 5.4 that if i: $\mathbb{Z}/2 \to SL(2,\mathbb{F}_q)$ is the inclusion, then

$$i_*(x_{4k}) = \sigma_k.$$

Lemma 5.6. $\tau_k^2 = 0$ in $H_*BSp(4,\mathbb{F}_q)$.

Proof: $\tau_k^2 = Q^{4n-1}\tau_k = 0$ by Chap. IX 1.5 and 3.1.

Finally we look at Bockstein operations and Steenrod operations in the (co-) homology of $SL(2,\mathbb{F}_q)$.

Proposition 5.7. (a) The differentials in the Bockstein spectral sequence of $H^*BSL(2,\mathbb{F}_q)$ are given by

$$d_k^*(P^m) = 0 \quad \text{for all } k \text{ and } m$$

$$d_k^*(\tilde{x}P^m) = 0 \quad \text{for all } m \text{ and all } k < \nu$$

$$d_\nu^*(\tilde{x}P^m) = y^{m+1} \quad \text{for all } m$$

where $\nu = t + 1 = \nu(q^2-1)$ is the largest integer for which 2^ν divides $q^2 - 1$.

(b) The Steenrod operations in $H^*BSL(2,\mathbb{F}_q)$ are given by

$$Sq^i(P^k) = Sq^i(\tilde{x}P^k) = 0 \text{ whenever } i \not\equiv 0 \pmod 4$$

$$Sq^{4i}(P^k) = (i,k-i)P^{k+i}$$

$$Sq^{4i}(\tilde{x}P^k) = (i,k-i)\tilde{x}P^{k+i}$$

for all $i,k \geq 0$.

(c) The differentials in the Bockstein spectral sequence of $H_*BSL(2,\mathbb{F}_q)$ are given by

$$d_k(\tau_m) = 0 \text{ for all } k \text{ and } m$$

$$d_k(\sigma_m) = 0 \text{ for all } m \text{ and all } k < \nu$$

$$d_\nu(\sigma_m) = \tau_m \text{ for all } m > 0$$

where ν is as in (a)

(d) The dual Steenrod operations in $H_*BSL(2,\mathbb{F}_q)$ are given by

$$Sq_*^i(\sigma_k) = Sq_*^i(\tau_k) = 0 \quad \text{if} \quad i \not\equiv 0 \pmod 4$$

$$Sq_*^{4i}(\sigma_k) = (i,k-2i)\sigma_{k-i}$$

$$Sq_*^{4i}(\tau_k) = (i,k-1-2i)\tau_{k-i}$$

for all $i,k \geq 0$.

<u>Proof</u>: Part (a) follows from Prop. 5.2(b) and the fact that

$$i^*: H^nBSL(2,\mathbb{F}_q) \to H^nBSL(2,\mathbb{F}_q)$$

induced by inclusion $Q_t \to SL(2,\mathbb{F}_q)$, is an isomorphism if $n \equiv 0$ or

3 (mod 4).

To prove part (b) we note, according to Remark 5.3, that, as a class in H^*BQ_t, P is the image of the generator of H^4BS^3 under the map induced by the representation $\rho: Q_t \to S^3$. It follows that

$$Sq^i(P) = 0 \quad i = 1,2,3 \qquad Sq^4(P) = P^2$$

We also note that by part (a)

$$Sq^1(\tilde{x}) = d_1^*(\tilde{x}) = 0$$

while $Sq^2(\tilde{x}) = Sq^3(\tilde{x}) = 0$ since $H^5BSL(2,\mathbb{F}_q) = H^6BSL(2,\mathbb{F}_q) = 0$. The rest of part (b) follows by repeated applications of the Cartan formula.

Parts (c) and (d) follow by dualization.

We now prove a lemma which we needed in Chapter IV. We use the notation of Def. III 6.6.

Lemma 5.8. Let $\overline{M}_*$ denote the $\mathbb{Z}/2$-submodule of $H_*BSL(2,\mathbb{F}_q)$ generated by $\{\tau_k | k \geq 2\}$ and let $\overline{N}_*$ denote the $\mathbb{Z}/2$-submodule generated by τ_2. Then $\overline{M}_*$ and $\overline{N}_*$ are $\mathcal{A}_*$-submodules of $H_*BSL(2,\mathbb{F}_q)$ and $CL(\overline{N}_*) = \overline{M}_*$.

Proof: The fact that $\overline{M}_*$ and $\overline{N}_*$ are $\mathcal{A}_*$-submodules follows directly from Prop. 5.7(d). To prove $CL(\overline{N}_*) = \overline{M}_*$ we argue as in Lemma 2.6.

As an induction hypothesis, suppose that $\tau_j \in CL(\overline{N}_*)$ for $1 < j < i$. If i is odd then

$$Sq_*^4(\tau_i) = (1,i-1-2)\tau_{i-1} = \tau_{i-1}$$

so that τ_i is Steenrod related to $CL(\overline{N}_*)$ and hence $\tau_i \in CL(\overline{N}_*)$.

If $i \equiv 0 \pmod 4$ say $i = 4k$ then

$$Sq_*^8(\tau_{i+1}) = (2,i-4)\tau_{i-1} = \tau_{i-1}$$

so that τ_{i+1} is Steenrod related to $CL(\overline{N}_*)$ and hence $\tau_{i+1} \in CL(N_*)$. Then $\tau_i = Sq_*^4(\tau_{i+1}) \in CL(\overline{N}_*)$.

If $i \equiv 2 \pmod 4$ then $i = 4k + 2$ $k \geq 0$

$$Sq_*^8(\tau_i) = (2,i-5)\tau_{i-2} = \tau_{i-2}$$

provided $k \geq 1$. Hence τ_i is Steenrod related to $CL(\overline{N}_*)$ and so $\tau_i \in CL(\overline{N}_*)$.

The only remaining case is $i = 2$. But then $\tau_i = \tau_2 \in \overline{N}_*$ by hypothesis. This concludes the induction and proof.

We conclude this section by discussing inclusions

$$j: SL(2,\mathbb{F}_q) \to SL(2,\mathbb{F}_{q^r})$$

obtained by extending scalars and passage to the direct limit $SL(2,\overline{\mathbb{F}}_q)$. These results were needed in Chapter III §7.

<u>Proposition</u> 5.9. Let $j: SL(2,\mathbb{F}_q) \to SL(2,\mathbb{F}_{q^r})$ be the inclusion induced by extension of scalars. Let us denote by $\hat{P}, \hat{x}$ (respectively $\hat{\sigma}_i, \hat{\tau}_i$) the generators of $H^*(BSL(2,\mathbb{F}_{q^r}))$ (respectively $H_*BSL(2,\mathbb{F}_{q^r})$). Then

(a) If r is odd, then

$$j^*: H^*BSL(2,\mathbb{F}_{q^r}) \to H^*BSL(2,\mathbb{F}_q)$$

$$j_*: H_*BSL(2,\mathbb{F}_q) \to H_*BSL(2,\mathbb{F}_{q^r})$$

are isomorphisms, so that

$$j^*(\hat{P}) = P \quad j^*(\hat{x}) = \tilde{x}$$

$$j_*(\sigma_i) = \hat{\sigma}_i \quad j_*(\tau_i) = \hat{\tau}_i \text{ for all } i$$

(b) If r is even, then

$$j^*(\hat{P}) = P \qquad j^*(\hat{x}) = 0$$

$$j_*(\sigma_i) = \hat{\sigma}_i \qquad j_*(\tau_i) = 0 \text{ for all } i$$

<u>Proof</u>: As in Prop. 5.7 we let $\nu = \nu(q^2-1)$, $\eta = \nu(q^{2r}-1)$ be the largest integers such that 2^ν, 2^η divide $q^2 - 1$, $q^{2r} - 1$ respectively. Clearly $\nu \leq \eta$. Moreover by VII 2.3, $\nu = \eta$ in case (a) while $\nu < \eta$ in case (b).

It follows from Remark 5.3 and Prop. 5.4 that we have a commutative diagram

$$\begin{array}{ccccc} SL(2,\mathbb{F}_q) & \xleftarrow{\ i\ } & Q_{\nu-1} & & \\ & & & \searrow^{\rho} & \\ \downarrow{\scriptstyle j} & & \downarrow & & S^3 \\ & & & \swarrow_{\rho} & \\ SL(2,\mathbb{F}_{q^r}) & \xleftarrow{\ i\ } & Q_{\eta-1} & & \end{array}$$

and that $j^*(\hat{P}) = P$. Dualizing we get $j_*(\sigma_i) = \hat{\sigma}_i$ for all i. We now apply Prop. 5.7(c).

In case (a) $\nu = \eta$ so that

$$j_*(\tau_i) = j_* d_\nu(\sigma_i) = d_\eta j_*(\sigma_i) = d_\eta(\hat{\sigma}_i) = \hat{\tau}_i$$

for all i. Dually $j^*(\hat{x}) = \tilde{x}$.

In case (b) $\nu < \eta$ so that

$$j_*(\tau_i) = j_* d_\nu(\sigma_i) = d_\nu j_*(\sigma_i) = d_\nu(\hat{\sigma}_i) = 0$$

for all i. Dually $j^*(\hat{x}) = 0$.

<u>Corollary</u> 5.10. Let $\overline{\mathbb{F}}_p$ denote the algebraic closure of $\mathbb{F}_p$, p an odd prime. Then: (a) $H^*BSL(2,\overline{\mathbb{F}}_p) = \mathbb{Z}/2[P]$ where degree $P = 4$

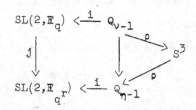

(b) $H_*BSL(2,\overline{\mathbb{F}}_p) = \begin{cases} \mathbb{Z}/2 \text{ on generator } \gamma_r & \text{if } k = 4r \\ 0 & \text{if } k \not\equiv 0 \pmod 4 \end{cases}$

(c) If $i\colon \mathbb{Z}/2 \to SL(2,\overline{\mathbb{F}}_p)$ is the inclusion, then $i_*\colon H_*B\mathbb{Z}/2 \to H_*BSL(2,\overline{\mathbb{F}}_p)$ is given by

$$i_*(x_k) = \begin{cases} \gamma_r & \text{if } k = 4r \\ 0 & \text{if } k \not\equiv 0 \pmod 4 \end{cases}$$

<u>Proof</u>: Since $SL(2,\overline{\mathbb{F}}_p) = \lim\limits_{\substack{\rightarrow \\ r}} SL(2,\mathbb{F}_{p^r})$, (a) and (b) follow immediately from Prop. 5.9. Finally (c) follows from Remark 5.5.

§6. $GL(2,\mathbb{F}_q)$ <u>and</u> $U(2,\mathbb{F}_q)$

In this section we compute the mod-2 homology of $GL(2,\mathbb{F}_q)$ and $U(2,\mathbb{F}_q)$.

In IV 7.1, we defined generators $\alpha_i \in H_{2i}BGL(1,\mathbb{F}_q)$ $i \geq 0$, $\beta_i \in H_{2i-1}BGL(1,\mathbb{F}_q)$ $i > 0$; $\rho_i \in H_{2i}BU(1,\mathbb{F}_{q^2})$ $i \geq 0$, $\eta_i \in H_{2i-1}BU(1,\mathbb{F}_{q^2})$ $i > 0$. We begin by establishing a relation

<u>Lemma</u> 6.1. (1) $\beta_i^2 = 0$ in $H_*BGL(2,\mathbb{F}_q)$

(2) $\eta_i^2 = 0$ in $H_*BU(2,\mathbb{F}_{q^2})$

<u>Proof</u>: Consider the following commutative diagram of homomorphisms

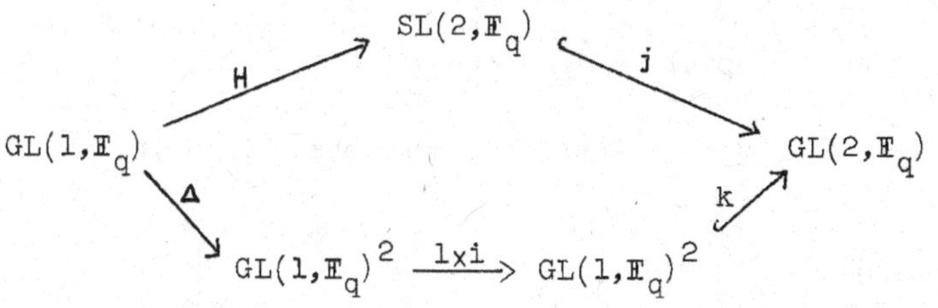

where j,k are the standard inclusions, Δ is the diagonal map, $H(x) = \begin{pmatrix} x & 0 \\ 0 & x^{-1} \end{pmatrix}$ and $i(x) = x^{-1}$. In the induced diagram in homology $i_* = \mathrm{id}$, hence tracing α_{2i-1} around the lower route we find $\alpha_{2i-1} \to \beta_i^2$ if $q \equiv 3 \pmod 4$. However $H_{4i-2}BSL(2,\mathbb{F}_q) = 0$ which proves (1). Still assuming $q \equiv 3 \bmod 4$ we find $\eta_i^2 = Q^{2i-1}\eta_i = 0$ by IX 4.2 which proves (2). Suppose $q \equiv 1 \bmod 4$. Then $\beta_i^2 = Q^{2i-1}\beta_i = 0$ by IX 4.1. Since $SU(2,\mathbb{F}_{q^2}) \approx SL(2,\mathbb{F}_q)$ a similar diagram chase as above proves that $\eta_i^2 = 0$ if $q \equiv 1 \bmod 4$.

Lemma 6.3 (a) The elements $\beta_1\alpha_0$, $\alpha_1\alpha_0$, $\alpha_1\beta_1$, $\beta_2\alpha_0$, α_1^2, $\alpha_2\alpha_0$, $\beta_1\beta_2$ are linearly independent in $H_*BGL(2,\mathbb{F}_q)$

(b) The elements $\eta_1\xi_0$, $\xi_1\xi_0$, $\xi_1\eta_1$, $\eta_2\xi_0$, ξ_1^2, $\xi_2\xi_0$, $\eta_1\eta_2$ are linearly independent in $H_*BU(2,\mathbb{F}_{q^2})$.

Proof: As in Chapter IV §7 let us denote by $\bar{\alpha}_i$, $\bar{\beta}_i$ the images of α_i, β_i respectively under the map

$$BGL(1,\mathbb{F}_q) \to BGL(\infty,\mathbb{F}_q) \to \Gamma_0 B\mathcal{GL}(\mathbb{F}_q)$$

Then the elements $\alpha_i\alpha_j$, $\beta_i\beta_j$, $\alpha_i\beta_j$ in $H_*BGL(2,\mathbb{F}_q)$ map under

$$BGL(2,\mathbb{F}_q) \to BGL(\infty,\mathbb{F}_q) \to \Gamma_0 B\mathcal{GL}(\mathbb{F}_q)$$

to the elements $\bar{\alpha}_i\bar{\alpha}_j$, $\bar{\beta}_i\bar{\beta}_j$, $\bar{\alpha}_i\bar{\beta}_j$ in $H_*\Gamma_0 B\mathcal{GL}(\mathbb{F}_q)$.

Now according to Lemma IV 7.5 and Lemma IV 7.6

$$\lambda_*(\bar{\alpha}_i) = \hat{a}_i \quad \forall i$$

$$\lambda_*(\bar{\beta}_k) = \hat{b}_k + \text{decomposables } k = 1,2$$

under the map

$H^*BGL(2,\mathbb{F}_q)$, it follows that k^* is an isomorphism onto S.

Part (b) is proved similarly.

$$\lambda_*: H_*\Gamma_0 B \mathcal{D}\mathcal{L}(\mathbb{F}_q) \to H_*JU(q) = \mathbb{Z}/2[\hat{a}_1, \hat{a}_2, \ldots] \otimes E[\hat{b}_1, \hat{b}_2, \ldots]$$

Consequently λ_* maps $\overline{S} = \{\overline{\beta}_1\overline{\alpha}_0, \overline{\alpha}_1\overline{\alpha}_0, \overline{\alpha}_1\overline{\beta}_1, \overline{\beta}_2\alpha_0, \overline{\alpha}_1^2, \overline{\alpha}_2\overline{\alpha}_0, \overline{\beta}_1\overline{\beta}_2\}$ into a linearly independent subset of $H_*JU(q)$. Consequently $\overline{S}$ must be linearly independent as must be $\beta_1\alpha_0$, $\alpha_1\alpha_0$, $\alpha_1\beta_1$, $\beta_2\alpha_0$, α_1^2, $\alpha_2\alpha_0$, $\beta_1\beta_2$. This proves (a). Part (b) is proved similarly.

Lemma 6.4 (a) The Serre spectral sequence in (co-) homology of the fibration

$$BSL(2,\mathbb{F}_q) \to BGL(2,\mathbb{F}_q) \xrightarrow{B\ \det} B\mathbb{F}_q^* = B\mathbb{Z}/q-1$$

collapses.

(b) The Serre spectral sequence in (co-) homology of the fibration

$$BSU(2,\mathbb{F}_{q^2}) \to BU(2,\mathbb{F}_{q^2}) \xrightarrow{B\ \det} BU(1,\mathbb{F}_{q^2}) = B\mathbb{Z}/q+1$$

(cf. II 6.7) collapses.

(c) The Poincaré series of $H^*BGL(2,\mathbb{F}_q)$, $H_*BGL(2,\mathbb{F}_q)$, $H^*BU(2,\mathbb{F}_{q^2})$, $H_*BU(2,\mathbb{F}_{q^2})$ is in all cases

$$\frac{1+t^3}{(1-t)(1-t^4)}$$

Proof: Recall that according to Prop. 5.4, $H^*BSL(2,\mathbb{F}_q) = \mathbb{Z}/2[P] \otimes E[\tilde{x}]$ where degree $\tilde{x} = 3$, degree $P = 4$. To prove (a) it suffices to show that P and $\tilde{x}$ are in the image of

$$i^*: H^*BGL(2,\mathbb{F}_q) \to H^*BSL(2,\mathbb{F}_q)$$

or equivalently that $1 \otimes P$ and $1 \otimes \tilde{x}$ are in the E_∞ term of the spectral sequence.

First we observe that $H^1BSL(2,\mathbb{F}_q) = 0$ or $\mathbb{Z}/2$ so the spectral sequence of (a) must have trivial local coefficients. Also

$$BGL(2,\mathbb{F}_q) \xrightarrow{B\det} B\mathbb{F}_q^*$$

must induce a monomorphism in cohomology since the map has a right inverse

$$B\mathbb{F}_q^* = BGL(1,\mathbb{F}_q) \to BGL(2,\mathbb{F}_q)$$

It follows that $1 \otimes \tilde{x}$ transgresses to zero and hence survives to E_∞.

It remains to show that $P \in \mathrm{Im}\, i^*$. To do this observe that the Brauer character of the representation

$$\mathbb{Z}/2 \to SL(2,\mathbb{F}_q) \to GL(2,\mathbb{F}_q)$$

is the character of the representation

$$\tau\colon \mathbb{Z}/2 \to U(2)$$

given by $1 \to \begin{pmatrix}1 & 0\\ 0 & 1\end{pmatrix}$, $-1 \to \begin{pmatrix}-1 & 0\\ 0 & -1\end{pmatrix}$. Hence there is a commutative diagram

$$\begin{array}{ccccccc}
B\mathbb{Z}/2 & \xrightarrow{k} & BSL(2,\mathbb{F}_q) & \xrightarrow{i} & BGL(2,\mathbb{F}_q) & \xrightarrow{j} & BGL(\infty,\mathbb{F}_q) \\
\downarrow{\scriptstyle\Delta} & & & & & & \downarrow{\scriptstyle\beta} \\
B\mathbb{Z}/2 \times B\mathbb{Z}/2 & \xrightarrow{f\times f} & BS^1 \times BS^1 = BU(1) \times BU(1) & & \xrightarrow{\oplus} & & BU(2) \to BU
\end{array}$$

Now chasing the Chern class $c_2 \in H^4BU$ around the diagram we obtain

$$k^* i^* j^* \beta^*(c_2) = \Delta^*(f^* \times f^*)(c_1 \otimes c_1) = \Delta^*(x^2 \otimes x^2) = x^4 \neq 0$$

Hence $i^* j^* \beta^*(c_2)$ is a nonzero class in $H^4 BSL(2,\mathbb{F}_q)$ and so

$$P = i^* j^* \beta^*(c_2)$$

is in the image of i^*. This proves (a).

Part (b) is proved similarly, first observing that by II 8.5($\overline{a}^*$) $SU(2,\mathbb{F}_{q^2}) \cong SL(2,\mathbb{F}_q)$. Part (c) follows immediately from (a) and (b) together with Lemmas 2.1 and 5.4.

Proposition 6.5 (a) The elements $\{\alpha_i\alpha_j | 0 \leq i \leq j\}$, $\{\beta_i\beta_j | 0 < i < j\}$ $\{\alpha_i\beta_j | 0 \leq i, 0 < j\}$ form a basis for $H_* BGL(2,\mathbb{F}_q)$.

(b) The elements $\{\xi_i\xi_j | 0 \leq i \leq j\}$, $\{\eta_i\eta_j | 0 < i < j\}$ $\{\xi_i\eta_j | 0 \leq i, 0 < j\}$ form a basis for $H_* BU(2,\mathbb{F}_{q^2})$.

Proof: Let $S \subseteq H^*(BGL(1,\mathbb{F}_q) \times BGL(1,\mathbb{F}_q))$ be the subalgebra generated by $\{s_1 = x \otimes 1 + 1 \otimes x$, $s_2 = y \otimes 1 + 1 \otimes y$, $s_3 = x \otimes y + y \otimes x$, $s_4 = y \otimes y\}$. We shall prove the dual of (a) in cohomology which by Lemma 6.1 and the commutativity of $\oplus$ becomes

$$k^*: H^* BGL(2,\mathbb{F}_q) \xrightarrow{\cong} S$$

where $k = B\oplus$: $BGL(1,\mathbb{F}_q) \times BGL(1,\mathbb{F}_q) \to BGL(2,\mathbb{F}_q)$.

Dualizing Lemma 6.3(a) we obtain that $s_1, s_2, s_3, s_4 \in \mathrm{Im}\, k^*$. Hence k^* maps epimorphically onto S. Since the Poincaré series of S is

$$\frac{1 + t^3}{(1-t)(1-t^4)}$$

which by Lemma 6.4(c) is the same as the Poincaré series of

Chapter VII

Detection Theorems at the Prime 2

§1. Introduction

In this chapter we find upper bounds for $H_*(BG(n))$ where $G(n) = O(n,\mathbb{F}_q)$, $O(\overline{n},\mathbb{F}_q)$, $Sp(2m,\mathbb{F}_q)$, $GL(n,\mathbb{F}_q)$, $U(n,\mathbb{F}_{q^2})$ and where homology is taken (throughout the chapter) with coefficients in $\mathbb{Z}/2$. Our results take the following form: For small integers a_i the direct sum homomorphism

$$G(a_1) \times G(a_2) \times \cdots \times G(a_r) \to G(n)$$

($n = \Sigma\, a_i$) induce epimorphisms in homology. In fact a_i is either 1 or 2 and so the $G(a_i)$ were analyzed in Chapter VI.

The plan of the calculations is to find wreath product subgroups $\mathcal{S}_m \wr G(a)$ which contain a Sylow 2-subgroup of $G(am)$. Quillen's Lemma (III 5.6) is then used to analyze $\mathcal{S}_m \wr G(a)$.

The orthogonal groups are treated first in §2 and serve as a pattern for the other cases. §3 is devoted to the symplectic groups and §4 to the general linear and unitary groups.

§2. $O(n,\mathbb{F}_q)$ and $O(\overline{n},\mathbb{F}_q)$

The purpose of this section is to prove

<u>Proposition 2.1.</u> The direct sum group homomorphisms

i) $O(2,\mathbb{F}_q) \times \cdots \times O(2,\mathbb{F}_q) \to O(2n,\mathbb{F}_q)$

ii) $O(2m,\mathbb{F}_q) \times O(1,\mathbb{F}_q) \to O(2m+1,\mathbb{F}_q)$

iii) $O(2m,\mathbb{F}_q) \times O(\overline{1},\mathbb{F}_q) \to O(\overline{2m+1},\mathbb{F}_q)$

and iv) $O(2m-1,\mathbb{F}_q) \times O(\overline{1},\mathbb{F}_q) \to O(\overline{2m},\mathbb{F}_q)$

induce epimorphisms on mod 2 homology.

We begin by showing that the wreath product $\mathcal{S}_m \wr O(2,\mathbb{F}_q)$ contains a Sylow 2-subgroup of $O(2m,\mathbb{F}_q)$. Let $\nu(m)$ be the largest integer k such that $2^k \mid m$.

<u>Lemma 2.2</u>. If $q \equiv \varepsilon = \pm 1 \pmod 4$ then $\nu(q^{2^n}-1) = n + \nu(q-\varepsilon)$, $n \geq 1$.

<u>Proof</u>: If $n \geq 2$ then $q^{2^{n-1}} + 1 \equiv 2 \pmod 4$. Hence

$$\nu(2^{2n} - 1) = \nu[(q^{2^{n-1}} - 1)(q^{2^{n-1}} + 1)]$$

$$= \nu(2^{2^{n-1}} - 1) + \nu(2^{2^{n-1}} + 1) = \nu(q^{2^{n-1}} - 1) + 1$$

Iterating this recursion formula we get

$$\nu(q^{2^n}-1) = \nu(q^2-1) + (n-1) \quad n \geq 1.$$

Since $q + \varepsilon \equiv 2 \pmod 4$ it follows that

$$\nu(q^{2^n}-1) = (n-1) + \nu(q^2-1) = (n-1) + \nu(q+\varepsilon) + \nu(q-\varepsilon) = n + \nu(q-\varepsilon)$$

<u>Corollary 2.3</u>. If $q \equiv \varepsilon = \pm 1 \pmod 4$ and m is even, $\nu(q^m-1) = \nu(m) + \nu(q-\varepsilon)$.

<u>Proof</u>. Let $s = \nu(m)$, $m = 2^s t$. Then by Lemma 2.2

$$\nu(q^m-1) = \nu((q^t)^{2^s}-1) = s+\nu(q^t-\varepsilon^t) = s+\nu(q-\varepsilon)+\nu(\Sigma_{i=0}^{t}\varepsilon^i q^{t-i}) = s+\nu(q-\varepsilon)$$

<u>Lemma 2.4</u>. $\mathcal{S}_m \wr O(2,\mathbb{F}_q)$ contains a 2-Sylow subgroup of $O(2m,\mathbb{F}_q)$

<u>Proof</u>. Since $q^m + \varepsilon^m \equiv 2 \pmod 4$ we have (by II 3.1)

$$\nu(|O(2m,\mathbb{F}_q)|) = 1 + \nu(q^m-\varepsilon^m) + \Sigma_{i=1}^{m-1}\nu(q^{2i}-1)$$

$$= \nu(q^m + \varepsilon^m) + \nu(q^m-\varepsilon^m) + \Sigma_{i=1}^{m-1}\nu(q^{2i}-1)$$

$$= \nu(q^{2m}-\epsilon^{2m}) + \Sigma_{i=1}^{m-1}\nu(q^{2i}-1)$$

$$= \Sigma_{i=1}^{m}\nu(q^{2i}-1) = \Sigma_{i=1}^{m}[1 + \nu(i) + \nu(q-\epsilon)]$$

On the other hand

$$\nu(|\mathcal{S}_m \wr O(2,\mathbb{F}_q)|) = \nu(m!\ (2(q-\epsilon))^m) = \nu(m!) + m + m\nu(q-\epsilon)$$

$$= \Sigma_{i=1}^{m}[1 + \nu(i) + \nu(q-\epsilon)]$$

Thus $\mathcal{S}_m \wr O(2,\mathbb{F}_q)$ contains a 2-Sylow subgroup of $O(2m,\mathbb{F}_q)$.

Proof of Proposition 2.1. (i) We first proceed by induction to show that

$$O(2,\mathbb{F}_q) \times O(2,\mathbb{F}_q) \times \cdots \times O(2,\mathbb{F}_q) \to O(2^n,\mathbb{F}_q)$$

detects $H^*(BO(2^n,\mathbb{F}_q))$. For $n = 1$ there is nothing to prove. Now assume the result for $n - 1$. We claim the subgroup $\mathcal{S}_2 \wr O(2^{n-1},\mathbb{F}_q)$ detects $H^*(BO(2^n,\mathbb{F}_q))$. To see this consider the diagram of inclusions

$$\begin{array}{ccc}
\mathcal{S}_{2^{n-1}} \wr O(2,\mathbb{F}_q) & \xrightarrow{\alpha_1} & O(2^n,\mathbb{F}_q) \\
\uparrow \beta_1 & & \uparrow \beta_2 \\
\mathcal{S}_2 \wr \mathcal{S}_{2^{n-2}} \wr O(2,\mathbb{F}_q) & \xrightarrow{\alpha_2} & \mathcal{S}_2 \wr O(2^{n-1},\mathbb{F}_q)
\end{array}$$

Now α_1^* is injective by Lemma 2.4. Since $\nu(|\mathcal{S}_2 \wr \mathcal{S}_{2^{n-1}}|) = \nu(|\mathcal{S}_{2^{n-1}}|)$ it follows that β_1^* is injective. It follows that β_2^* is injective.

Thus by III 5.6, $H^*(BO(2^n,\mathbb{F}_q))$ is detected by

$\mathcal{S}_2 \times O(2^{n-1},\mathbb{F}_q)$ and $O(2^{n-1},\mathbb{F}_q)^2$. We will show that $\mathcal{S}_2 \times O(2^{n-1},\mathbb{F}_q)$ is conjugate to a subgroup of $O(2^{n-1},\mathbb{F}_q)^2$.

The elements

$$\{B \otimes I_{2^{n-1}}, I_2 \otimes M \mid B = \begin{pmatrix} 0 & 1 \\ 1 & 0 \end{pmatrix}, \ M \in O(2^{n-1},\mathbb{F}_q)\}$$

generate $\mathcal{S}_2 \times O(2^{n-1},\mathbb{F}_q)$. Let $\gamma_1 = \frac{1}{\sqrt{2}}\begin{pmatrix} -1 & 1 \\ 1 & 1 \end{pmatrix}$. Let $\overline{\gamma} = \gamma_1 \oplus \cdots \oplus \gamma_1$. Then $\overline{\gamma}\, O(2^{n-1},\mathbb{F}_q)\overline{\gamma}^{-1} \subseteq O(2^{n-1},\mathbb{F}_q)$. If $\gamma = \gamma_1 \otimes \overline{\gamma}$ then $\gamma \in O(2^n,\mathbb{F}_q)$ and

$$\gamma(B \otimes I_{2^{n-1}})\gamma^{-1} = (\gamma_1 B \gamma_1^{-1}) \otimes I_{2^{n-1}} = \begin{pmatrix} -1 & 0 \\ 0 & 1 \end{pmatrix} \otimes I_{2^{n-1}}$$

$$= \begin{pmatrix} -I_{2^{n-1}} & 0 \\ 0 & I_{2^{n-1}} \end{pmatrix} \in O(2,\mathbb{F}_q)^{2^{n-1}}$$

$$\gamma(I_2 \otimes M)\gamma^{-1} = I_2 \otimes (\overline{\gamma} M \overline{\gamma}^{-1}) \in O(2^{n-1},\mathbb{F}_q)^2$$

Hence $H^*(BO(2^n,\mathbb{F}_q))$ is detected by $O(2^{n-1},\mathbb{F}_q)^2$ and therefore by induction hypothesis it is detected by $O(2,\mathbb{F}_q)^{2^{n-1}}$. This completes the induction and the proof for $m = 2^{n-1}$.

Finally we consider the general case $2m$. We express m in the 2-adic form $m = \Sigma_{i=0}^{r} a_i 2^i$, $a_i = 0$ or 1. Now consider the diagram of inclusions

$$\begin{array}{ccc}
(\mathcal{S}_{a_0 2^0} \times \mathcal{S}_{a_1 2^1} \times \cdots \times \mathcal{S}_{a_r 2^r}) \wr O(2,\mathbb{F}_q) & & \\
\| & & \\
(\mathcal{S}_{a_0 2^0} \wr O(2,\mathbb{F}_q)) \times \cdots \times (\mathcal{S}_{a_r 2^r} \wr O(2,\mathbb{F}_q)) & \xrightarrow{f} & O(a_0 2^1,\mathbb{F}_q) \times \cdots \times O(a_r 2^{r+1},\mathbb{F}_q) \\
\downarrow u & & \downarrow g \\
\mathcal{S}_m \wr O(2,\mathbb{F}_q) & \xrightarrow{h} & O(2m,\mathbb{F}_q)
\end{array}$$

Since $\nu(|\mathcal{S}_{a_0 2^0} \times \mathcal{S}_{a_1 2^1} \times \cdots \times \mathcal{S}_{a_r 2^r}|) = \nu(|\mathcal{S}_m|)$, u* is injective. By Lemma 2.4, h* is injective. Hence g* is injective. By the previous cases $H^*(BO(2^i, \mathbb{F}_q))$ is detected by $O(2,\mathbb{F}_q)^{2^{i-1}}$. Hence

$$O(2,\mathbb{F}_q)^m = O(2,\mathbb{F}_q)^{a_0 2^0} \times O(2,\mathbb{F}_q)^{a_1 2^1} \times \cdots \times O(2,\mathbb{F}_q)^{a_r 2^r}$$

detects $H^*(BO(2m,\mathbb{F}_q))$. This completes the proof of (i).

To prove (ii) we observe that (by II 3.1)

$$\nu(|O(2m+1,\mathbb{F}_q)|) = \nu(N_{m+1}) + 1 = \nu(q^{2m} - 1) + \nu(N_m) + 1$$

$$\nu(|O(2m,\mathbb{F}_q)|) = \nu(q^m - \epsilon^m) + \nu(N_m) + 1$$

and that

$$\nu(q^{2m}-1) = \nu(q^{2m}-\epsilon^{2m}) = \nu(q^m-\epsilon^m) + \nu(q^m + \epsilon^m) = \nu(q^m-\epsilon^m) + 1$$

since $q^m + \epsilon^m \equiv 2 \pmod 4$. Hence

$$\begin{aligned}\nu(|O(2m+1,\mathbb{F}_q)|) &= \nu(|O(2m,\mathbb{F}_q)|) + 1 \\ &= \nu(|O(2m,\mathbb{F}_q) \times O(1,\mathbb{F}_q)|)\end{aligned}$$

which completes the proof of (ii). The proof of (iii) is similar.

From II 4.5 we have

$$\begin{aligned}\nu(|O(2m-1,\mathbb{F}_q) \times O(\overline{1},\mathbb{F}_q)|) &= \nu(|O(2m-1,\mathbb{F}_q)|) + \nu(|O(\overline{1},\mathbb{F}_q)|) \\ &= 2 + \nu(N_m)\end{aligned}$$

and

$$\nu(|O(\overline{2m},\mathbb{F}_q)|) = 1 + \nu(q^m + \epsilon^m) + \nu(N_m) = 2 + \nu(N_m)$$

since

$$q^m + \epsilon^m \equiv 2\epsilon^m \equiv 2 (\text{mod } 4)$$

Hence $O(2m-1,\mathbb{F}_q) \times O(\overline{1},\mathbb{F}_q)$ contains a Sylow 2-subgroup of $O(\overline{2m},\mathbb{F}_q)$ which completes the proof of (iv).

Finally we prove a result that we needed in Chapter VI §4. Recall that a 2-primary abelian group is said to be elementary if every non-trivial element has order 2.

In what follows we will denote by $A_{m,n}$ the group

$$O(1,\mathbb{F}_q)^m \times O(\overline{1},\mathbb{F}_q)^n$$

regarded as a subgroup of $O(m + \overline{n},\mathbb{F}_q)$ by the direct sum homomorphism

$$A_{m,n} = O(1,\mathbb{F}_q)^m \times O(\overline{1},\mathbb{F}_q)^n \xrightarrow{\oplus} O(m + \overline{n},\mathbb{F}_q).$$

Obviously $A_{m,n}$ is an elementary abelian 2-subgroup of $O(m + \overline{n},\mathbb{F}_q)$.

Proposition 2.5(a) An elementary abelian 2-subgroup of $O(a,\mathbb{F}_q)$, $a \in \tilde{\overline{N}}$ is conjugate to a subgroup of $A_{m,n}$ for some m,n such that $m + \overline{n} = a$. (In other words the $A_{m,n}$'s are maximal elementary abelian 2-subgroups of $O(a,\mathbb{F}_q)$.)

(b) A_{m_1,n_1} is conjugate to A_{m_2,n_2} in $O(a,\mathbb{F}_q)$ iff $m_1 = m_2$ and $n_1 = n_2$.

Proof (a) Let A be an elementary abelian 2-subgroup of $O(a,\mathbb{F}_q)$. Then

$$A \cong (\mathbb{Z}/2)^r \quad r \leq a$$

Let $\{S_i\}_{i=1}^r$ be a basis for A. Let (V,Q) be the quadratic space on which $O(a,\mathbb{F}_q)$ acts. We will construct inductively a sequence of subspaces $\{V_i\}_{i=1}^t$ such that

(1) $V = \oplus_{i=1}^t V_i$

(2) Each V_i is an ± 1 eigenspace of each S_j

We begin by observing that since $S_1^2 = I$ we have

$$V = W_1 \oplus W_2$$

where W_1 is the ± 1 eigenspace of S_1 and W_2 is the -1 eigenspace of S_1. Since $S_2S_1 = S_1S_2$ it follows that S_2 maps W_1 and W_2 to themselves. By the same argument

$$W_1 = W_{11} \oplus W_{12} \quad W_2 = W_{21} \oplus W_{22}$$

where W_{i1} is the $+1$ eigenspace of $S_2|W_i$ and W_{i2} is the -1 eigenspace of $S_2|W_i$. Continuing in this way we split V as

$$V = \oplus^2_{i_1,i_2,i_3,\ldots,i_r=1} W_{i_1 i_2 \ldots i_r}$$

where each $W_{i_1 i_2 \ldots i_r}$ is a $-(-1)^{i_k}$ eigenspace of S_k. This produces a decomposition having properties (1) and (2).

By discarding the zero-dimensional subspaces among the V_i's and splitting up the multidimensional subspaces, we may assume that all the V_i's are one dimensional subspaces. By rearranging the V_i's if necessary we may assume that

$$(V_i,Q) \cong (\mathbb{F}_q,Q_+) \quad i = 1,2,\ldots,m$$
$$(V_i,Q) \cong (\mathbb{F}_q,Q_-) \quad i = m+1,m+2,\ldots,m+n$$

Taking the direct sum of these isomorphisms, we get an isomorphism

$$\tau\colon (V,Q) = \oplus_i(V_i,Q) \cong (\mathbb{F}_q,Q_+)^m \oplus (\mathbb{F}_q,Q_-)^n$$

Combining this with the standard isomorphism

$$\sigma\colon (\mathbb{F}_q,Q_+)^m \oplus (\mathbb{F}_q,Q_-)^n \cong (V,Q)$$

which gives the imbedding $A_{m,n} \to O(a,\mathbb{F}_q)$ via $P \to \sigma P \sigma^{-1}$ we see that

$$(\sigma\tau)A(\sigma\tau)^{-1} \subseteq A_{m,n}$$

with $\sigma\tau \in O(a,\mathbb{F}_q)$. This proves part (1).

(b) Suppose that $\sigma A_{m_2,n_2}\sigma^{-1} = A_{m_1,n_1}$ with $\sigma \in O(a,\mathbb{F}_q)$. Let (V,Q) denote the quadratic space on which $O(a,\mathbb{F}_q)$ acts. Then there is a unique decomposition of V into one dimensional subspaces $V = \oplus_{i=1}^{a} V_{1i}$ such that for any $S \in A_{m_1,n_1}$, V_{1i} is a ± 1 eigenspace of S. Similarly there is a unique decomposition $V = \oplus_{j=1}^{a} V_{2j}$ into one dimensional subspaces such that for any $T \in A_{m_2,n_2}$, V_{2j} is a ± 1 eigenspace of T. Moreover σ must map each V_{1i} isomorphically onto some V_{2j}.

Now there are m_1 indexes i for which $(V_{1i},Q) \cong (\mathbb{F}_q,Q_+)$. Similarly there are m_2 indexes j for which $(V_{2j},Q) \cong (\mathbb{F}_q,Q_+)$. Since σ sets up a 1-1 correspondence between these spaces (V_{1i},Q) and (V_{2j},Q), it follows that $m_1 = m_2$. Similarly we obtain $n_1 = n_2$.

§3. $Sp(2m,\mathbb{F}_q)$

In this section we prove

Proposition 3.1. The direct sum group homomorphism

$$Sp(2,\mathbb{F}_q) \times \cdots \times Sp(2,\mathbb{F}_q) \to Sp(2m,\mathbb{F}_q)$$

induces an epimorphism in mod 2 homology.

First we observe that $\mathcal{S}_m \wr Sp(2,\mathbb{F}_q)$ contains a Sylow 2-subgroup of $Sp(2m,\mathbb{F}_q)$. Let $\nu(m)$ be the largest integer k such that $2^k | m$.

Lemma 3.2. $\nu(|\mathcal{S}_m \wr Sp(2,\mathbb{F}_q)|) = \nu(|Sp(2m,\mathbb{F}_q)|)$

Proof. From II 6.3 and II 3.1 we have

$$\nu(|Sp(2m,\mathbb{F}_q)|) = \nu(\tfrac{1}{2}|O(2m+1,\mathbb{F}_q)|) = \nu(|O(2m,\mathbb{F}_q)|)$$

Thus by Lemma 2.4

$$\nu(|Sp(2m,\mathbb{F}_q)|) = \nu(|O(2m,\mathbb{F}_q)|) = \nu(|\mathcal{S}_m \wr O(2,\mathbb{F}_q)|)$$
$$= \nu(|\mathcal{S}_m \wr Sp(2,\mathbb{F}_q)|) .$$

Proof of Proposition 3.1: Identical to the proof of Proposition 2.1. We need only observe that $\gamma_1 \otimes \gamma_1 \in Sp(4,\mathbb{F}_q)$.

§4. $GL(n,\mathbb{F}_q)$ and $U(n,\mathbb{F}_{q^2})$

The purpose of this section is to prove the following propositions

Proposition 4.1. The direct sum group homomorphisms

$$GL(2,\mathbb{F}_q) \times \cdots \times GL(2,\mathbb{F}_q) \to GL(2m,\mathbb{F}_q)$$

$$GL(2m,\mathbb{F}_q) \times GL(1,\mathbb{F}_q) \to GL(2m+1,\mathbb{F}_q)$$

induce epimorphisms in mod 2 homology.

Proposition 4.2. The direct sum group homomorphisms

$$U(2,\mathbb{F}_{q^2}) \times \cdots \times U(2,\mathbb{F}_{q^2}) \to U(2m,\mathbb{F}_{q^2})$$

$$U(2m,\mathbb{F}_{q^2}) \times U(1,\mathbb{F}_{q^2}) \to U(2m+1,\mathbb{F}_{q^2})$$

induce epimorphisms in mod 2 homology.

We begin by showing that the wreath products

$$\mathcal{S}_m \wr GL(2,\mathbb{F}_q) \subset GL(2m,\mathbb{F}_q)$$

$$\mathcal{S}_m \wr U(2,\mathbb{F}_{q^2}) \subset U(2m,\mathbb{F}_{q^2})$$

contain Sylow 2-subgroups. Let $\nu(m)$ be the largest integer k such that $2^k | m$.

Lemma 4.3. (i) $\nu(|\mathcal{S}_m \wr GL(2,\mathbb{F}_q)|) = \nu(|GL(2m,\mathbb{F}_q)|)$

(ii) $\nu(|\mathcal{S}_m \wr U(2,\mathbb{F}_{q^2})|) = (|U(2m,\mathbb{F}_{q^2})|)$

Proof: We prove (ii), (i) is similar. From II 6.7 we have

$$\nu(|U(2m,\mathbb{F}_{q^2})|) = \nu(\Pi_{i=1}^{2m}(q^i - (-1)^i)q^{i-1}) = \Sigma_{i=1}^{2m}\nu(q^i - (-1)^i).$$

By Corollary 2.3 we have for i even

$$\nu(q^i - (-1)^i) = \nu(q^i - 1) = \nu(i) + \nu(q - \epsilon)$$

where $\epsilon = \pm 1$ according as $q \equiv \pm 1 \pmod 4$. If i is odd then

$$\nu(q^i - (-1)^i) = \nu(q^i + 1) = \nu(q + 1) + \nu(\Sigma_{j=0}^{i-1}(-1)^i q^j) = \nu(q + 1)$$

$$= \nu(i) + \nu(q + 1)$$

Hence

$$\nu(|U(2m,\mathbb{F}_{q^2})|) = \Sigma_{i=1}^{m}(\nu(2i) + \nu(q-\epsilon)) + \Sigma_{i=1}^{m}(\nu(2i-1) + \nu(q + 1))$$

$$= \Sigma_{i=1}^{2m}\nu(i) + m[\nu(q-\epsilon) + \nu(q + 1)]$$

$$= \nu((2m)!) + m[\nu(q-\epsilon) + \nu(q + 1)]$$

In particular

$$\nu(|U(2,\mathbb{F}_{q^2})|) = 1 + \nu(q-\epsilon) + \nu(q + 1)$$

Hence

$$\nu(|\mathcal{S}_m \wr U(2,\mathbb{F}_{q^2})|) = \nu(m!) + m\nu(|U(2,\mathbb{F}_{q^2})|)$$

$$= \nu(m!) + m + m[\nu(q-\epsilon) + \nu(q + 1)]$$

But

$$\nu((2m)!) = \Sigma_{i=1}^{\infty}[\frac{2m}{2^i}] = \Sigma_{i=1}^{\infty}[\frac{m}{2^{i-1}}] = m + \Sigma_{i=1}^{\infty}[\frac{m}{2^i}] = m + \nu(m!)$$

Hence it follows that

$$\nu(|\mathcal{S}_m \wr U(2,\mathbb{F}_{q^2})|) = \nu(|U(2m,\mathbb{F}_{q^2})|).$$

Proof of Propositions 4.1 and 4.2: Identical to the proof of Proposition 2.1. We need only check that $\gamma_1 \otimes \gamma_1 \in GL(4,\mathbb{F}_q) \cap U(4,\mathbb{F}_{q^2})$.

Chapter VIII

Detection Theorems at Odd Primes

§1. Introduction

Let ℓ be an odd prime. In this chapter we give the mod ℓ versions of the detection results of Chapter VII. The idea of the proofs is closely analogous to the mod 2 case (see Introduction, Chapter VII) but differs in enough details to make a separate treatment much clearer.

In §2 we begin by collecting number theoretic preliminaries associated with the orders of the classical groups. In §3 we simultaneously treat $GL(m,\mathbb{F}_q)$ $q \equiv 1 \pmod{\ell}$ and $U(m,\mathbb{F}_{q^2})$ $q \equiv -1 \pmod{\ell}$. These results are then used in §4 to handle the more difficult orthogonal groups. Sections 5, 6, and 7 are devoted to $Sp(2m,\mathbb{F}_q)$, $GL(m,\mathbb{F}_q)$ and $U(m,\mathbb{F}_{q^2})$. Throughout this chapter homology will be taken with $\mathbb{Z}/\ell$ coefficients.

§2. Number theoretic preliminaries.

In this section we will collect several number theoretic results which will be used in the calculations of the succeeding sections.

Definition 2.1. Let b be an integer. Then $\nu_\ell(b)$ will denote the largest exponent k for which ℓ^k divides b.

Lemma 2.2. If $b \equiv 1 \pmod{\ell}$ then

$$\nu_\ell(b^\ell - 1) = 1 + \nu_\ell(b - 1)$$

Proof. Let $b = s\ell^k + 1$, where g.c.d.$(\ell,s) = 1$. Then

$$b^\ell - 1 = (s\ell^k + 1)^\ell - 1 = \ell s\ell^k + \frac{\ell(\ell-1)}{2}(s\ell^k)^2 + \Sigma_{j=3}^{\ell}\binom{\ell}{j}(s\ell^k)^j$$

$$= \ell^{k+1}[s + \ell(\frac{\ell-1}{2}\, s\ell^{k-1} + \Sigma_{j=3}^{\ell}\binom{\ell}{j} s^j \ell^{(j-1)k-2})]$$

Consequently

$$\nu_\ell(b^\ell - 1) = k + 1 = \nu_\ell(b - 1) + 1.$$

Corollary 2.3. If $b \equiv 1 \pmod{\ell}$ then

$$\nu_\ell(b^{\ell^k} - 1) = k + \nu_\ell(b - 1), \ k \geq 0$$

Proof. We proceed by induction on k, using Lemma 2.2. For $k = 0$ the statement is obviously true. Assuming this for $k - 1$ we get

$$\begin{aligned}\nu_\ell(b^{\ell^k} - 1) &= \nu_\ell((b^{\ell^{k-1}})^\ell - 1) = 1 + \nu_\ell(b^{\ell^{k-1}} - 1)\\ &= 1 + k - 1 + \nu_\ell(b - 1)\\ &= k + \nu_\ell(b - 1)\end{aligned}$$

which concludes the induction and proof.

Lemma 2.4. Let m be an integer relatively prime to ℓ. Let c be the smallest positive integer such that $m^c \equiv 1 \pmod{\ell}$. Let a be a positive integer. Then

$$\nu_\ell(m^a - 1) = \begin{cases} 0 & \text{if } c \text{ does not divide } a \\ \nu_\ell(a) + \nu_\ell(m^c - 1) & \text{if } c \text{ divides } a \end{cases}$$

Proof. If c does not divide a then $a = sc + t$, where $0 < t < c$. Hence

$$m^a = (m^c)^s m^t \equiv m^t \not\equiv 1 \pmod{\ell}$$

by the minimality of c. Hence $\nu_\ell(m^a - 1) = 0$.

Now suppose $c|a$. Then write $a = s\ell^k c$ where $k = \nu_\ell(a)$ so g.c.d.$(s,\ell) = 1$. Hence

$$\nu_\ell(m^a - 1) = \nu_\ell((m^{\ell^k c})^s - 1) = \nu_\ell[(m^{\ell^k c} - 1)\Sigma_{i=0}^{s-1}(m^c)^{i\ell^k}]$$

$$= \nu_\ell(m^{\ell^k c} - 1) + \nu_\ell(\Sigma_{i=0}^{s-1}(m^c)^{i\ell^k})$$

$$= \nu_\ell(m^{\ell^k c} - 1)$$

since $\Sigma_{i=0}^{s-1}(m^c)^{i\ell^k} \equiv \Sigma_{i=0}^{s-1} 1 \equiv s \not\equiv 0 \pmod{\ell}$. Hence by Corollary 2.3

$$\nu_\ell(m^a - 1) = \nu_\ell((m^c)^{\ell^k} - 1) = k + \nu_\ell(m^c - 1)$$

$$= \nu_\ell(a) + \nu_\ell(m^c - 1).$$

Definition 2.5. Let q be a fixed integer relatively prime to ℓ. We will denote by $u(+)$, $u(-)$, d the minimal positive integers such that

$$q^{u(+)} \equiv 1 \pmod{\ell}$$

$$(-q)^{u(-)} \equiv 1 \pmod{\ell}$$

$$(q^2)^d = q^{2d} \equiv 1 \pmod{\ell}$$

respectively.

Lemma 2.6. The integers $u(+)$ and $u(-)$ are divisors of $\ell-1$. The integer d is a divisor of $\frac{1}{2}(\ell-1)$. Moreover d divides both $u(+)$ and $u(-)$

Proof: According to Fermat's theorem

$$q^{\ell-1} \equiv 1 \pmod{\ell}$$

$$(-q)^{\ell-1} \equiv 1 \pmod{\ell}$$

The first statement follows immediately. The second statement follows from

$$(q^2)^{1/2(\ell-1)} = q^{\ell-1} \equiv 1 \pmod{\ell}$$

Finally we have

$$(q^2)^{u(+)} = q^{2u(+)} \equiv 1 \pmod{\ell}$$

$$(q^2)^{u(-)} = (-q)^{2u(-)} \equiv 1 \pmod{\ell}$$

Hence d divides both $u(+)$ and $u(-)$.

Lemma 2.7. Let q, $u(+)$, $u(-)$, d be as in Def. 2.5. Then

$$\nu_\ell(q^{u(+)} - 1) = \nu_\ell((-q)^{u(-)} - 1) = \nu_\ell(q^{2d} - 1)$$

Proof. According to Lemma 2.6, $u(+) = sd$, $u(-) = vd$. Similarly from

$$q^{2d} \equiv 1 \pmod{\ell} \qquad q^{2d} = (-q)^{2d} \equiv 1 \pmod{\ell}$$

it follows that $u(+)$ and $u(-)$ are divisors of $2d$. Hence $2d = ku(+)$, $2d = tu(-)$. Consequently

$$2d = ku(+) = ksd \Longrightarrow 2 = ks$$

$$2d = tu(-) = tvd \Longrightarrow 2 = tv$$

and k,s,t,v are all relatively prime to ℓ.

Applying Lemma 2.4 we obtain

$$\nu_\ell(q^{2d}-1) = \nu_\ell(q^{ku(+)}-1) = \nu_\ell(ku(+)) + \nu_\ell(q^{u(+)}-1) = \nu_\ell(q^{u(+)}-1)$$

$$\nu_\ell(q^{2d}-1) = \nu_\ell((-q)^{2d}-1) = \nu_\ell((-q)^{tu(-)}-1)$$

$$= \nu_\ell(tu(-)) + \nu_\ell((-q)^{u(-)}-1) = \nu_\ell((-q)^{u(-)}-1)$$

Definition 2.8. Let q, u(+), u(-), d be as in Def. 2.5. Then we will denote by ν the common value

$$\nu = \nu_\ell(q^{u(+)}-1) = \nu_\ell((-q)^{u(-)}-1) = \nu_\ell(q^{2d}-1)$$

§3. $GL(m,\mathbb{F}_q)$ $q \equiv 1 \pmod{\ell}$ and $U(m,\mathbb{F}_{q^2})$ $q \equiv -1 \pmod{\ell}$

In this section we will use Quillen's detection methods to prove the following result.

Theorem 3.1. (a) If $q \equiv 1 \pmod{\ell}$ then the direct sum homomorphism

$$GL(1,\mathbb{F}_q) \times GL(1,\mathbb{F}_q) \times \cdots \times GL(1,\mathbb{F}_q) \xrightarrow{\oplus} GL(m,\mathbb{F}_q)$$

induces an epimorphism in mod-ℓ homology.

(b) If $q \equiv -1 \pmod{\ell}$ then the direct sum homomorphism

$$U(1,\mathbb{F}_{q^2}) \times U(1,\mathbb{F}_{q^2}) \times \cdots \times U(1,\mathbb{F}_{q^2}) \xrightarrow{\oplus} U(m,\mathbb{F}_{q^2})$$

induces an epimorphism in mod-ℓ homology.

As a first step we show that

Proposition 3.2. (a) If $q \equiv 1 \pmod{\ell}$ then $\mathcal{S}_m \wr GL(1,\mathbb{F}_q)$ contains a Sylow ℓ-subgroup of $GL(m,\mathbb{F}_q)$.

(b) If $q \equiv -1 \pmod{\ell}$ then $\mathcal{S}_m \wr U(1,\mathbb{F}_{q^2})$ contains a Sylow ℓ-subgroup of $U(m,\mathbb{F}_{q^2})$.

Proof. We shall confine ourselves to proving (b). Case (a) is proved similarly. Since $q \equiv -1 \pmod{\ell}$ we have

$$\begin{aligned}\nu_\ell(|\mathcal{S}_m \wr U(1,\mathbb{F}_{q^2})|) &= \nu_\ell(|\mathcal{S}_m| \times |U(1,\mathbb{F}_{q^2})|^m) \\ &= \nu_\ell(|\mathcal{S}_m|) + m\nu_\ell(|U(1,\mathbb{F}_{q^2})|) \\ &= \nu_\ell(m!) + m\nu_\ell(q+1)\end{aligned}$$

On the other hand (cf. II §6.7) by Lemma 2.4

$$\begin{aligned}\nu_\ell(|U(m,\mathbb{F}_{q^2})|) &= \nu_\ell(\Pi_{i=1}^m (q^i-(-1)^i)q^{i-1}) \\ &= \Sigma_{i=1}^m \nu_\ell(q^i-(-1)^i) = \Sigma_{i=1}^m \nu_\ell((-q)^i-1) \\ &= \Sigma_{i=1}^m [\nu_\ell(i) + \nu_\ell(-q-1)] \\ &= \Sigma_{i=1}^m \nu_\ell(i) + m\nu_\ell(q+1) \\ &= \nu_\ell(m!) + m\nu_\ell(q+1)\end{aligned}$$

Consequently $\mathcal{S}_m \wr U(1,\mathbb{F}_{q^2})$ contains a Sylow ℓ-subgroup of $U(m,\mathbb{F}_{q^2})$.

Lemma 3.3. Let $\mathbb{Z}/\ell \subseteq \mathcal{S}_\ell$ be the subgroup generated by the ℓ-cycle $(1,2,3,\ldots,\ell)$.

(a) If $q \equiv 1 \pmod{\ell}$, then $\mathbb{Z}/\ell \wr GL(\ell^n,\mathbb{F}_q)$ contains a Sylow ℓ-subgroup of $GL(\ell^{n+1},\mathbb{F}_q)$

(b) If $q \equiv -1 \pmod{\ell}$, then $\mathbb{Z}/\ell \wr U(\ell^n,\mathbb{F}_{q^2})$ contains a Sylow ℓ-subgroup of $U(\ell^{n+1},\mathbb{F}_{q^2})$

Proof. Suppose $q \equiv 1 \pmod{\ell}$. Consider the commutative diagram of inclusions

$$\begin{array}{ccc} \mathcal{S}_m \wr \mathcal{S}_{\ell^n} \wr GL(1,\mathbb{F}_q) & \to & \mathcal{S}_\ell \wr GL(\ell^n,\mathbb{F}_q) \\ \downarrow & & \downarrow \\ \mathcal{S}_{\ell^{n+1}} \wr GL(1,\mathbb{F}_q) & \to & GL(\ell^{n+1},\mathbb{F}_q) \end{array}$$

Since $\nu_\ell(|\mathcal{S}_\ell \wr \mathcal{S}_{\ell^n}|) = (\ell^{n+1}-1)/(\ell-1) = \nu_\ell(|\mathcal{S}_{\ell^{n+1}}|)$ it follows from Prop. 3.2 that

$$\begin{aligned} \nu_\ell(|\mathcal{S}_\ell \wr \mathcal{S}_{\ell^{n+1}} \wr GL(1,\mathbb{F}_q)|) &= \nu_\ell(|\mathcal{S}_{\ell^{n+1}} \wr GL(1,\mathbb{F}_q)|) \\ &= \nu_\ell(|GL(\ell^{n+1},\mathbb{F}_q)|) \end{aligned}$$

Consequently we have

$$\nu_\ell(|\mathcal{S}_\ell \wr GL(\ell^n,\mathbb{F}_q)|) = \nu_\ell(|GL(\ell^{n+1},\mathbb{F}_q)|)$$

Since $\nu_\ell(|\mathbb{Z}/\ell|) = 1 = \nu_\ell(|\mathcal{S}_\ell|)$ it follows that

$$\nu_\ell(|\mathbb{Z}/\ell \wr GL(\ell^n,\mathbb{F}_q)|) = \nu_\ell(|GL(\ell^{n+1},\mathbb{F}_q)|)$$

This proves case (a). Case (b) is proved similarly.

Lemma 3.4 (a) If $q \equiv 1 \pmod{\ell}$, then the direct sum homomorphism

$$GL(\ell^n,\mathbb{F}_q) \times GL(\ell^n,\mathbb{F}_q) \times \cdots \times GL(\ell^n,\mathbb{F}_q) \xrightarrow{\oplus} GL(\ell^{n+1},\mathbb{F}_q)$$

induces an epimorphism in mod-ℓ homology.

(b) If $q \equiv -1 \pmod{\ell}$, then the direct sum homomorphism

$$U(\ell^n,\mathbb{F}_{q^2}) \times U(\ell^n,\mathbb{F}_{q^2}) \times \cdots \times U(\ell^n,\mathbb{F}_{q^2}) \xrightarrow{\oplus} U(\ell^{n+1},\mathbb{F}_{q^2})$$

induces an epimorphism in mod-ℓ homology.

Proof: We consider case (b): Let $q \equiv -1 \pmod{\ell}$. By Lemma 3.3 $\mathbb{Z}/\ell \wr U(\ell^n, \mathbb{F}_{q^2})$ contains a Sylow ℓ-subgroup of $U(\ell^{n+1}, \mathbb{F}_{q^2})$. It follows from Quillen's Lemma (cf. III 5.6) that $\mathbb{Z}/\ell \times U(\ell^n, \mathbb{F}_{q^2})$ and $U(\ell^n, \mathbb{F}_{q^2})^\ell$ detect $H_* BU(\ell^{n+1}, \mathbb{F}_{q^2})$. We shall show that $\mathbb{Z}/\ell \times U(\ell^n, \mathbb{F}_{q^2})$ is conjugate in $U(\ell^{n+1}, \mathbb{F}_{q^2})$ to a subgroup of $U(\ell^n, \mathbb{F}_{q^2})^\ell$.

We observe that $\mathbb{Z}/\ell \times U(\ell^n, \mathbb{F}_{q^2})$ is generated by the elements

$$\{\tau \otimes I_{\ell^n}, \{I_\ell \otimes M \mid M \in U(\ell^n, \mathbb{F}_{q^2})\}\}$$

where τ is the permutation matrix corresponding to the ℓ-cycle $(1\ 2\ 3\ \dots\ \ell)$. Let $\{e_i\}_{i=1}^{\ell}$ denote the standard basis for $F^{\ell}_{q^2}$. Let α denote the matrix which corresponds to the change of basis

$$e_i \to f_i = \Sigma_{j=0}^{\ell-1} \lambda^{\ell-ij} e_{j+1}$$

where λ denotes a primitive ℓ-th root of unity (which exists in $\mathbb{F}_{q^2}$ since $q^2 \equiv 1 \pmod{\ell}$). Observe that

$$H(f_i, f_k) = \Sigma_{j=0}^{\ell-1} \lambda^{\ell-ij} \lambda^{(\ell-kj)q} = \Sigma_{j=0}^{\ell-1} \lambda^{(k-i)j}$$

$$= \begin{cases} 1 + \lambda + \lambda^2 + \cdots + \lambda^{\ell-1} = 0 & \text{if } k \neq i \\ 1 + 1 + 1 + \cdots + 1 = \ell & \text{if } k = i. \end{cases}$$

Hence if we let $\alpha_n = \oplus_{i=1}^{\ell^{n-1}} \alpha$ then $\frac{1}{\ell}\alpha \otimes \alpha_n \in U(\ell^{n+1}, \mathbb{F}_{q^2})$. Also observe that $\alpha^{-1}\tau\alpha$ is the diagonal matrix

$$\alpha^{-1}\tau\alpha = [1, \lambda, \lambda^2, \dots, \lambda^{\ell-1}]$$

which is unitary since $\lambda\bar{\lambda} = \lambda^{q+1} = 1$. Then

$$(\tfrac{1}{\ell}\alpha \otimes \alpha_n)^{-1}(\tau \otimes I_{\ell^n})(\tfrac{1}{\ell}\alpha \otimes \alpha_n) = (\alpha^{-1}\tau\alpha) \otimes I_{\ell^n} = \oplus_{i=0}^{\ell-1}\lambda^i I_{\ell^n} \subseteq U(\ell^n, \mathbb{F}_{q^2})^\ell$$

$$(\tfrac{1}{\ell}\alpha \otimes \alpha_n)^{-1}(I_\ell \otimes M)(\tfrac{1}{\ell}\alpha \otimes \alpha_n) = I_\ell \otimes (\alpha_n^{-1}M\alpha_n) \subseteq U(\ell^n, \mathbb{F}_{q^2})^\ell$$

Therefore $\mathbb{Z}_\ell \times U(\ell^n, \mathbb{F}_{q^2})$ is conjugate to a subgroup of $U(\ell^n, \mathbb{F}_{q^2})^\ell$.

It follows that $U(\ell^n, \mathbb{F}_{q^2})^\ell$ detects $H_*BU(\ell^{n+1}, \mathbb{F}_{q^2})$. This proves case (b).

The proof in case (a) is similar but even simpler.

<u>Corollary</u> 3.5(a) If $q \equiv 1 \pmod{\ell}$ then the direct sum homomorphism

$$GL(1,\mathbb{F}_q) \times GL(1,\mathbb{F}_q) \times \cdots \times GL(1,\mathbb{F}_q) \xrightarrow{\oplus} GL(\ell^n,\mathbb{F}_q)$$

induces an epimorphism in mod-ℓ homology.

(b) If $q \equiv -1 \pmod{\ell}$ then the direct sum homomorphism

$$U(1,\mathbb{F}_{q^2}) \times U(1,\mathbb{F}_{q^2}) \times \cdots \times U(1,\mathbb{F}_{q^2}) \xrightarrow{\oplus} U(\ell^n,\mathbb{F}_{q^2})$$

induces an epimorphism in mod-ℓ homology.

<u>Proof</u>: This follows from Lemma 3.4 by induction on n.

<u>Lemma</u> 3.6. If $0 \leq k \leq \ell - 1$ and

(a) If $q \equiv 1 \pmod{\ell}$ then the direct sum homomorphism

$$GL(\ell^n,\mathbb{F}_q)^k \xrightarrow{\oplus} GL(k\ell^n,\mathbb{F}_q)$$

detects $H_*BGL(k\ell^n,\mathbb{F}_q)$

(b) If $q \equiv -1 \pmod{\ell}$ then the direct sum homomorphism

$$U(\ell^n,\mathbb{F}_{q^2})^k \to U(k\ell^n,\mathbb{F}_{q^2})$$

detects $H_*BU(k\ell^n,\mathbb{F}_{q^2})$.

<u>Proof</u>: In case (a), $q \equiv 1 \pmod{\ell}$ and we have a commutative diagram of inclusions

$$\begin{array}{ccc} \mathcal{S}_k \wr \mathcal{S}_{\ell^n} \wr GL(1,\mathbb{F}_q) & \to & \mathcal{S}_k \wr GL(\ell^n,\mathbb{F}_q) \\ \downarrow & & \downarrow \\ \mathcal{S}_{k\ell^n} \wr GL(1,\mathbb{F}_q) & \longrightarrow & GL(k\ell^n,\mathbb{F}_q) \end{array}$$

Since $\nu_\ell(|\mathcal{S}_k \wr \mathcal{S}_{\ell^n}|) = \nu_\ell(|\mathcal{S}_{k\ell^n}|)$, it follows from Prop. 3.2 that

$$\nu_\ell(|\mathcal{S}_k \wr \mathcal{S}_{\ell^n} \wr GL(1,\mathbb{F}_q)|) = \nu_\ell(|\mathcal{S}_{k\ell^n} \wr GL(1,\mathbb{F}_q)|) = \nu_\ell(|GL(k\ell^n,\mathbb{F}_q)|)$$

Therefore

$$\nu_\ell(|GL(k\ell^n,\mathbb{F}_q)|) = \nu_\ell(|\mathcal{S}_k \wr GL(\ell^n,\mathbb{F}_q)|) = \nu_\ell(|GL(\ell^n,\mathbb{F}_q)^k|)$$

so $GL(\ell^n,\mathbb{F}_q)^k$ detects $H_*BGL(k\ell^n,\mathbb{F}_q)$.

Case (b) is proved similarly.

<u>Proof of Theorem</u> 3.1. Suppose $q \equiv 1 \pmod{\ell}$. In Corollary 3.5 we showed that

$$GL(1,\mathbb{F}_q)^m \xrightarrow{\oplus} GL(m,\mathbb{F}_q)$$

detects $H_*BGL(m,\mathbb{F}_q)$ if m is of the form $m = \ell^n$. For general m, we write m in ℓ-adic form

$$m = \Sigma_{i=0}^{i} a_i \ell^i$$

where $0 \leq a_i \leq \ell - 1$. We then consider the commutative diagram of inclusions

$$\begin{array}{ccc}
\Pi_{i=0}^{k}(\mathcal{S}_{a_i \ell^i} \wr GL(1,\mathbb{F}_q)) & \xrightarrow{f} & \Pi_{i=0}^{k} GL(a_i \ell^i, \mathbb{F}_q) \\
\| & & \\
(\Pi_{i=0}^{k} \mathcal{S}_{a_i \ell^i}) \wr GL(1,\mathbb{F}_q) & & \downarrow g \\
\downarrow u & & \\
\mathcal{S}_m \wr GL(1,\mathbb{F}_q) & \xrightarrow{h} & GL(m,\mathbb{F}_q)
\end{array}$$

Since

$$\nu_\ell(|\Pi_{i=0}^{k} \mathcal{S}_{a_i \ell^i}|) = \nu_\ell(|\mathcal{S}_m|),$$

u_* is surjective. By Prop. 3.2, h_* is surjective. Hence g_* is surjective. By Lemma 3.6 and Cor. 3.5, $H_* BGL(a_i \ell^i, \mathbb{F}_q)$ is detected by $GL(1,\mathbb{F}_q)^{a_i \ell^i}$. Hence

$$GL(1,\mathbb{F}_q)^m = \Pi_{i=0}^{k} GL(1,\mathbb{F}_q)^{a_i \ell^i}$$

detects $H_* BGL(m,\mathbb{F}_q)$. This proves case (a).

Case (b) is proved by an identical argument.

§4. $O(m,\mathbb{F}_q)$ and $O(\bar{m},\mathbb{F}_q)$

In this section we will use Quillen's detection methods to construct generators for the mod-ℓ homology algebras

$$H_*B\bar{\mathcal{O}}(\mathbb{F}_q) = \oplus_{n\in\bar{\mathbb{N}}} H_*BO(n,\mathbb{F}_q) \quad q \text{ odd}$$

$$H_*B\bar{\mathcal{O}}^{ev}(\mathbb{F}_q) = \oplus_{n\in\tilde{\mathbb{N}}^{ev}} H_*BO(n,\mathbb{F}_q) \quad q \text{ even}$$

(Here as throughout this section we will use the notation of II 4.7. cf. also II 7.18). The two cases $\bar{\mathcal{O}}(\mathbb{F}_q)$ q odd and $\bar{\mathcal{O}}^{ev}(\mathbb{F}_q)$ q even are so similar that they will be treated in tandem, the second case being subsumed under the former.

<u>Definition</u> 4.1. Let d be the minimal positive integer such that $q^{2d} \equiv 1 \pmod{\ell}$ (as in Def. 2.3). If q is even we denote $\epsilon = 1$. If q is odd we denote $\epsilon = \pm 1$ according as $q \equiv \pm 1 \pmod 4$. Since

$$q^{2d} - 1 = q^{2d} - \epsilon^{2d} = (q^d - \epsilon^d)(q^d + \epsilon^d).$$

ℓ must divide exactly one of the factors $q^d - \epsilon^d$, $q^d + \epsilon^d$. Accordingly we define an element $c = c_q \in \tilde{\mathbb{N}}^{ev}$ given by

$$c = \begin{cases} 2d & \text{if } q^d \equiv \epsilon^d \pmod{\ell} \\ \overline{2d} & \text{if } q^d \equiv -\epsilon^d \pmod{\ell} \end{cases}$$

<u>Lemma</u> 4.2. The element c is the smallest index in $\tilde{\mathbb{N}}$ such that ℓ divides the order of $O(c,\mathbb{F}_q)$. More precisely ℓ divides $|O(m,\mathbb{F}_q)|$ iff $c \le m$.

<u>Proof</u>. According to II.4.5 and II.7.8

$$|O(2m-1,\mathbb{F}_q)| = |O(\overline{2m-1},\mathbb{F}_q)| = 2N_m \quad q \text{ odd}$$

$$\left.\begin{aligned} |O(2m,\mathbb{F}_q)| &= 2(q^m - \varepsilon^m)q^{m-1}N_m \\ |O(\overline{2m},\mathbb{F}_q)| &= 2(q^m + \varepsilon^m)q^{m-1}N_m \end{aligned}\right\} q \quad \text{even or odd}$$

where $N_m = \Pi_{i=1}^{m-1}(q^{2i} - 1)q^{2i-1}$ and $\varepsilon = \pm 1$ as defined in 4.1.

If ℓ divides $|O(2m,\mathbb{F}_q)|$ or $|O(\overline{2m},\mathbb{F}_q)|$, then either ℓ divides $q^{2i} - 1$ for some $i < m$ or ℓ divides $q^m \pm \varepsilon^m$. In the first case $d \leq i < m$ by minimality so that c is $\leq 2m$ or $\overline{2m}$ as the case may be. In the second case ℓ divides $q^{2m} - 1 = (q^m + \varepsilon^m)(q^m - \varepsilon^m)$ so again by minimality $d \leq m$. If $d < m$ then c is $\leq 2m$ or $\overline{2m}$ as the case may be. If $d = m$, then $c = 2m$ or $\overline{2m}$.

If ℓ divides $|O(2m-1,\mathbb{F}_q)|$ or $|O(\overline{2m-1},\mathbb{F}_q)|$, then ℓ divides N_m. Consequently ℓ must divide either $|O(2m-2,\mathbb{F}_q)|$ or $|O(\overline{2m-2},\mathbb{F}_q)|$. Hence by the previous argument $c \leq 2m - 2$ or $c \leq \overline{2m-2}$, which implies $c \leq 2m - 1$ or $c \leq \overline{2m - 1}$.

We now state the main results of this section.

Theorem 4.3(a). The direct sum homomorphism

$$O(c,\mathbb{F}_q) \times O(c,\mathbb{F}_q) \times \cdots \times O(c,\mathbb{F}_q) \xrightarrow{\oplus} O(mc,\mathbb{F}_q)$$

induces an epimorphism in mod-ℓ homology.

(b) Let $k \in \overline{\mathbb{N}}$. Let mc be the largest integral multiple of c such that $mc \leq k$. Then the standard inclusion

$$O(mc,\mathbb{F}_q) \to O(k,\mathbb{F}_q)$$

induces an epimorphism in mod-ℓ homology.

Theorem 4.4. Any Sylow ℓ-subgroup of $O(c,\mathbb{F}_q)$ is a cyclic subgroup $\mathbb{Z}/\ell^\nu$ where $\nu = \nu_\ell(q^{2d} - 1)$. An inclusion $\mathbb{Z}/\ell^\nu \subseteq O(c,\mathbb{F}_q)$ is given by the following chains of inclusions

(a) If $q^d \equiv -1 \pmod{\ell}$ then

$$\mathbb{Z}/\ell^\nu \to \mathbb{Z}/q^d+1 = U(1,\mathbb{F}_{q^{2d}}) \xrightarrow{i} O(c_{q^d},\mathbb{F}_{q^d}) \xrightarrow{j} O(c,\mathbb{F}_q)$$

where i is the monomorphism II 8.5($\bar{e}$) or ($\bar{f}$) given by the functor defined in II.8.5 and j is the monomorphism II.8.3($\bar{a}$) or ($\bar{f}$) defined by the forgetful functor II.8.3(a) or (f).

(b) If $q^d \equiv 1 \pmod{\ell}$ then

$$\mathbb{Z}/\ell^\nu \to \mathbb{Z}/q^d-1 = GL(1,\mathbb{F}_{q^d}) \xrightarrow{h} O(c_{q^d},\mathbb{F}_{q^2}) \xrightarrow{j} O(c,\mathbb{F}_q)$$

where h is the monomorphism II.8.4($\bar{a}$) or ($\bar{e}$) defined by the hyperbolic functor II.8.4(a) or (e) and j is as in (a).

Consequently the inclusion $\mathbb{Z}/\ell^\nu \to O(c,\mathbb{F}_q)$ induces an epimorphism in mod-ℓ homology. Moreover the mod-ℓ homology of $O(c,\mathbb{F}_q)$ is given by

$$H_nBO(c,\mathbb{F}_q) = \begin{cases} \text{cyclic on generator } \gamma_k \text{ if } n = 4kd \\ \text{cyclic on generator } \delta_k \text{ if } n = 4kd-1 \\ 0 \quad \text{otherwise} \end{cases}$$

where the generators γ_k, δ_k are images of generators of $H_*B\mathbb{Z}/\ell^\nu$ in corresponding degrees.

In what follows we use the notation of II 2.16(viii) and denote by [n] the generator of $H_0BO(n,\mathbb{F}_q)$.

Theorem 4.5(a) The homology algebra $H_*B\bar{\mathcal{O}}(\mathbb{F}_q) = \oplus_{n\in\bar{\mathbb{N}}} H_*BO(n,\mathbb{F}_q)$ q odd is generated by γ_i, δ_i $i > 0$, [1] and [$\bar{1}$].

(b) The homology algebra $H_*B\bar{\mathcal{O}}^{ev}(\mathbb{F}_q) = \oplus_{n\in\bar{\mathbb{N}}^{ev}} H_*BO(n,\mathbb{F}_q)$ q even is generated by γ_i, δ_i $i > 0$, [2] and [$\bar{2}$].

(c) The homology algebra $H_*BO(\infty,\mathbb{F}_q) = \lim_{n\to\infty} H_*BO(n,\mathbb{F}_q)$ q even or odd is generated by the images of γ_i, δ_i $i > 0$, under the natural inclusion $O(c,\mathbb{F}_q) \to O(\infty,\mathbb{F}_q)$.

As a first step towards proving Theorem 4.3 we show that

Proposition 4.6. $\mathcal{S}_m \wr O(c,\mathbb{F}_q)$ contains a Sylow ℓ-subgroup of $O(mc,\mathbb{F}_q)$ with

$$\nu_\ell(|\mathcal{S}_m \wr O(c,\mathbb{F}_q)|) = \nu_\ell(|O(cm,\mathbb{F}_q)|) = \Sigma_{i=1}^{m}\nu_\ell(q^{2di} - 1)$$

Proof: Define $\delta = \pm 1$ according as $q^d \equiv \pm\varepsilon^d \pmod{\ell}$. Since $\ell | (q^d - \delta\varepsilon^d)$, it follows that $\ell | (q^{dm} - (\delta\varepsilon^d)^m)$ and that $\ell \nmid (q^{dm} + (\delta\varepsilon^d)^m)$. Consequently,

$$\nu_\ell(q^{dm}-(\delta\varepsilon^d)^m) = \nu_\ell(q^{dm}-(\delta\varepsilon^d)^m) + \nu_\ell(q^{dm} + (\delta\varepsilon^d)^m) = \nu_\ell(q^{2dm}-(\delta\varepsilon^d)^{2m})$$

$$= \nu_\ell(q^{2dm} - 1)$$

In particular $\nu_\ell(q^d - \delta\varepsilon^d) = \nu_\ell(q^{2d} - 1) = \nu$.

Next we see that

$$\begin{aligned}\nu_\ell(|\mathcal{S}_m \wr O(c,\mathbb{F}_q)|) &= \nu_\ell(|\mathcal{S}_m| \times |O(c,\mathbb{F}_q)|^m) \\ &= \nu_\ell(|\mathcal{S}_m|) + m\,\nu_\ell(|O(c,\mathbb{F}_q)|) \\ &= \nu_\ell(m!) + m\nu\end{aligned}$$

while

$$\begin{aligned}\nu_\ell(|O(mc,\mathbb{F}_q)|) &= \nu_\ell((q^{dm}-(\delta\varepsilon^d)^m)\,q^{dm-1}\Pi_{i=1}^{dm-1}(q^{2i}-1)\,q^{2i-1}) \\ &= \nu_\ell(q^{dm}-(\delta\varepsilon^d)^m) + \Sigma_{i=1}^{dm-1}\nu_\ell(q^{2i}-1) \\ &= \nu_\ell(q^{2dm}-1) + \Sigma_{i=1}^{m-1}\nu_\ell(q^{2di} - 1)\end{aligned}$$

$$= \Sigma_{i=1}^{m} \nu_{\ell}(q^{2di} - 1)$$

$$= \Sigma_{i=1}^{m} [\nu_{\ell}(i) + \nu_{\ell}(q^{2d} - 1)]$$

$$= \nu_{\ell}(m!) + m\nu.$$

Consequently $\mathcal{S}_m \wr O(c,\mathbb{F}_q)$ contains a Sylow ℓ-subgroup of $O(mc,\mathbb{F}_q)$ and $\nu_{\ell}(|O(mc,\mathbb{F}_q)|)$ is as claimed.

Lemma 4.7. Theorem 4.3(a) holds if d = 1.

Proof If d = 1 then c = 2 or $\overline{2}$ and we have

$$q^2 - 1 = (q - 1)(q + 1) \equiv 0 \pmod{\ell}$$

Hence $q \equiv \pm 1 \pmod{\ell}$

Case 1 (Unitary Case). $q \equiv -1 \pmod{\ell}$.

Consider the monomorphism

$$U(m,\mathbb{F}_{q^2}) \to O_{\pm}(2m,\mathbb{F}_q) \tag{4.8}$$

defined in II 8.5($\overline{e}$) or ($\overline{f}$). Since $|U(1,\mathbb{F}_{q^2})| = q + 1$ and $|O(c,\mathbb{F}_q)| = 2(q + 1)$ are divisible by ℓ while $|O(c + \gamma,\mathbb{F}_q)| = 2(q-1)$ is not divisible by ℓ, it follows that in case m = 1 the inclusion (4.8) is given by

$$U(1,\mathbb{F}_{q^2}) \to O(c,\mathbb{F}_q)$$

Since the inclusions (4.8) arise from a permutative functor, these inclusions are compatible with respect to direct sums and wreath products. Consequently the inclusion (4.8) takes the form

$$U(m, \mathbb{F}_{q^2}) \to O(mc, \mathbb{F}_q)$$

and we get a commutative diagram of inclusions

$$\begin{array}{ccc} \mathcal{S}_m \wr U(1, \mathbb{F}_{q^2}) & \to & U(m, \mathbb{F}_{q^2}) \\ \downarrow & & \downarrow \\ \mathcal{S}_m \wr O(c, \mathbb{F}_q) & \to & O(mc, \mathbb{F}_q) \end{array}$$

Since

$$\nu_\ell(|U(1, \mathbb{F}_{q^2})|) = \nu_\ell(q + 1) = \nu_\ell(|O(c, \mathbb{F}_q)|)$$

it follows from Prop. 4.6 that

$$\nu_\ell(|\mathcal{S}_m \wr U(1, \mathbb{F}_{q^2})|) = \nu_\ell(|\mathcal{S}_m \wr O(c, \mathbb{F}_q)|) = \nu_\ell(|O(mc, \mathbb{F}_q)|)$$

Hence

(4.9) $$\nu_\ell(|U(m, \mathbb{F}_{q^2})|) = \nu_\ell(|O(mc, \mathbb{F}_q)|)$$

We now consider the commutative diagram of inclusions

$$\begin{array}{ccc} U(1, \mathbb{F}_{q^2})^m & \xrightarrow{f} & U(m, \mathbb{F}_{q^2}) \\ \downarrow & & \downarrow g \\ O(c, \mathbb{F}_q)^m & \xrightarrow{u} & O(mc, \mathbb{F}_q) \end{array}$$

By (4.9) g_* is an epimorphism. By Theorem 3.1(b), f_* is an epimorphism. It follows that u_* is an epimorphism. This completes the proof of Theorem 4.3(a) in Case 1.

Case 2 (Hyperbolic Case) $q \equiv 1 \pmod{\ell}$.

Consider the hyperbolic monomorphism

$$GL(m,\mathbb{F}_q) \to O_{\pm}(2m,\mathbb{F}_q)$$

defined in II 8.4($\bar{a}$) or ($\bar{e}$). By the same argument as in Case 1 these inclusions take the form

$$GL(m,\mathbb{F}_q) \to O(mc,\mathbb{F}_q)$$

and we have

$$\nu_{\ell}(|GL(m,\mathbb{F}_q)|) = \nu_{\ell}(|O(mc,\mathbb{F}_q)|) \tag{4.10}$$

Analyzing the analogous diagram

$$\begin{array}{ccc} GL(1,\mathbb{F}_q)^m & \to & GL(m,\mathbb{F}_q) \\ \downarrow & & \downarrow \\ O(c,\mathbb{F}_q)^m & \xrightarrow{u} & O(mc,\mathbb{F}_q) \end{array}$$

in a similar way as in Case 1 using Theorem 3.1(a), we obtain that u_* is an epimorphism, which completes the proof.

4.11. Proof of Theorem 4.3. Observe that it follows from Lemma 4.7 that

$$O(c_{q^d},\mathbb{F}_{q^2})^m \xrightarrow{\oplus} O(mc_{q^d},\mathbb{F}_{q^2})$$

induces an epimorphism in mod-ℓ homology.

Now consider the "forgetful" monomorphisms

$$O_\sigma(m,\mathbb{F}_{q^d}) \to O_\tau(md,\mathbb{F}_q)$$

$\sigma = \pm$, $\tau = \pm$ defined in II.8.3($\overline{b}$) or ($\overline{f}$). By the same reasoning as in the proof of Lemma 4.7, these inclusions must take the form

$$O(mc_{q^d},\mathbb{F}_{q^d}) \to O(mc, \mathbb{F}_q)$$

and we must have

$$\nu_\ell(|O(mc_{q^d},\mathbb{F}_{q^d})|) = \nu_\ell(|O(mc, \mathbb{F}_q)|) \tag{4.12}$$

Arguing as in Lemma 4.7 with the analogous diagram

$$\begin{array}{ccc} O(c_{q^d},\mathbb{F}_{q^d})^m & \longrightarrow & O(mc_{q^d},\mathbb{F}_{q^d}) \\ \downarrow & & \downarrow \\ O(c,\mathbb{F}_q)^m & \xrightarrow{\ u\ } & O(mc, \mathbb{F}_{q^d}) \end{array}$$

we see that u_* is an epimorphism. This proves part (a) of Theorem 4.3.

To prove part (b) observe that if m is the largest positive integer such that $mc \leq k$, then

$$\nu_\ell(|O(mc,\mathbb{F}_q)|) = \nu_\ell(|O(k,\mathbb{F}_q)|)$$

so that $O(mc,\mathbb{F}_q)$ contains a Sylow ℓ-subgroup of $O(k,\mathbb{F}_q)$ and hence detects $H_*BO(k,\mathbb{F}_q)$.

4.13. <u>Proof of Theorem</u> 4.4. The first part of Theorem 4.4 concerning the structure of the Sylow ℓ-subgroup of $O(c,\mathbb{F}_q)$ follows from the proofs of Lemma 4.7 and Theorem 4.3 especially 4.12, 4.9 and 4.10.

Let $w_i \in H_i BO(c,\mathbb{F}_q)$ be the image of a generator of $H_i B\mathbb{Z}/\ell^\nu$ under the inclusion

$$\mathbb{Z}/\ell^\nu \to O(c,\mathbb{F}_q)$$

The proof of Theorem 4.4 will be complete once we show that $w_i = 0$ unless i is of the form $4kd$ or $4kd - 1$.

We first consider the hyperbolic case where $q^d \equiv 1 \pmod{\ell}$ and $\mathbb{Z}/\ell^\nu$ includes via

$$\mathbb{Z}/\ell^\nu \to \mathbb{Z}/q^d-1 = GL(1,\mathbb{F}_{q^d}) \to O(c_{q^d},\mathbb{F}_{q^d}) \to O(c,\mathbb{F}_q)$$

Recall that according to Chap. II 8.4, this inclusion is given by considering $\mathbb{F}_{q^d} \oplus \mathbb{F}_{q^d}$ as a vector space over $\mathbb{F}_q$ with quadratic form

$$Q((x,y)) = tr(xy)$$

and by considering an element $u \in \mathbb{F}_q^* = GL(1,\mathbb{F}_{q^d})$ as acting on $\mathbb{F}_{q^d} \oplus \mathbb{F}_{q^d}$ by

$$u(x,y) = (ux,u^{-1}y)$$

Now let $\alpha\colon F_{q^d} \oplus F_{q^d} \to F_{q^d} \oplus F_{q^d}$ be the map given by $\alpha(x,y) = (x^q,y^q)$ and let $\beta\colon \mathbb{F}_{q^d} \oplus \mathbb{F}_{q^d} \to \mathbb{F}_{q^d} \oplus \mathbb{F}_{q^d}$ be the map given by $\beta(x,y) = (y,x)$. Then $\alpha \in O(c,\mathbb{F}_q)$ and $\beta \in O(c,\mathbb{F}_q)$. Now if a is a generator of $\mathbb{Z}_{\ell^\nu}$ then

$$\alpha a \alpha^{-1}(x,y) = \alpha a(x^{q^{d-1}},y^{q^{d-1}}) = \alpha(ax^{q^{d-1}},a^{-1}y^{q^{d-1}}) = (a^q x^{q^d},a^{-q}y^{q^d})$$

$$= (a^q x,a^{-q}y) = a^q(x,y)$$

$$\beta a \beta^{-1}(x,y) = \beta a(y,x) = \beta(ay,a^{-1}x) = (a^{-1}x,ay) = a^{-1}(x,y)$$

In other words $\alpha a\alpha^{-1} = a^q$ and $\beta a\beta^{-1} = a^{-1}$. Homologically this tells us that the following diagrams commute

$$\begin{array}{ccc} H_*(B\mathbb{Z}_{\ell^\nu}) & \xrightarrow{(q)_*} & H_*(B\mathbb{Z}_{\ell^\nu}) \\ & \searrow \quad \swarrow & \\ & H_*(BO(c,\mathbb{F}_q)) & \end{array} \qquad \begin{array}{ccc} H_*(B\mathbb{Z}_{\ell^\nu}) & \xrightarrow{(-1)_*} & H_*(B\mathbb{Z}_{\ell^\nu}) \\ & \searrow \quad \swarrow & \\ & H_*(BO(c,\mathbb{F}_q)) & \end{array}$$

This tells us that $q^i w_{2i} = w_{2i}$, $q^i w_{2i-1} = w_{2i-1}$, $(-1)^i w_{2i} = w_{2i}$ $(-1)^i w_{2i-1} = w_{2i-1}$ or equivalently

(a) $\quad (q^i - 1)w_{2i} = 0 \qquad (q^i - 1)w_{2i-1} = 0$

(b) $\quad (1-(-1)^i)w_{2i} = 0 \qquad (1-(-1)^i)w_{2i-1} = 0.$

From (a) we get $w_i = 0$ unless i is of the form $i = 2kd$ or $i = 2kd-1$. From (b) we get $w_i = 0$ unless i is of the form $i = 4k$ or $i = 4k-1$.

Now since ℓ divides $q^d - 1$, it follows that d is odd. (Otherwise $\frac{1}{2}d$ would be the smallest integer such that $q^{2(1/2d)} \equiv 1 \pmod{\ell}$.) Hence (a) and (b) imply that $w_i = 0$ unless $i = 4kd$ or $4kd - 1$.

The proof in the unitary case is similar.

We now relabel the elements w_i as follows: $w_{4kd} = \gamma_k$ and $w_{4kd-1} = \delta_k$. It follows that

$$H_n BO(c,\mathbb{F}_q) = \begin{cases} \text{cyclic on generator } \gamma_k \text{ if } n = 4kd \\ \text{cyclic on generator } \delta_k \text{ if } n = 4kd-1 \\ 0 \quad \text{otherwise} \end{cases}$$

Theorem 4.5 is an immediate consequence of Theorems 4.3 and 4.4.

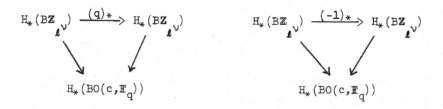

§5. $Sp(2m,\mathbb{F}_q)$

In this section we will use Quillen's detection methods to construct generators for the mod-ℓ homology algebra

$$H_*B\mathcal{S}p(\mathbb{F}_q) = \oplus_{n\in\mathbb{N}} Sp(2n,\mathbb{F}_q)$$

In most cases our proofs will be so analogous to those in §4 that little or no comment need be made.

As in §4 we begin by defining

Definition 5.1. Throughout this section (as in §4) d will denote the smallest positive integer such that $q^{2d} \equiv 1 \pmod{\ell}$.

As in Lemma 4.2 we show that

Lemma 5.2. The integer d is the smallest positive integer for which ℓ divides the order of $Sp(2d,\mathbb{F}_q)$: that is ℓ divides $|Sp(2m,\mathbb{F}_q)|$ iff $d \le m$.

We now state the main results of this section

Theorem 5.3(a) The direct sum homomorphism

$$Sp(2d,\mathbb{F}_q) \times Sp(2d,\mathbb{F}_q) \times \cdots \times Sp(2d,\mathbb{F}_q) \xrightarrow{\oplus} Sp(2md,\mathbb{F}_q)$$

induces an epimorphism in mod-ℓ homology

(b) Let $k\in\mathbb{N}$. Let md be the largest integral multiple of d such that $md \le k$. Then the standard inclusion

$$Sp(2md,\mathbb{F}_q) \to Sp(2k,\mathbb{F}_q)$$

induces an epimorphism in mod-ℓ homology.

Theorem 5.4. Any Sylow ℓ-subgroup of $Sp(2d,\mathbb{F}_q)$ is a cyclic subgroup $\mathbb{Z}/\ell^\nu$ where $\nu = \nu_\ell(q^{2d}-1)$. An inclusion $\mathbb{Z}/\ell^\nu \subseteq Sp(2d,\mathbb{F}_q)$ is given by the following chains of inclusions

(a) If $q^d \equiv -1 \pmod{\ell}$ then

$$\mathbb{Z}/\ell^\nu \to \mathbb{Z}/q^d + 1 = U(1,\mathbb{F}_{q^{2d}}) \xrightarrow{i} Sp(2,\mathbb{F}_{q^d}) \xrightarrow{j} Sp(2d,\mathbb{F}_q)$$

where i is the monomorphism II 8.5($\bar{d}$) given by the functor defined in II.8.5 and j is the monomorphism II 8.3($\bar{c}$) given by the forgetful functor II 8.3(c)

(b) If $q^d \equiv 1 \pmod{\ell}$ then

$$\mathbb{Z}/\ell^\nu \to \mathbb{Z}/q^d-1 = GL(1,\mathbb{F}_{q^d}) \xrightarrow{h} Sp(2,\mathbb{F}_{q^d}) \xrightarrow{j} Sp(2d,\mathbb{F}_q)$$

where h is the monomorphism II 8.4($\bar{b}$) defined by the hyperbolic functor II 8.4(b) and j is as in (a).

Consequently the inclusion $\mathbb{Z}/\ell^\nu \to Sp(2d,\mathbb{F}_q)$ induces an epimorphism in mod-ℓ homology. Moreover the mod-ℓ homology of $Sp(2d,\mathbb{F}_q)$ is given by

$$H_n BSp(2d,\mathbb{F}_q) = \begin{cases} \text{cyclic on generator } \sigma_k \text{ if } n = 4kd \\ \text{cyclic on generator } \tau_k \text{ if } n = 4kd-1 \\ 0 \quad \text{otherwise} \end{cases}$$

where the generators σ_k, τ_k are images of generators of $H_*B\mathbb{Z}/\ell^\nu$ in corresponding degrees.

In what follows we use the notation of II 2.16(viii) and denote by [2n] the generator of $H_0BSp(\overline{2n},\mathbb{F}_q)$.

Theorem 5.5(a). The homology algebra $H_*B\mathcal{Sp}(\mathbb{F}_q) = \oplus_{n \in \mathbb{N}} H_*BSp(2n,\mathbb{F}_q)$ is generated by σ_i, τ_i $i > 0$ and [2]

(b) The homology algebra $H_*BSp(\infty, \mathbb{F}_q) = \lim_{n\to\infty} H_*BSp(2n, \mathbb{F}_q)$ is generated by the images of σ_i, τ_i $i > 0$ under the natural inclusion $Sp(2d, \mathbb{F}_q) \to Sp(\infty, \mathbb{F}_q)$.

The proofs of Theorem 5.3, 5.4, and 5.5 proceed in an analogous fashion as the proofs of Theorem 4.3, 4.4 and 4.5. We list the necessary steps, commenting on proofs only in where they differ from the proofs of the corresponding results in the orthogonal case.

Proposition 5.6. $\mathcal{S}_m \wr Sp(2d, \mathbb{F}_q)$ contains a Sylow ℓ-subgroup of $Sp(2md, \mathbb{F}_q)$.

Lemma 5.7. Theorem 5.3(a) holds if $d = 1$.

Proof: The proof is completely analogous to that of Lemma 4.7 except that in the unitary case $q \equiv -1 \pmod{\ell}$, we use the monomorphism

$$U(m, \mathbb{F}_{q^2}) \to Sp(2m, \mathbb{F}_q)$$

defined in II 8.5($\overline{d}$); and in the hyperbolic case $q \equiv 1 \pmod{\ell}$, we use the hyperbolic monomorphism

$$GL(m, \mathbb{F}_q) \to Sp(2m, \mathbb{F}_q)$$

defined in II 8.4(b)

Proof of Theorem 5.3. This proceeds in a fashion completely analogous to that of Theorem 4.3, except that we use the "forgetful" monomorphisms

$$Sp(2m, \mathbb{F}_{q^d}) \to Sp(2md, \mathbb{F}_q)$$

defined in II 8.3($\bar{c}$).

Proof of Theorem 5.4. Again this proceeds in a fashion completely analogous to that of Theorem 4.4 except that in the hyperbolic case the map $\beta: \mathbb{F}_{q^d} \oplus \mathbb{F}_{q^d} \to \mathbb{F}_{q^d} \oplus \mathbb{F}_{q^d}$ is defined to be $\beta(x,y) = (y,-x)$ in order to have $\beta \in Sp(2d, \mathbb{F}_q)$.

Theorem 5.5 is then an immediate consequence of Theorems 5.3 and 5.4.

§6. $GL(m, \mathbb{F}_q)$

In this section we will use Quillen's detection methods to construct generators for the mod-ℓ homology algebra

$$H_* B\mathcal{GL}(\mathbb{F}_q) = \oplus_{n \in \mathbb{N}} GL(n, \mathbb{F}_q)$$

In most cases our proofs will be analogous to those in §4 and §5, and so we will be as brief as possible.

As in §4 and §5 we begin by defining

Definition 6.1. Throughout this section (cf Def. 2.5) $u(+)$ will denote the smallest positive integer such that $q^{u(+)} \equiv 1 \pmod{\ell}$.

As in Lemma 4.2 or 5.2 we show that

Lemma 6.2. The integer $u(+)$ is the smallest positive integer for which ℓ divides the order of $GL(u(+), \mathbb{F}_q)$: that is ℓ divides $|GL(m, \mathbb{F}_q)|$ iff $u(+) \leq m$.

We now state the main results of this section

Theorem 6.3(a) The direct sum homomorphism

$$GL(u(+),\mathbb{F}_q) \times GL(u(+),\mathbb{F}_q) \times \cdots \times GL(u(+),\mathbb{F}_q) \xrightarrow{\oplus} GL(mu(+),\mathbb{F}_q)$$

induces an epimorphism in mod-ℓ homology

(b) Let $k \in \mathbb{N}$. Let $mu(+)$ be the largest integral multiple of $u(+)$ such that $mu(+) \leq k$. Then the standard inclusion

$$GL(mu(+),\mathbb{F}_q) \to GL(k,\mathbb{F}_q)$$

induces an epimorphism in mod-ℓ homology.

<u>Theorem</u> 6.4. Any Sylow ℓ-subgroup of $GL(u(+),\mathbb{F}_q)$ is a cyclic subgroup $\mathbb{Z}/\ell^\nu$ where $\nu = \nu_\ell(q^{u(+)}-1)$. An inclusion $\mathbb{Z}/\ell^\nu \subseteq GL(u(+),\mathbb{F}_q)$ is given by the following chains of inclusions

$$\mathbb{Z}/\ell^\nu \to \mathbb{Z}/q^{u(+)}-1 = GL(1,\mathbb{F}_{q^{u(+)}}) \xrightarrow{j} GL(u(+),\mathbb{F}_q)$$

where j is the monomorphism II 8.3($\bar{a}$) given by the forgetful functor II 8.3(a).

Consequently the inclusion $\mathbb{Z}/\ell^\nu \to GL(u(+),\mathbb{F}_q)$ induces an epimorphism in mod-ℓ homology. Moreover the mod-ℓ homology of $GL(u(+),\mathbb{F}_q)$ is given by

$$H_n BGL(u(+),\mathbb{F}_q) = \begin{cases} \text{cyclic on generator } \alpha_k \text{ if } n = 2ku(+) \\ \text{cyclic on generator } \beta_k \text{ if } n = 2ku(+)-1 \\ 0 \quad \text{otherwise} \end{cases}$$

where the generators α_k, β_k are images of generators of $H_* B\mathbb{Z}/\ell^\nu$ in corresponding degrees.

In what follows we use the notation of II 2.16(viii) and denote by $[n]$ the generator of $H_0 BGL(n,\mathbb{F}_q)$.

Theorem 6.5(a). The homology algebra $H_*B\mathcal{GL}(\mathbb{F}_q) = \oplus_{n\in\mathbb{N}} H_*BGL(n,\mathbb{F}_q)$ is generated by α_i, β_i $i > 0$ and [1]

(b) The homology algebra $H_*BGL(\infty,\mathbb{F}_q) = \lim_{n\to\infty} H_*BGL(n,\mathbb{F}_q)$ is generated by the images of α_i, β_i $i > 0$ under the natural inclusion $GL(u(+),\mathbb{F}_q) \to GL(\infty,\mathbb{F}_q)$.

The proofs of Theorems 6.3, 6.4 and 6.5 proceed in an analogous manner as the proofs of Theorem 4.3, 4.4 and 4.5. We list the necessary steps, commenting on proofs only in where they differ from the proofs of the corresponding results in the orthogonal case.

Proposition 6.6. $\mathcal{S}_m \wr GL(u(+),\mathbb{F}_q)$ contains a Sylow ℓ-subgroup of $GL(mu(+),\mathbb{F}_q)$.

Lemma 6.7. Theorem 6.3(a) holds if $u(+) = 1$.

Proof. This is equivalent to Theorem 3.1(a)

Proof of Theorem 6.3. This proceeds in a fashion completely analogous to that of Theorem 4.3, except that we use the "forgetful" monomorphisms

$$GL(m,\mathbb{F}_{q^{u(+)}}) \to GL(mu(+),\mathbb{F}_q)$$

defined in II 8.3($\bar{a}$).

The proof of Theorem 6.4 proceeds in an entirely analogous fashion as that of Theorem 4.4 except that it is simpler.

Again Theorem 6.5 is an immediate consequence of Theorems 6.3 and 6.4.

§7. $U(m,\mathbb{F}_{q^2})$

In this section we will use Quillen's detection methods to

construct generators for the mod-ℓ homology algebra

$$H_* B\mathcal{U}(\mathbb{F}_{q^2}) = \oplus_{n\in\mathbb{N}} H_* BU(n,\mathbb{F}_{q^2})$$

In most cases our proofs will be so analogous to those in §4, §5 and §6, that little or no comment need be made.

As in §4 we begin by defining

Definition 7.1. Throughout this section (cf. Def. 2.5) $u(-)$ will denote the smallest positive integer such that $(-q)^{u(-)} \equiv 1 \pmod{\ell}$.

As in Lemma 4.2 we show that

Lemma 7.2. The integer $u(-)$ is the smallest positive integer for which ℓ divides the order of $U(m,\mathbb{F}_{q^2})$: that is ℓ divides $|U(m,\mathbb{F}_{q^2})|$ iff $m \leq u(-)$.

We now state the main results of this section

Theorem 7.3(a) The direct sum homomorphism

$$U(u(-),\mathbb{F}_{q^2}) \times U(u(-),\mathbb{F}_{q^2}) \times \cdots \times U(u(-),\mathbb{F}_{q^2}) \xrightarrow{\oplus} U(mu(-),\mathbb{F}_{q^2})$$

induces an epimorphism in mod-ℓ homology

(b) Let $k\in\mathbb{N}$. Let $mu(-)$ be the largest integral multiple of $u(-)$ such that $mu(-) \leq k$. Then the standard inclusion

$$U(mu(-),\mathbb{F}_{q^2}) \to U(k,\mathbb{F}_{q^2})$$

induces an epimorphism in mod-ℓ homology.

Theorem 7.4. Any Sylow ℓ-subgroup of $U(u(-),\mathbb{F}_{q^2})$ is a cyclic subgroup

$\mathbb{Z}/\ell^{\nu}$ where $\nu = \nu_{\ell}((-q)^{u(-)}-1)$. An inclusion $\mathbb{Z}/\ell^{\nu} \subseteq U(u(-),\mathbb{F}_{q^2})$ is given by the following chains of inclusions

(a) If u(-) is odd then

$$\mathbb{Z}/\ell^{\nu} \to \mathbb{Z}/q^{u(-)}+1 = U(1,\mathbb{F}_{q^{2u(-)}}) \xrightarrow{j} U(u(-),\mathbb{F}_{q^2})$$

where j is the monomorphism II 8.3($\overline{d}$) given by the forgetful functor II 8.3(d)

(b) If u(-) is even then u(-) = 2d where d is as in Def. 5.1 and

$$\mathbb{Z}/\ell^{\nu} \to Sp(2d,\mathbb{F}_q) \xrightarrow{i} U(u(-),\mathbb{F}_{q^2})$$

where the first inclusion is that given in Theorem 5.4 and i is the monomorphism II 8.5($\overline{a}$) given by the functor II 8.5(a).

Consequently the inclusion $\mathbb{Z}/\ell^{\nu} \to U(u(-),\mathbb{F}_{q^2})$ induces an epimorphism in mod-ℓ homology. Moreover the mod-ℓ homology of $U(u(-),\mathbb{F}_{q^2})$ is given by

$$H_nBU(u(-),\mathbb{F}_{q^2}) = \begin{cases} \text{cyclic on generator } \xi_k & \text{if } n = 2ku(-) \\ \text{cyclic on generator } \eta_k & \text{if } n = 2ku(-)-1 \\ 0 & \text{otherwise} \end{cases}$$

where the generators ξ_k, η_k are images of generators of $H_*B\mathbb{Z}/\ell^{\nu}$ in corresponding degrees.

In what follows we use the notation of II 2.16(viii) and denote by [n] the generator of $H_0BU(n,\mathbb{F}_{q^2})$.

<u>Theorem</u> 7.5(a). The homology algeba $H_*B\mathcal{U}(\mathbb{F}_{q^2}) = \oplus_{n\in\mathbb{N}}H_*BU(n,\mathbb{F}_{q^2})$ is generated by ξ_i, η_i $i > 0$ and [1]

(b) The homology algebra $H_*BU(\infty,\mathbb{F}_{q^2}) = \lim_{n\to\infty} H_*BU(n,\mathbb{F}_{q^2})$ is generated by the images of ξ_i, η_i $i > 0$ under the natural inclusion $U(u(-),\mathbb{F}_{q^2}) \to U(\infty,\mathbb{F}_{q^2})$.

The proofs of Theorems 7.3, 7.4, and 7.5 proceed in an analogous fashion as the proofs of Theorems 4.3, 4.4 and 4.5. We list the necessary steps, commenting on proofs only in where they differ from the proofs of the corresponding results in the orthogonal case.

Proposition 7.6. $\mathcal{S}_m \wr U(u(-),\mathbb{F}_{q^2})$ contains a Sylow ℓ-subgroup of $U(mu(-),\mathbb{F}_{q^2})$.

Lemma 7.7. Theorem 7.3(a) holds if $u(-) = 1$.

Proof. This is equivalent to Theorem 3.1(b).

Proof of Theorem 7.3. There are two cases to consider

Case 1. $u(-)$ is odd.

We then proceed as in the proof of Theorem 4.3 except that we use the "forgetful" monomorphism

$$U(m,\mathbb{F}_{q^{2u(-)}}) \to U(mu(-),\mathbb{F}_{q^2})$$

defined in II 8.3($\overline{d}$).

Case 2. $u(-)$ is even.

It is then easy to see that $u(-) = 2d$ where d is as in Def. 5.1. It then follows that

$$\nu_\ell(|Sp(2d,\mathbb{F}_q)|) = \nu_\ell(q^{2d}-1) = \nu_\ell((-q)^{u(-)}-1) = \nu_\ell(|U(u(-),\mathbb{F}_{q^2})|)$$

and consequently by Prop. 5.6 and 7.6

$$\nu_\ell(|Sp(2md,\mathbb{F}_q)|) = \nu_\ell(|\mathcal{S}_m \wr Sp(2d,\mathbb{F}_q)|) = \nu_\ell(|\mathcal{S}_m \wr U(u(-),\mathbb{F}_{q^2})|)$$

$$= \nu_\ell(|U(mu(-),\mathbb{F}_{q^2})|)$$

It follows that the monomorphism II 8.5($\bar{a}$)

$$Sp(2md,\mathbb{F}_q) \to U(mu(-),\mathbb{F}_{q^2})$$

maps epimorphically in mod-ℓ homology. We now argue as in the proof of Lemma 4.7 using the diagram

$$\begin{array}{ccc} Sp(2d,\mathbb{F}_q)^m & \to & Sp(2md,\mathbb{F}_q) \\ \downarrow & & \downarrow \\ U(u(-),\mathbb{F}_q)^m & \to & U(mu(-),\mathbb{F}_{q^2}) \end{array}$$

to show that the bottom arrow maps epimorphically in mod ℓ-homology.

Thus we prove part (a) of Theorem 7.3. Part (b) is proved exactly in the manner of Theorem 4.3(b).

The proof of Theorem 7.4 proceeds in the manner of the proof of Theorem 4.4 if u(-) is odd. If u(-) is even, then Theorem 7.4 is an immediate consequence of Theorem 5.4.

Again Theorem 7.5 follows from Theorems 7.3 and 7.4.

Chapter IX

Homology Operations Associated with the Classical Groups

§1. Introduction

In this chapter we calculate the (additive) mod-2 homology operations on the classifying spaces $B\mathcal{A}$ of the various permutative categories treated in Chapter IV and on their associated infinite loop spaces $\Gamma B\mathcal{A}$. These results were used at several crucial points in Chapter IV and in Chapter VI. Throughout this section homology will always be taken with coefficients in $Z/2$. The finite fields $\mathbb{F}_q$, $\mathbb{F}_{q^2}$ considered in this chapter will all have odd characteristic.

The basic method used in our computation is that employed in Priddy [33]. We shall treat the orthogonal category $\mathcal{O}(\mathbb{F}_q)$ in §2, the symplectic category $\mathcal{S}p(\mathbb{F}_q)$ in §3, and the general linear and unitary categories $\mathcal{GL}(\mathbb{F}_q)$ and $\mathcal{U}(\mathbb{F}_{q^2})$ in §4.

For reference throughout this chapter we recall some basic notions about homology operations. For details and proofs the reader is referred to May [27].

Let $E\mathcal{S}_2$ denote a contractible space on which the symmetric group $\mathcal{S}_2$ acts freely. Let $\mathcal{A} = \coprod G(n)$ denote one of the permutative categories of Chapter II. Then the wreath product construction gives a Dyer-Lashof map

(1.1) $$\theta\colon E\mathcal{S}_2 \times_{\mathcal{S}_2} (B\mathcal{A})^2 \to \coprod E\mathcal{S}_2 \times_{\mathcal{S}_2} (BG(n))^2$$

$$= \coprod B(\mathcal{S}_2 \wr G(n)) \to \coprod BG(2n) \to B\mathcal{A}$$

Furthermore there is a compatible Dyer-Lashof map

$$\theta\colon E\mathcal{S}_2 \times_{\mathcal{S}_2} (\Gamma B\mathcal{A})^2 \to \Gamma B\mathcal{A}$$

which makes the following diagram commute

$$\begin{array}{ccc} E\mathcal{S}_2 \times_{\mathcal{S}_2} (B\mathcal{A})^2 & \xrightarrow{\theta} & B\mathcal{A} \\ \downarrow & & \downarrow \\ E\mathcal{S}_2 \times_{\mathcal{S}_2} (\Gamma B\mathcal{A})^2 & \xrightarrow{\theta} & \Gamma B\mathcal{A} \end{array} \tag{1.2}$$

where $B\mathcal{A} \to \Gamma B\mathcal{A}$ is the natural map (cf. II 2.17).

Now given a Dyer-Lashof operation

$$\theta\colon E\mathcal{S}_2 \times_{\mathcal{S}_2} X^2 \to X$$

on an E_∞ space X we can construct homology operations in H_*X as follows: Let W_* denote the standard $\mathbb{Z}/2[\mathbb{Z}/2]$ free resolution of $\mathbb{Z}/2$ and let $s_*\colon W_* \to C_*(E\mathcal{S}_2)$ be a map of $\mathbb{Z}/2[\mathbb{Z}/2]$ complexes. We define $\theta_*\colon H_*(W_* \otimes (C_*(X))^2) \to H_*(X)$ to be the map induced in homology by the composite

$$W_* \otimes (C_*(X))^2 \xrightarrow{s_*\otimes\eta} C_*(E\mathcal{S}_2) \otimes C_*(X^2) \xrightarrow{\eta} C_*(E\mathcal{S}_2 \times X^2) \xrightarrow{C_*(\theta)} C_*(X)$$

Then the homology operations $Q^t\colon H_i(X) \to H_{i+t}(X)$ are defined by the formulas

$$Q^t(x) = 0 \quad \text{if} \quad t < i \tag{1.3}$$

$$Q^t(x) = \theta_*(e_{t-i} \otimes x \otimes x) \quad \text{if} \quad t \geq i \tag{1.4}$$

where $e_j \in W_j$ denotes the standard generator.

For future reference we list some basic properties of homology operations in the homology algebra $H_*(X)$

(1.5) $$Q^t(x) = x^2 \quad \text{if} \quad t = \text{degree}(x)$$

(1.6) $Q^t([0]) = 0$ if $t > 0$ (where $[0] \in H_0(X)$ denotes the generator of the zero component, cf. II 2.16(viii))

(1.7) (Cartan formula) $Q^t(xy) = \Sigma_{a+b=t} Q^a(x) Q^b(y)$

(1.8) The Q^t are stable, ie. $Q^t \sigma_* = \sigma_* Q^t$ where

$$\sigma_*: \tilde{H}_*(\Omega X) \to H_{*+1}(X) \text{ is the homology suspension}$$

(1.9) (Adem relations) If $r > 2t$, then

$$Q^r Q^t = \Sigma_i (2i-r, r-i-t-1) Q^{r+t-i} Q^i$$

(1.10) The Q^t are natural with respect to infinite loop maps

(1.11) If X is an infinite loop space and $\chi: H_*X \to H_*X$ is the automorphism induced by $-1: X \to X$, then $Q^t \chi = \chi Q^t$. (This is a special case of 1.10 since $-1: X \to X$ is an infinite loop map)

In the cases $X = B\mathcal{O} = \coprod BG(n)$ and $X = \Gamma B\mathcal{O}$ that we consider in this chapter it follows from 1.1 and 1.2 that

$$Q^t: H_i BG(n) \to H_{i+t} BG(2n)$$

$$Q^t: H_i \Gamma_n B\mathcal{O} \to H_{i+t} \Gamma_{2n} B\mathcal{O}$$

Moreover it follows from diagram 1.2 that the natural map

$$(1.12) \qquad H_*B\mathcal{A} = \oplus H_*BG(n) \to H_*\Gamma B\mathcal{A}$$

preserves homology operations. Our technique will be to evaluate the wreath product maps

$$B(\mathcal{S}_2 \wr G(n)) \to BG(2n)$$

and use (1.1) to compute homology operations in $H_*B\mathcal{A}$ and then use (1.12) and properties (1.5) - (1.11) to extend these computations to $H_*\Gamma B\mathcal{A}$.

§2. Homology operations in $H_*B\bar{\mathcal{O}}(\mathbb{F}_q)$ and $H_*\Gamma B\bar{\mathcal{O}}(\mathbb{F}_q)$

In this section we prove the following results on homology operations in the mod-2 homology algebras $H_*B\bar{\mathcal{O}}(\mathbb{F}_q)$ and $H_*\Gamma B\bar{\mathcal{O}}(\mathbb{F}_q)$:

Theorem 2.1. The homology operations in $H_*B\bar{\mathcal{O}}(\mathbb{F}_q) = \oplus_{n\in\tilde{\mathbb{N}}} H_*BO(n,\mathbb{F}_q)$ are given in the standard generators of Theorem IV 2.1 by

(a) If $q \equiv \pm 1 \pmod 8$

$$Q^n(v_k) = \Sigma_{t=0}^k (t, n-k-1)\, v_{n+t}v_{k-t} \qquad n > k$$

$$Q^n(y_k) = \Sigma_{t=0}^k (t, n-k-1)\, y_{n+t}y_{k-t} \qquad n > k$$

(b) If $q \equiv \pm 3 \pmod 8$

$$Q^n(v_k) = \Sigma_{t=0}^k (t, n-k-1)\, y_{n+t}y_{k-t} \qquad n > k$$

$$Q^n(y_k) = \Sigma_{t=0}^k (t, n-k-1)\, v_{n+t}v_{k-t} \qquad n > k$$

while in all cases

$$Q^n(v_n) = v_n^2 = y_n^2 = Q^n(y_n)$$

$$Q^n(v_k) = 0 = Q^n(y_k) \qquad n < k$$

<u>Theorem</u> 2.2. The homology operations in $H_*\Gamma B\overline{O}(\mathbb{F}_q)$ are given on the standard generators of Theorem IV 2.3 by

(a) If $q \equiv \pm 1 \pmod 8$ and $n > k$

$$Q^n(\overline{v}_k * [1]) = \Sigma_{a+b=n+k}(a-n, n-k-1)\overline{v}_a\overline{v}_b * [2]$$

$$Q^n(\overline{v}_k) = \Sigma_{a+b+c=n+k}(k-c, n-a-k-1)\,\chi(\overline{v}_a)\overline{v}_b\overline{v}_c$$

(b) If $q \equiv \pm 3 \pmod 8$ and $n > k$

$$Q^n(\overline{v}_k * [1]) = \Sigma_{a+b+c+d=n+k}(a+c-n, n-k-1)\overline{v}_a\overline{v}_b\overline{u}_c\overline{u}_d * [2]$$

$$Q^n(\overline{v}_k) = \Sigma_{a+b+c+d+e+f=n+k}(k-c-f, n-a-d-k-1)\,\chi(\overline{v}_a)\overline{v}_b\overline{v}_c\overline{u}_d\overline{u}_e\overline{u}_f$$

(c) In all cases

$$Q^n(\overline{v}_n) = v_n^2 \qquad Q^n(\overline{v}_k) = 0 \quad \text{if} \quad n < k$$

$$Q^n(\overline{u}_k) = \Sigma_{a+b+c=n+k}(k-a, n-k-b-1)\overline{u}_a\overline{u}_b\overline{u}_c$$

$$Q^n([\gamma]) = \overline{u}_n \quad \text{where} \quad \gamma = 1 - \overline{1}.$$

As a first step we prove that

<u>Lemma</u> 2.3. The following formula holds for all $n, k \geq 0$

$$\Sigma_{s=0}^{[k/2]}(s, k-2s)\,Q^{n+s}(v_s) = \Sigma_{t=0}^{k}(t, n)\,\rho_*(v_{n+1}v_{k-t})$$

where $\rho\colon B\bar{\mathcal{O}}(\mathbb{F}_q) \to B\bar{\mathcal{O}}(\mathbb{F}_q)$ is the identity map if $q \equiv \pm 1 \pmod 8$ and $\rho = \Phi\colon B\bar{\mathcal{O}}(\mathbb{F}_q) \to B\bar{\mathcal{O}}(\mathbb{F}_q)$ is the automorphism of II 4.4 if $q \equiv \pm 3 \pmod 8$.

Proof. We begin with the commutative diagram

$$\begin{array}{ccccccc}
 & \mathcal{S}_2 \times \mathcal{S}_2 & \xrightarrow{1\times\Delta} & \mathcal{S}_2 \wr \mathcal{S}_2 \to \mathcal{S}_2 \wr O(1,\mathbb{F}_q) & \xrightarrow{j} & O(2,\mathbb{F}_q) \\
 & \Big\downarrow 1\times\Delta & & & & \Big\uparrow D(\cdot)D^{-1} \\
\mathcal{S}_2 \times \mathcal{S}_2 \times \mathcal{S}_2 & \xrightarrow{\mu\times 1} & \mathcal{S}_2 \times \mathcal{S}_2 = O(1,\mathbb{F}_q) \times O(1,\mathbb{F}_q) & & \xrightarrow{\oplus} & O(2,\mathbb{F}_q)
\end{array}$$

where

$$D = \begin{pmatrix} -1 & 1 \\ 1 & 1 \end{pmatrix}$$

μ = group multiplication for $\mathcal{S}_2$

Let us now apply $H_*B(\cdot)$ to the diagram. The map $1 \times \Delta\colon \mathcal{S}_2 \times \mathcal{S}_2 \to \mathcal{S}_2 \wr \mathcal{S}_2$ has been evaluated in homology by Adem (see May [27]). Consider the resulting diagram for the element $x_n \otimes x_k$

$$\begin{array}{ccc}
x_n \otimes x_k & \xrightarrow{B(1\times\Delta)_*} & \Sigma_{s=0}^{[k/2]}(s,k-2s)\, x_{n+2s-k} \otimes v_{k-s}^2 \\
 & & \Big\downarrow (Bj)_* \\
\Big\downarrow & & \Sigma_{s=0}^{[k/2]}(s,k-2s)\, Q^{n+s}(v_{k-s}) \\
 & & \Big\uparrow \rho_* \\
\Sigma_{t=0}^{k} x_n \otimes x_t \otimes x_{k-t} & \xrightarrow{B(\mu\times 1)_*} & \Sigma_{t=0}^{k}(t,n)\, v_{n+t}v_{k-t} \\
 & & \Big\downarrow B\oplus_* \\
 & & \Sigma_{t=0}^{k}(t,n)\, v_{n+t}v_{k-t}
\end{array}$$

where ρ_* is the map induced on homology by conjugation by the matrix D. This gives the required formula

$$\Sigma_{s=0}^{[k/2]}(s,k-2s)Q^{n+s}(v_{k-s}) = \Sigma_{t=0}^{k}(t,n)\rho_*(v_{n+t}v_{k-t})$$

If $q \equiv \pm 1 \pmod 8$ then 2 is a square in $\mathbb{F}_q$ and conjugation by D is equivalent to conjugation by $\frac{1}{\sqrt{2}}D \in O(2,\mathbb{F}_q)$ and hence ρ_* induces the identity on homology.

If $q \equiv \pm 3 \pmod 8$ then 2 is a nonsquare in $\mathbb{F}_q$ and in Theorem II 3.12 we may take conjugation by

$$\begin{pmatrix} -a & b \\ b & a \end{pmatrix} = \begin{pmatrix} -1 & 1 \\ 1 & 1 \end{pmatrix}$$

to define the automorphism $\Phi: \bar{O}(\mathbb{F}_q) \to \bar{O}(\mathbb{F}_q)$. Consequently in this case $\rho_* = \Phi_*$.

2.4 <u>Proof of Theorem</u> 2.1. By Prop. IV 3.1 it is enough to show that

$$(*) \qquad Q^n(v_k) = \Sigma_{t=0}^{k}(t,n-k-1)\rho_*(v_{n+t}v_{k-t})$$

for $n > k$. (The statements about $Q^n(v_k)$ $n \leq k$ are immediate from 1.3 and 1.5.) We proceed by induction on k using Lemma 2.3.

For $k = 0$ we have

$$Q^n(v_0) = \rho_*(v_n v_0)$$

in agreement with Lemma 2.3. Assume we have proved (*) for $k' < k$. Then Lemma 2.3 gives

$$Q^n(v_k) = \Sigma_{s=1}^{k}(s,k-2s)Q^{n+s}(v_{k-s}) + \Sigma_{t=0}^{k}(t,n)\rho_*(v_{n+t}v_{k-t})$$

$$= \Sigma_{s=1}^{k}\Sigma_{\ell=0}^{k-s}(s,k-2s)(\ell,n-k+2s-1)\rho_*(v_{n+s+\ell}v_{k-s-\ell})$$

$$+ \Sigma_{t=0}^{k}(t,n)\rho_*(v_{n+t}v_{k-t})$$

$$= \Sigma_{r=1}^{k}\Sigma_{s=1}^{r}(s,k-2s)(r-s,n-k+2s-1)\rho_*(v_{n+r}v_{k-r})$$

$$+ \Sigma_{t=0}^{k}(t,n)\rho_*(v_{n+t}v_{k-t})$$

We now apply Adem's binomial coefficient identity for integers a,b,c

$$\Sigma_{s=0}^{c}(s,a-2s)(c-s,b-c+2s) \equiv (c,a+b-c+1) \pmod 2$$

Setting $a = k$, $b = n-k+r-1$ and $c = r$ we obtain

$$Q^n(v_k) = \Sigma_{r=1}^{k}[(r,n) + (r,n-k-1)]\rho_*(v_{n+r}v_{k-r})$$

$$+ \Sigma_{t=0}^{k}(t,n)\rho_*(v_{n+t}v_{k-t})$$

$$= \Sigma_{r=1}^{k}(r,n-k-1)\rho_*(v_{n+r}v_{k-r}) + \rho_*(v_n v_k)$$

$$= \Sigma_{t=0}^{k}(t,n-k-1)\rho_*(v_{n+t}v_{k-t})$$

This completes the induction and proof.

2.5. <u>Proof of Theorem</u> 2.2. Since $\overline{v}_k = v_k*[-1]$ the statements about $Q^n(\overline{v}_k*[1])$ are merely restatements of Theorem 2.1 using IV.2.8. We then calculate $Q^n(\overline{v}_k)$ using the Cartan formula and 1.11 and IV.2.10

$$Q^n(\overline{v}_k) = Q^n((\overline{v}_k*[1])*\chi(v_0)) = \Sigma_{i=0}^n Q^i(\overline{v}_k*[1])\chi Q^{n-i}(v_0)$$

The formula for $Q^n(\overline{u}_k)$ is obtained from the formulas for $Q^n(\overline{v}_k)$ by applying the map $(\Phi-1)_*$ using Prop. IV.3.2 and Cor. IV.2.10. If $q \equiv \pm 3 \pmod 8$

$$\begin{aligned}
Q^n(\overline{u}_k) &= Q^n(\Phi-1)_*(\overline{v}_k) = (\Phi-1)_*Q^n(\overline{v}_k)\\
&= (\Phi-1)_*[\Sigma_{a+b+c+d+e+f=n+k}(k-c-f,n-a-d-k-1)\,\chi(\overline{v}_a)\overline{v}_b\overline{v}_c\overline{u}_d\overline{u}_e\overline{u}_f]\\
&= \Sigma_{a+b+c=n+k}(k-c,n-a-k-1)\,\chi(\overline{u}_a)\overline{u}_b\overline{u}_c\\
&= \Sigma_{a+b+c=n+k}(k-c,n-a-k-1)\overline{u}_a\overline{u}_b\overline{u}_c\\
&= \Sigma_{a+b+c=n+k}(k-a,n-b-k-1)\overline{u}_a\overline{u}_b\overline{u}_c
\end{aligned}$$

A similar argument works if $q \equiv \pm 1 \pmod 8$

We also have

$$Q^n([\gamma]) = Q^n(y_0\chi(v_0)) = \Sigma_{i=0}^n Q^i(y_0)\chi Q^{n-i}(v_0)$$

$$= \begin{cases} \Sigma_{i=0}^n y_i y_0 \chi(v_{n-i}v_0) = \Sigma_{i=0}^n \overline{y}_i\chi(\overline{v}_{n-i}) = \overline{u}_n & q \equiv \pm 1 \pmod 8 \\ \Sigma_{i=0}^n v_i v_0 \chi(y_{n-i}y_0) = \chi(\Sigma_{i=0}^n \chi(\overline{v}_i)\overline{y}_{n-i}) = \chi(\overline{u}_n) = \overline{u}_n & \text{if } q \equiv \pm 3 \pmod 8 \end{cases}$$

This completes the proof.

We conclude with an application of homology operations

<u>Theorem</u> 2.6. The 2-primary components of $\Gamma_0 B\overline{\mathcal{O}}(\mathbb{F}_q) \simeq JO(q)$ and $\Gamma_0 B\mathcal{N}^{ev}(\mathbb{F}_q) \simeq \Gamma_0 B\mathcal{N}\mathcal{D}^{ev}(\mathbb{F}_q) \simeq J(q)$ are indecomposable as infinite loop spaces if $q \equiv \pm 3 \pmod 8$.

Proof. Assume that

$$\Gamma_0 B\bar{\mathcal{O}}(\mathbb{F}_q)_{(2)} \simeq X \times Y$$

as an infinite loop space. We may assume that $\bar{v}_1 \in H_* X$.

Now label the primitive elements of $H_* \Gamma_0 B\bar{\mathcal{O}}(\mathbb{F}_q)$ as follows:

$$P_* \mathbb{Z}/2[\bar{v}_1, \bar{v}_2, \ldots] = \{\bar{p}_1, \bar{p}_2, \ldots\}$$

$$P_* E[\bar{u}_i \mid i \geq 1] = \{\bar{q}_1, \bar{q}_3, \ldots, \bar{q}_{2n+1}, \ldots\}$$

Define $\bar{q}_n = 0$ if n is even.

Since if k is odd

$$\bar{p}_k = \bar{v}_k + \text{decomposables}$$

$$\bar{q}_k = \bar{u}_k + \text{decomposables}$$

and Q^n sends decomposable elements to decomposable elements, it follows from Theorem 2.2. that

$$Q^n \bar{p}_k = (k, n-k-1)[\bar{p}_{n+k} + \bar{q}_{n+k}]$$

$$Q^n \bar{q}_k = (k, n-k-1)\bar{q}_{n+k}$$

if n is even and k is odd. Consequently for all $n \geq 1$

$$\bar{p}_{2n+1} + \bar{q}_{2n+1} = Q^{2n} \bar{v}_1 \in H_* X$$

and hence also

$$\bar{p}_{2n} = \bar{p}_n^2 = (\bar{p}_n + \bar{q}_n)^2 \in H_* X$$

Now $\bar{v}_1 + \bar{u}_1$ cannot lie in H_*Y since then we would have

$$\bar{v}_1^2 = (\bar{v}_1 + \bar{u}_1)^2 \in H_*Y$$

while on the other hand $\bar{v}_1^2 \in H_*X$.

We cannot have $u_1 \in H_*Y$ for then we would have

$$\bar{q}_7 = (3,0)\bar{q}_7 = Q^4(\bar{q}_3) = Q^4Q^2(\bar{u}_1) \in H_*Y$$

while on the other hand

$$\bar{q}_7 = [\bar{p}_7 + \bar{q}_7] - \bar{p}_7 = [\bar{p}_7 + \bar{q}_7] - \Phi_*[(3,0)(\bar{p}_7 + \bar{q}_7)]$$

$$= [\bar{p}_7 + \bar{q}_7] - \Phi_*Q^4(\bar{p}_3) = [\bar{p}_7 + \bar{q}_7] - Q^4(\bar{p}_3 + \bar{q}_3) \in H_*X$$

Therefore $\bar{u}_1, \bar{v}_1 + \bar{u}_1 \in H_*X$ and

$$Q^{2n}\bar{u}_1 = (1,2n-2)\bar{q}_{2n+1} = \bar{q}_{2n+1} \in H_*X \qquad n \geq 1$$

Consequently

$$\bar{p}_{2n+1} = (\bar{p}_{2n+1} + \bar{q}_{2n+1}) - \bar{q}_{2n+1} \in H_*X \qquad n \geq 1$$

Therefore

$$P_*H_*X = P_*H_*\Gamma_0 B\mathcal{O}^{ev}(\mathbb{F}_q)$$

$$P_*H_*Y = 0$$

It follows that $H_*Y = 0$ and Y is contractible. Hence $X \simeq \Gamma_0 B\bar{\mathcal{O}}(\mathbb{F}_q)$.

A similar argument shows that $\Gamma_0 B\,\mathcal{N}^{ev}(\mathbb{F}_q)_{(2)} \simeq \Gamma_0 B\,\mathcal{N}\mathcal{D}^{ev}(\mathbb{F}_q)_{(2)} \simeq J(q)_{(2)}$ is an indecomposable infinite loop space.

§3. Homology operations in $H_*B\mathcal{Sp}(\mathbb{F}_q)$ and $H_*\Gamma B\mathcal{Sp}(\mathbb{F}_q)$

In this section we prove the following results on homology operations in the mod-2 homology algebras $H_*B\mathcal{Sp}(\mathbb{F}_q)$ and $H_*\Gamma B\mathcal{Sp}(\mathbb{F}_q)$.

Theorem 3.1. The homology operations in $H_*B\mathcal{Sp}(\mathbb{F}_q) = \oplus_{n\in\mathbb{N}} H_*BSp(2n,\mathbb{F}_q)$ are given on the standard generators of Theorem IV.5.2 by

(i) $Q^m(\sigma_k) = Q^m(\tau_k) = 0$ if $m \not\equiv 0 \pmod 4$

(ii) $Q^{4n}(\tau_k) = \Sigma_{u=0}^{k-1}(u,n-k)(\tau_{n+u+1}\sigma_{k-1-u} + \sigma_{n+u}\tau_{k-u})$

(iii) $Q^{4n}(\sigma_k) = \Sigma_{u=0}^{k}(u,n-k-1)\sigma_{n+u}\sigma_{k-u}$ $\quad n > k$

(iv) $Q^{4n}(\sigma_n) = \sigma_n^2 \qquad Q^{4n}(\sigma_k) = 0 \quad n < k$

Theorem 3.2. The homology operations in $H_*\Gamma B\mathcal{Sp}(\mathbb{F}_q)$ are given on the standard generators of Theorem IV.5.3 by

(i) $Q^m(\bar\sigma_k*[2]) = Q^m(\bar\sigma_k) = Q^m(\bar\tau_k) = Q^m(\bar\tau_k*[2]) = 0$ if $m \not\equiv 0 \pmod 4$

(ii) $Q^{4n}(\bar\tau_k*[2]) = \Sigma_{a+b=n+k}[(a-n-1,n-k)+(k-a,n-k)]\bar\tau_a\bar\sigma_b*[4]$

(iii) $Q^{4n}(\bar\tau_k) = \Sigma_{a+b+c=n+k}[(k-c-1,n-a-k)+(k-b,n-a-k]\chi(\bar\sigma_a)\bar\tau_b\bar\sigma_c$

(iv) $Q^{4n}(\bar\sigma_k*[2]) = \Sigma_{a+b=n+k}(a-n,n-k-1)\bar\sigma_a\bar\sigma_b*[4] \quad n > k$

(v) $Q^{4n}(\bar\sigma_k) = \Sigma_{a+b+c=n+k}(k-c,n-a-k-1)\chi(\bar\sigma_a)\bar\sigma_b\bar\sigma_c \quad n > k$

(vi) $Q^{4n}(\bar\sigma_n) = \bar\sigma_n^2 \qquad Q^{4n}(\bar\sigma_k) = 0 \quad n < k$

As a first step we prove the following

Lemma 3.3. Let $\mu\colon \mathbb{Z}/2 \times Sp(2,\mathbb{F}_q) \to Sp(2,\mathbb{F}_q)$ be the homomorphism defined by $\mu(\pm 1,w) = \pm w$. Then

$$\mu_*\colon H_*B(\mathbb{Z}/2 \times Sp(2,\mathbb{F}_q)) \to H_*BSp(2,\mathbb{F}_q)$$

is given by

(i) $\mu_*(x_{4n} \otimes \sigma_k) = (k,n)\sigma_{n+k}$

(ii) $\mu_*(x_m \otimes \sigma_k) = 0$ if $m \not\equiv 0 \pmod 4$

(iii) $\mu_*(x_{4n} \otimes \tau_k) = (k-1,n)\tau_{n+k}$

(iv) $\mu_*(x_m \otimes \tau_k) = 0$ if $m \not\equiv 0 \pmod 4$)

<u>Proof</u>: Consider the commutative diagram

$$\begin{array}{ccc}
\mathbb{Z}/2 \times \mathbb{Z}/2 & \xrightarrow{\mu} & \mathbb{Z}/2 \\
\downarrow{\scriptstyle 1\times i} & & \downarrow{\scriptstyle i} \\
\mathbb{Z}/2 \times Sp(2,\mathbb{F}_q) & \xrightarrow{\mu} & Sp(2,\mathbb{F}_q)
\end{array}$$

In homology $i_*(x_{4k}) = \sigma_k$ and $i_*(x_m) = 0$ if $m \not\equiv 0 \pmod 4$ (cf. diagram of VI.5.3 and Prop. VI.5.7(c).). Consequently (i) and (ii) follow from Prop. VI.2.3(c).

According to Props. VI.2.1 and VI.5.6 we have

$$H^*BSp(2,\mathbb{F}_q) = \mathbb{Z}/2[P] \otimes E[\tilde{x}] \qquad H^*B\mathbb{Z}/2 = \mathbb{Z}/2[x]$$

where $\deg x = 1$, $\deg \tilde{x} = 3$ and $\deg P = 4$. Thus to prove (iii) and (iv) it suffices to show that

$$\mu^*(P) = 1 \otimes P + x^4 \otimes 1$$

$$\mu^*(\tilde{x}) = 1 \otimes \tilde{x}$$

The first statement follows from (i) and (ii). To prove the second statement write

$$\mu^*(\tilde{x}) = 1 \otimes \tilde{x} + \alpha x^3 \otimes 1$$

Then by Prop. VI.5.7(b)

$$0 = \mu^*(Sq^1\tilde{x}) = Sq^1\mu^*(\tilde{x}) = Sq^1(1 \otimes \tilde{x}) + \alpha Sq^1(x^3 \otimes 1) = \alpha x^4 \otimes 1$$

so that $\alpha = 0$ and $\mu^*(\tilde{x}) = 1 \otimes \tilde{x}$.

3.4. Proof of Theorem 3.1. We consider the following commutative diagram of homomorphisms

$$\begin{array}{ccccccc}
\mathbb{Z}/2 \times Sp(2,\mathbb{F}_q) \cong \mathcal{S}_2 \times Sp(2,\mathbb{F}_q) & \xrightarrow{1\times\Delta} & \mathcal{S}_2 \wr Sp(2,\mathbb{F}_q) & \xrightarrow{j} & & & Sp(4,\mathbb{F}_q) \\
\Big\downarrow 1\times\Delta & & & & & & \Big\downarrow \delta \\
\mathbb{Z}/2 \times Sp(2,\mathbb{F}_q)^2 & \xrightarrow{\mu\times 1} & Sp(2,\mathbb{F}_q)^2 & \xrightarrow{\epsilon\times\epsilon} & Sp(2,\mathbb{F}_q)^2 & \xrightarrow{\oplus} & Sp(4,\mathbb{F}_q)
\end{array}$$

where $\delta = A(\cdot)A^{-1}$ is conjugation by $A = \frac{1}{2}\begin{pmatrix}1 & -1\\ 1 & 1\end{pmatrix} \otimes \begin{pmatrix}1 & -1\\ 1 & 1\end{pmatrix}$ and $\epsilon = C(\cdot)C^{-1}$ is conjugation by $C = \begin{pmatrix}1 & -1\\ 1 & 1\end{pmatrix}$. We apply $H_*B(\cdot)$ to the diagram. Since $A \in Sp(4,\mathbb{F}_q)$, δ_* is the identity map. Since $H_*BSp(2,\mathbb{F}_q)$ has at most one non trivial element in each dimension, ϵ_* is the identity map. The map

$$1 \times \Delta\colon \mathcal{S}_2 \times Sp(2,\mathbb{F}_q) \to \mathcal{S}_2 \wr Sp(2,\mathbb{F}_q)$$

is evaluated in May [28, Lemma 9.1]. The map

$$\mu\colon \mathbb{Z}/2 \times Sp(2,\mathbb{F}_q) \to Sp(2,\mathbb{F}_q)$$

has been evaluated in Lemma 3.3. The map

$$\Delta\colon Sp(2,\mathbb{F}_q) \to Sp(2,\mathbb{F}_q)^2$$

is evaluated in VI.5.5.

Hence chasing the elements $x_{4n} \otimes \sigma_k$, $x_{4n} \otimes \tau_k$, $x_m \otimes \sigma_k$ and $x_m \otimes \tau_k$, $m \not\equiv 0 \pmod 4$ around the above diagram we obtain

$$\begin{array}{ccccc}
x_{4n} \otimes \sigma_k & \to & \Sigma_{v=0}^{[k/2]} x_{4n+8v-4k} \otimes (Sq_*^{4v}\, \sigma_k)^2 & \to & \Sigma_{v=0}^{[k/2]} Q^{4n+4v}(Sq_*^{4v}\, \sigma_k) \\
\downarrow & & & & \| \\
x_{4n} \otimes \Sigma_{u=0}^{k} \sigma_u \otimes \sigma_{k-u} & \to & \Sigma_{u=0}^{k} (u,n)\, \sigma_{u+n}\sigma_{k-u} & \to & \Sigma_{u=0}^{k} (u,n)\, \sigma_{u+n}\sigma_{k-u}
\end{array}$$

$$\begin{array}{ccccc}
x_{4n} \otimes \tau_k & \to & \Sigma_{v=0}^{[k/2]} x_{4n+8v-4k+1} \otimes (Sq_*^{4v}\, \tau_k)^2 & \to & \Sigma_{v=0}^{[k/2]} Q^{4n+4v}(Sq_*^{4v}\, \tau_k) \\
\downarrow & & & & \| \\
x_{4n} \otimes \Sigma_{u=0}^{k-1} (\tau_{u+1} \otimes \sigma_{k-1-u} + \sigma_u \otimes \tau_{k-u}) & \to & \Sigma_{u=0}^{k-1} (u,n)\, [\tau_{n+u+1}\sigma_{k-1-u} + \sigma_{n+u}\tau_{k-u}] & &
\end{array}$$

$$\begin{array}{ccccc}
x_m \otimes \sigma_k & \to & \Sigma_{v=0}^{[k/2]} x_{m+8v-4k} \otimes (Sq_*^{4v}\, \sigma_k)^2 & \to & \Sigma_{v=0}^{[k/2]} Q^{m+4v}(Sq_*^{4v}\, \sigma_k) \\
\downarrow & & & & \| \\
x_m \otimes \Sigma_{u=0}^{k} \sigma_u \otimes \sigma_{k-u} & \longrightarrow & 0 & \longrightarrow & 0
\end{array}$$

$$\begin{array}{ccccc}
x_m \otimes \tau_k & \to & \Sigma_{v=0}^{[k/2]} x_{m+8v-4k+1} \otimes (Sq_*^{4v}\, \tau_k)^2 & \to & \Sigma_{v=0}^{[k/2]} Q^{m+4v}(Sq_*^{4v}\, \tau_k) \\
\downarrow & & & & \| \\
x_m \otimes \Sigma_{u=0}^{k-1} (\tau_{u+1} \otimes \sigma_{k-1-u} + \sigma_u \otimes \tau_{k-u}) & \longrightarrow & 0 & \longrightarrow & 0
\end{array}$$

According to Prop. VI 5.7(d), $Sq_*^{4v}\, \sigma_k = (v,k-2v)\sigma_{k-v}$ and $Sq_*^{4v}(\tau_k) = (v,k-1-2v)\tau_{k-v}$. Hence we obtain

$$\Sigma_{v=0}^{[k/2]} (v,k-2v)\, Q^{4n+4v}(\sigma_{k-v}) = \Sigma_{u=0}^{k} (u,n)\, \sigma_{u+n}\sigma_{k-u}$$

$$\Sigma_{v=0}^{[k/2]} (v,k-1-2v)\, Q^{4n+4v}(\tau_{k-v}) = \Sigma_{u=0}^{k-1} (u,n)\, [\tau_{n+u+1}\sigma_{k-1-u} + \sigma_{n+u}\tau_{k-u}]$$

$$\Sigma_{v=0}^{[k/2]} (v,k-2v)\, Q^{m+4v}(\sigma_{k-v}) = 0$$

$$\Sigma_{v=0}^{[k/2]} (v,k-1-2v)\, Q^{m+4v}(\tau_{k-v}) = 0$$

Now arguing inductively as in the proof of Theorem 2.1 we obtain the indicated formulas.

3.5. Proof of Theorem 3.2. Since $\bar{\sigma}_k = \sigma_k*[-2]$ and $\bar{\tau}_k = \tau_k*[-2]$ the statements about $Q^r(\bar{\sigma}_k*[2])$ and $Q^r(\bar{\tau}_k*[2])$ are merely restatements of Theorem 3.1. The statements about $Q^r(\bar{\sigma}_k)$ and $Q^r(\bar{\tau}_k)$ follow from the Cartan formula and 1.11:

$$Q^r(\bar{\sigma}_k) = Q^r(\sigma_k \chi(\sigma_0)) = \Sigma_{i=0}^{r} Q^i(\sigma_k)\,\chi Q^{r-i}(\sigma_0)$$

$$Q^r(\bar{\tau}_k) = Q^r(\tau_k \chi(\sigma_0)) = \Sigma_{i=0}^{r} Q^i(\tau_k)\,\chi Q^{r-i}(\sigma_0).$$

§4. Homology operations in $H_*B\mathcal{GL}(\mathbb{F}_q)$, $H_*\Gamma B\mathcal{GL}(\mathbb{F}_q)$, $H_*B\mathcal{U}(\mathbb{F}_{q^2})$ and $H_*\Gamma B\mathcal{U}(\mathbb{F}_{q^2})$

In this section we compute the homology operations in the mod-2 homology algebras listed in the title. The results we obtain and their proofs are virtually identical for the general linear and unitary cases. Since the proofs are analogous to those in the preceding sections and because the general linear case has already been treated in Priddy [33], we shall confine ourselves to giving the barest outlines of the proofs.

Theorem 4.1. The homology operations in $H_*B\mathcal{GL}(\mathbb{F}_q) = \oplus_{n\in N} H_*BGL(n,\mathbb{F}_q)$ are given on the standard generators of Theorem IV.7.2 by the following formulas (where we denote $w_{2k} = \alpha_k$, $w_{2k-1} = \beta_k$).

(1) If $q \equiv 1 \pmod 4$ then

(i) $Q^{2n+1}(w_k) = 0$

(ii) $Q^{2n}(w_{2k+1}) = \Sigma_{u=0}^{2k+1}([u/2], n-k-1)\, w_{2n+u} w_{2k+1-u}$

(iii) $Q^{2n}(w_{2n}) = \Sigma_{u=0}^{k}(u, n-k-1)\, w_{2n+2u} w_{2k-2u}$ for $n > k$

(iv) $Q^{2n}(w_{2n}) = w_{2n}^2 \quad Q^{2n}(w_{2k}) = 0$ for $n < k$

(2) If $q \equiv -1 \pmod 4$ then

(i) $Q^n(w_k) = \Sigma_{u=0}^{k}(u,n-k-1)\,w_{n+u}w_{k-u}$ for $n > k$

(ii) $Q^n(w_n) = w_n^2 \quad Q^n(w_k) = 0$ for $n < k$.

Theorem 4.2. The homology operations in $H_*B\mathcal{U}(\mathbb{F}_{q^2}) = \oplus_{n\in\mathbb{N}} H_*BU(n,\mathbb{F}_{q^2})$ are given on the standard generators of Theorem IV 7.2 by the following formulas (where we denote $w_{2k} = \xi_k$, $w_{2k-1} = \eta_k$)

(1) If $q \equiv 1 \pmod 4$ then

(i) $Q^n(w_k) = \Sigma_{u=0}^{k}(u,n-k-1)\,w_{n+u}w_{k-u}$ for $n > k$

(ii) $Q^n(w_n) = w_n^2 \quad Q^n(w_k) = 0$ for $n < k$

(2) If $q \equiv -1 \pmod 4$ then

(i) $Q^{2n+1}(w_k) = 0$ for $n,k \geq 0$

(ii) $Q^{2n}(w_{2k+1}) = \Sigma_{u=0}^{2k+1}([u/2],n-k-1)\,w_{2n+u}w_{2k+1-u}$

(iii) $Q^{2n}(w_{2k}) = \Sigma_{u=0}^{k}(u,n-k-1)\,w_{2n+2u}w_{2k-2u}$ for $n > k$

(iv) $Q^{2n}(w_{2n}) = w_{2n}^2 \quad Q^{2n}(w_{2k}) = 0$ for $n < k$.

Theorem 4.3. The homology operations in $H_*\Gamma B\mathcal{O}\mathcal{L}(\mathbb{F}_q)$ are given on the standard generators of Theorem IV 7.3 by the following formulas (where we denote $\overline{w}_{2k} = \overline{\alpha}_k$, $\overline{w}_{2k-1} = \overline{\beta}_k$)

(1) If $q \equiv 1 \pmod 4$ then

(i) $Q^{2n+1}(\overline{w}_k*[1]) = Q^{2n+1}(\overline{w}_k) = 0$

(ii) $Q^{2n}(\overline{w}_{2k+1}*[1]) = \Sigma_{a+b=2n+2k+1}([a/2]-n,n-k-1)\,\overline{w}_a\overline{w}_b*[2]$

(iii) $Q^{2n}(\overline{w}_{2k+1}) = \Sigma_{2a+b+c=2n+2k+1}([b/2]+a-n,n-a-k-1)\,\chi(\overline{w}_a)\overline{w}_b\overline{w}_c$

(iv) $Q^{2n}(\overline{w}_{2k}*[1]) = \Sigma_{a+b=n+k}(a-n,n-k-1)\,\overline{w}_{2a}\overline{w}_{2b}*[2]$ for $n > k$

(v) $Q^{2n}(\overline{w}_{2k}) = \Sigma_{a+b+c=n+k}(k-c,n-a-k-1)\,\chi(\overline{w}_{2a})\overline{w}_{2b}\overline{w}_{2c}$ for $n > k$

(vi) $Q^{2n}(\overline{w}_{2n}) = \overline{w}_{2n}^2 \quad Q^{2n}(\overline{w}_{2k}) = 0$ for $n < k$

(2) If $q \equiv -1 \pmod 4$ then

(i) $Q^n(\overline{w}_k*[1]) = \Sigma_{a+b=n+k}(a-n,n-k-1)\,\overline{w}_a\overline{w}_b*[2]$ for $n > k$

(ii) $Q^n(\overline{w}_k) = \Sigma_{a+b+c=n+k}(k-c,n-a-k-1)\chi(\overline{w}_a)\overline{w}_b\overline{w}_c$ for $n > k$

(iii) $Q^n(\overline{w}_n) = \overline{w}_n^2 \quad Q^n(\overline{w}_k) = 0$ for $n < k$

Theorem 4.4. The homology operations in $H_*\Gamma B\mathcal{U}(\mathbb{F}_{q^2})$ are given on the standard generators of Theorem IV 7.3 by the following formulas (where we denote $\overline{w}_{2k} = \overline{\xi}_k$ and $\overline{w}_{2k-1} = \overline{\eta}_k$).

(1) If $q \equiv 1 \pmod 4$ then

(i) $Q^n(\overline{w}_k*[1]) = \Sigma_{a+b=n+k}(a-n,n-k-1)\overline{w}_a\overline{w}_b*[2]$ for $n > k$

(ii) $Q^n(\overline{w}_k) = \Sigma_{a+b+c=n+k}(k-c,n-a-k-1)\chi(\overline{w}_a)\overline{w}_b\overline{w}_c$ for $n > k$

(iii) $Q^n(\overline{w}_n) = \overline{w}_n^2 \quad Q^n(\overline{w}_k) = 0$ for $n < k$

(2) If $q \equiv -1 \pmod 4$ then

(i) $Q^{2n+1}(\overline{w}_k*[1]) = Q^{2n+1}(\overline{w}_k) = 0$

(ii) $Q^{2n}(\overline{w}_{2k+1}*[1]) = \Sigma_{a+b=2n+2k+1}([a/2]-n,n-k-1)\overline{w}_a\overline{w}_b*[2]$

(iii) $Q^{2n}(\overline{w}_{2k+1}) = \Sigma_{2a+b+c=2n+2k+1}([b/2]+a-n,n-a-k-1)\chi(\overline{w}_a)\overline{w}_b\overline{w}_c$

(iv) $Q^{2n}(\overline{w}_{2k}*[1]) = \Sigma_{a+b=n+k}(a-n,n-k-1)\overline{w}_{2a}\overline{w}_{2b}*[2]$ for $n > k$

(v) $Q^{2n}(\overline{w}_{2k}) = \Sigma_{a+b+c=n+k}(k-c,n-a-k-1)\chi(\overline{w}_{2a})\overline{w}_{2b}\overline{w}_{2c}$ for $n > k$

(vi) $Q^{2n}(\overline{w}_{2n}) = \overline{w}_{2n}^2 \quad Q^{2n}(\overline{w}_{2k}) = 0$ for $n < k$.

As a first step we prove the following

Lemma 4.5. Let r be an even positive integer. Let

$$\mu\colon \mathbb{Z}/2 \times \mathbb{Z}/r \to \mathbb{Z}/r$$

be the multiplication map $\mu(\pm 1,w) = \pm w$. Then

$$\mu_*\colon H_*B(\mathbb{Z}/2 \times \mathbb{Z}/r) \to H_*B\mathbb{Z}/r$$

is given by

(1) if $r \equiv 2 \pmod 4$, then

$$\mu_*(x_n \otimes x_k) = (k,n)\, x_{n+k}$$

(2) if $r \equiv 0 \pmod 4$, then

$$\mu_*(x_{2n} \otimes x_k) = ([\tfrac{k}{2}],n)\, x_{2n+k}$$

$$\mu_*(x_{2n-1} \otimes x_k) = 0$$

Proof. This follows immediately from Prop. VI 2.1, the commutative diagram

$$\begin{array}{ccc} \mathbb{Z}/2 \times \mathbb{Z}/r & & \\ \big\downarrow{\scriptstyle i\times 1} & \overset{\mu}{\searrow} \ \ \underset{\mu}{\nearrow} & \mathbb{Z}/r \\ \mathbb{Z}/r \times \mathbb{Z}/r & & \end{array}$$

and the well-known fact that the induced map $i_*\colon H_*B\mathbb{Z}/2 \to H_*B\mathbb{Z}/r$ is a isomorphism in all degrees if $r \equiv 2 \pmod 4$; while if $r \equiv 0 \pmod 4$ then i_* is an isomorphism in even degrees and the zero map in odd degrees [Proposition VI, 2.7].

4.6. Sketch of Proof of Theorem 4.1. We consider the commutative diagram of homomorphisms

$$\begin{array}{ccccc} \mathbb{Z}/2 \times \mathbb{Z}/q\text{-}1 = \mathcal{S}_2 \times GL(1,\mathbb{F}_q) & \xrightarrow{1\times\Delta} & \mathcal{S}_2 \wr GL(1,\mathbb{F}_q) & \xrightarrow{j} & GL(2,\mathbb{F}_q) \\ \big\downarrow{\scriptstyle 1\times\Delta} & & & & \big\downarrow{\scriptstyle A(\ )A^{-1}} \\ \mathbb{Z}/2 \times (\mathbb{Z}/q\text{-}1)^2 & \xrightarrow{\mu\times 1} & (\mathbb{Z}/q\text{-}1)^2 = GL(1,\mathbb{F}_q)^2 & \xrightarrow{\oplus} & GL(2,\mathbb{F}_q) \end{array}$$

where

$$A = \begin{pmatrix} -1 & 1 \\ 1 & 1 \end{pmatrix}.$$

We now apply $H_*B(\cdot)$ to the diagram. Since $A \in GL(2,\mathbb{F}_q)$, conjugation by A induces the identity map on homology. According to May [28; Prop. 9.1] the map $1 \times \Delta\colon \mathcal{S}_2 \times GL(1,\mathbb{F}_q) \to \mathcal{S}_2 \wr GL(1,\mathbb{F}_q)$ is given in homology by

$$(1 \times \Delta)_*(x_m \otimes x_k) = \Sigma_{v=0}^{[k/2]} x_{m+2v-k} \otimes (Sq_*^v(x_k))^2,$$

while $Sq_*^v(x_k)$ has been evaluated in Prop. VI 2.4. The map induced by $\mu\colon \mathbb{Z}/2 \times \mathbb{Z}/q-1 \to \mathbb{Z}/q-1$ has been evaluated in Lemma 4.5. One then proceeds as in the proofs of Theorems 2.1 and 3.1.

4.7. <u>Sketch of Proof of Theorem</u> 4.2. We consider the commutative diagram of homomorphisms

$$\begin{array}{ccccc}
\mathbb{Z}/2 \times \mathbb{Z}/q+1 = \mathcal{S}_2 \times U(1,\mathbb{F}_{q^2}) & \xrightarrow{1\times\Delta} & \mathcal{S}_2 \wr U(1,\mathbb{F}_{q^2}) & \xrightarrow{j} & U(2,\mathbb{F}_{q^2}) \\
\big\downarrow{\scriptstyle 1\times\Delta} & & & & \big\downarrow{\scriptstyle A(\ )A^{-1}} \\
\mathbb{Z}/2 \times (\mathbb{Z}/q+1)^2 & \xrightarrow{\mu\times 1} & (\mathbb{Z}/q+1)^2 = U(1,\mathbb{F}_{q^2})^2 & \xrightarrow{\oplus} & U(2,\mathbb{F}_{q^2})
\end{array}$$

where

$$A = \frac{1}{\sqrt{-2}} \begin{pmatrix} -1 & 1 \\ 1 & 1 \end{pmatrix} \in U(2,\mathbb{F}_{q^2})$$

Note that -1 is a square in $\mathbb{F}_{q^2}$ since $q^2 \equiv 1 \pmod 4$ and 2 is a square in $\mathbb{F}_{q^2}$ since $q^2 \equiv 1 \pmod 8$. One then proceeds as in the proof of Theorem 4.1.

Theorems 4.3 and 4.4 follow from Theorems 4.1 and 4.2 by the same reasoning that was used to derive Theorems 2.2 and 3.2 from Theorems 2.1 and 3.1 respectively.

Appendix: Multiplicative Homology Operations in $H_*(B\bar{\mathcal{O}}(\mathbb{F}_q);\ \mathbb{Z}/2)$ and $H_*(\Gamma_0 B\bar{\mathcal{O}}(\mathbb{F}_q);\ \mathbb{Z}/2)$

§1. Introduction.

In the various permutative categories $\mathcal{A}$ considered in Chapter II, the infinite loop structure on $\Gamma B\mathcal{A}$ arose from the direct sum functor

$$\oplus:\ \mathcal{A} \times \mathcal{A} \to \mathcal{A}.$$

Moreover this functor gives rise to the homology operations in $H_*(B\mathcal{A};\ \mathbb{Z}/2)$ and $H_*(\Gamma B\mathcal{A};\mathbb{Z}/2)$ which we calculated in Chapter IX.

In many cases (eg. for $\mathcal{A} = \mathcal{GL}(\mathbb{F}_q)$, $\bar{\mathcal{O}}(\mathbb{F}_q)$ or $\mathcal{U}(\mathbb{F}_{q^2})$) there is an equally important functor: the tensor product

$$\otimes:\ \mathcal{A} \times \mathcal{A} \to \mathcal{A}.$$

This endows the category $\mathcal{A}$ and its associated spaces $B\mathcal{A}$, $\Gamma B\mathcal{A}$ with a rich additional structure: It makes $\mathcal{A}$ into a bipermutative category, $B\mathcal{A}$ into an E_∞ ring space and $\Gamma B\mathcal{A}$ into the zeroth space of an E_∞ ring spectrum. Also it endows the 1-component, $\Gamma_1 B\mathcal{A}$ with a different infinite loop structure and produces a new set of homology operations on $H_*(\Gamma B\mathcal{A};\mathbb{Z}/2)$.

In this appendix we analyze this additional structure induced by the tensor product on $H_*(B\bar{\mathcal{O}}(\mathbb{F}_q);\ \mathbb{Z}/2)$ and $H_*(\Gamma B\bar{\mathcal{O}}(\mathbb{F}_q);\ \mathbb{Z}/2)$ q odd. This is the case of primary interest to topologists because of its intimate connection to the J-homomorphism (cf. Introduction).

Throughout this appendix q will denote a fixed <u>odd</u> prime power. All homology will be taken with $\mathbb{Z}/2$-coefficients.

§2. Bipermutative categories and their associated infinite loop spaces.

In this section we recall some of the basic results on bipermutative categories and their associated infinite loop spaces and fix notation.

Definition 2.1. A bipermutative category $\mathcal{A}$ is a small category which is a permutative category separately under each of two monoidal operations

$$\oplus: \mathcal{A} \times \mathcal{A} \to \mathcal{A}$$

$$\otimes: \mathcal{A} \times \mathcal{A} \to \mathcal{A}$$

and such that $\otimes$ distributes strictly over $\oplus$ from the left and distributes from the right up to coherent natural isomorphism. For details the reader is referred to May [25, Chap 6]. We shall denote by 0,1 the identity objects of $\oplus$ and $\otimes$ respectively.

We say that $(\mathcal{A},\oplus,\otimes)$ is a symmetric bimonoidal category if $(\mathcal{A},\oplus)$, $(\mathcal{A},\otimes)$ are symmetric monoidal categories and $\otimes$ distributes from both right and left up to coherent natural isomorphisms. There is a natural way to convert symmetric bimonoidal categories into equivalent bipermutative categories (cf. May [25]).

2.2. Examples of Bipermutative and Symmetric Bimonoidal Categories.

(a) Let R be a commutative ring. Then the symmetric monoidal category $\mathcal{P}(A)$ of finitely generated projective modules over R and all isomorphisms between them (cf. II 2.4) has a symmetric bimonoidal structure, with $\oplus$ denoting direct sum and $\otimes$ denoting tensor product over R.

(b) Let $\mathbf{F}$ be a field. Then the permutative category $\mathcal{GL}(\mathbf{F})$ defined in II 2.6 is a bipermutative category with $\otimes$ given by

$$\otimes(m,n) = mn$$

on objects and with

$$\otimes\colon \ GL(m,\mathbf{F}) \times GL(n,\mathbf{F}) \to GL(mn,\mathbf{F})$$

given by the usual tensor product of matrices

$$A \otimes B = \begin{pmatrix} a_{11}B & a_{12}B & \cdots & a_{1m}B \\ a_{21}B & a_{22}B & \cdots & a_{2m}B \\ \cdots & \cdots & \cdots & \cdots \\ a_{m1}B & a_{m2}B & \cdots & a_{mm}B \end{pmatrix}$$

(c) Let $\mathbf{F}_q$ denote the field with q elements, q odd. Then tensor product endows the symmetric monoidal category $\bar{\mathcal{O}}(\mathbf{F}_q)$ of non-degenerate quadratic spaces over $\mathbf{F}_q$ with a symmetric bimonoidal structure. For if (V_1,Q_1), (V_2,Q_2) are quadratic spaces, then $V_1 \otimes_{\mathbf{F}_q} V_2$ has a natural quadratic form given by

$$(Q_1 \otimes Q_2)(\Sigma\, \alpha_{ij} e_i \otimes f_j) = \Sigma\, \alpha_{ij} Q_1(e_i) Q_2(f_j)$$

where $\{e_i\}_{i=1}^m$, $\{f_j\}_{j=1}^n$ are orthogonal bases for V_1 and V_2 respectively.

2.3. If $\mathcal{A}$ is a bipermutative category, then its classifying space carries two associative operations with units

$$B\oplus: \ B\mathcal{A} \times B\mathcal{A} \to B\mathcal{A}$$

$$B\otimes: \ B\mathcal{A} \times B\mathcal{A} \to B\mathcal{A}$$

with $B\otimes$ distributing from the left over $B\oplus$. This gives $\pi_0 B\mathcal{A}$ the structure of a commutative semiring with unit.

May has shown that $B\mathcal{A}$ has the structure of an E_∞-ring space and that this E_∞-ring structure carries over to the space $\Gamma B\mathcal{A}$ obtained by taking the group completion of $B\mathcal{A}$ with respect to $B\oplus$. In particular $\pi_0 \Gamma B\mathcal{A}$ is a commutative ring with unit. Moreover this E_∞ ring structure is reflected in the infinite loop structure of $\Gamma B\mathcal{A}$; for $\Gamma B\mathcal{A}$ becomes the zeroth space of an E_∞ ring spectrum and the path component $\Gamma_1 B\mathcal{A}$ corresponding to the multiplicative unit in $\pi_0 \Gamma B\mathcal{A}$ carries an infinite loop structure induced by $\otimes$. (cf. May [25]).

2.4. The tensorial wreath products and Dyer-Lashof operations.

In the categories $\mathcal{A} = \coprod_{\alpha \in \pi_0 \mathcal{A}} G(\alpha)$ of Example 2.2 we can define a tensorial wreath product by defining inclusions

$$\mathcal{S}_n \wr G(\alpha) \to G(\alpha^n)$$

by having $(\tau, f_1, f_2, \ldots, f_n)$ act on $V^{\otimes n}$ where V is a representative object of α via the formula

$$(\tau, f_1, f_2, \ldots, f_n)(\Sigma\, v_{i_1} \otimes v_{i_2} \otimes \cdots \otimes v_{i_n})$$

$$= \Sigma\, f_1(v_{i_{\tau^{-1}(1)}}) \otimes f_2(v_{i_{\tau^{-1}(2)}}) \otimes \cdots \otimes f_n(v_{i_{\tau^{-1}(n)}})$$

Using this tensorial wreath product construction, we obtain a multiplicative Dyer-Lashof map

$$\rho: E\mathcal{S}_2 \times_{\mathcal{S}_2} (B\mathcal{A})^2 \to \coprod E\mathcal{S}_2 \times_{\mathcal{S}_2} (BG(\alpha))^2 = \coprod B(\mathcal{S}_2 \wr G(\alpha))$$
$$\to \coprod BG(\alpha^2) \to B\mathcal{A}$$

Passing to group completions we obtain a compatible Dyer-Lashof map

$$\rho: E\mathcal{S}_2 \times_{\mathcal{S}_2} (\Gamma B\mathcal{A})^2 \to \Gamma B\mathcal{A}$$

which makes the following diagram commute

$$\begin{array}{ccc} E\mathcal{S}_2 \times_{\mathcal{S}_2} (B\mathcal{A})^2 & \xrightarrow{\rho} & B\mathcal{A} \\ \downarrow & & \downarrow \\ E\mathcal{S}_2 \times_{\mathcal{S}_2} (\Gamma B\mathcal{A})^2 & \xrightarrow{\rho} & \Gamma B\mathcal{A} \end{array}$$

where $B\mathcal{A} \to \Gamma B\mathcal{A}$ is the natural map (cf. II 2.17). As sketched in Chapter IX §1, ρ defines multiplicative homology operations

$$\tilde{Q}^t: H_i(B\mathcal{A}; \mathbb{Z}/2) \to H_{i+t}(B\mathcal{A}; \mathbb{Z}/2)$$

$$\tilde{Q}^t: H_i(\Gamma B\mathcal{A}; \mathbb{Z}/2) \to H_{i+t}(\Gamma B\mathcal{A}; \mathbb{Z}/2)$$

2.5. Properties of the Multiplicative Homology Operations.

Let $\mathcal{A} = \coprod G(\alpha)$ be one of the symmetric bimonoidal categories of Example 2.2. Then both $B\mathcal{A}$ and $\Gamma B\mathcal{A}$ and their mod-2 homology have two product structures, one induced by $\oplus$, the other by $\otimes$. We shall denote by $+$ the H-space operation induced by $\oplus$ on $B\mathcal{A}$ or $\Gamma B\mathcal{A}$ and will denote by $*$ the corresponding Pontrjagin product on $H_*(B\mathcal{A}; \mathbb{Z}/2)$ or $H_*(\Gamma B\mathcal{A}; \mathbb{Z}/2)$. We shall denote by $\#$ the product induced by $\otimes$ both on the level of spaces and the level of homology. We observe that $*$ takes $H_i(BG(\alpha);\ \mathbb{Z}/2) \otimes H_j(BG(\beta);\ \mathbb{Z}/2)$ to $H_{i+j}(BG(\alpha+\beta);\ \mathbb{Z}/2)$ and $H_i(\Gamma_\alpha B\mathcal{A};\ \mathbb{Z}/2) \otimes H_j(\Gamma_\beta B\mathcal{A};\ \mathbb{Z}/2)$ to $H_{i+j}(\Gamma_{\alpha+\beta}B\mathcal{A};\ \mathbb{Z}/2)$. Similarly $\#$ takes $H_i(BG(\alpha);\ \mathbb{Z}/2) \otimes H_j(BG(\beta);\ \mathbb{Z}/2)$ to $H_{i+j}(BG(\alpha\beta);\ \mathbb{Z}/2)$ and $H_i(\Gamma_\alpha B\mathcal{A};\ \mathbb{Z}/2) \otimes H_j(\Gamma_\beta B\mathcal{A};\ \mathbb{Z}/2)$ to $H_{i+j}(\Gamma_{\alpha\beta}B\mathcal{A};\ \mathbb{Z}/2)$. As in II 2.16(viii), we denote by $[\alpha]$ the generator of $H_0(BG(\alpha);\ \mathbb{Z}/2)$ or $H_0(\Gamma_\alpha B\mathcal{A};\ \mathbb{Z}/2)$. Then $[\alpha]*[\beta] = [\alpha+\beta]$ and $[\alpha]\ \#\ [\beta] = [\alpha\beta]$.

We will denote by Q^s the homology operations coming from $\oplus$; Q^s takes $H_m(BG(\alpha),\ \mathbb{Z}/2)$ to $H_{m+s}(BG(2\alpha);\ \mathbb{Z}/2)$ and $H_m(\Gamma_\alpha B\mathcal{A};\ \mathbb{Z}/2)$ to $H_{m+s}(\Gamma_{2\alpha}B\mathcal{A};\ \mathbb{Z}/2)$. We will denote by $\tilde{Q}^s$ the homology operations induced by $\otimes$; $\tilde{Q}^s$ takes $H_m(BG(\alpha);\ \mathbb{Z}/2)$ to $H_{m+s}(BG(\alpha^2);\ \mathbb{Z}/2)$ and $H_m(\Gamma_\alpha B\mathcal{A};\ \mathbb{Z}/2)$ to $H_{m+s}(\Gamma_{\alpha^2}B\mathcal{A};\ \mathbb{Z}/2)$. Also we have

$$Q^0([\alpha]) = [2\alpha], \qquad \tilde{Q}^0([\alpha]) = [\alpha^2]$$

Let $\epsilon\colon H_*(B\mathcal{A};\ \mathbb{Z}/2) \to \mathbb{Z}/2$, $\epsilon\colon H_*(\Gamma B\mathcal{A};\ \mathbb{Z}/2) \to \mathbb{Z}/2$ denote the augmentations. Note that $\epsilon([\alpha]) = 1$. Let $\Delta\colon H_*(B\mathcal{A};\ \mathbb{Z}/2) \to H_*(B\mathcal{A};\ \mathbb{Z}/2) \otimes H_*(B\mathcal{A};\ \mathbb{Z}/2)$ and $\Delta\colon H_*(\Gamma B\mathcal{A};\ \mathbb{Z}/2) \to H_*(\Gamma B\mathcal{A};\ \mathbb{Z}/2) \otimes H_*(\Gamma B\mathcal{A};\ \mathbb{Z}/2)$ denote the coproduct induced by the diagonal. We observe that $\Delta([\alpha]) = [\alpha] \otimes [\alpha]$. Let $\chi\colon H_*(\Gamma B\mathcal{A};\ \mathbb{Z}/2) \to H_*(\Gamma B\mathcal{A};\ \mathbb{Z}/2)$ denote the conjugation with respect to $*$, i.e. the automorphism induced by the map $-1\colon \Gamma B\mathcal{A} \to \Gamma B\mathcal{A}$ (cf. IV 2.7). Then χ takes $H_i(\Gamma_\alpha B\mathcal{A};\ \mathbb{Z}/2)$ to $H_i(\Gamma_{-\alpha}B\mathcal{A};\ \mathbb{Z}/2)$ and $\chi([\alpha]) = [-\alpha]$. Moreover $*(1 \otimes \chi)\Delta = \eta\epsilon$, where $\eta\colon \mathbb{Z}/2 \to H_*(\Gamma B\mathcal{A};\ \mathbb{Z}/2)$ is the unit for $*$, i.e. $\eta(1) = [0]$.

We now list for future reference some basic properties of $*$, Q^s and $\tilde{Q}^s$, in addition to those for Q^s already listed in

Chapter IX §1 and the corresponding properties for $\tilde{Q}^s$.

Let X denote one of the spaces $B\mathcal{O}$ or $\Gamma B\mathcal{O}$. Let $x,y,z \in H_*(X; \mathbf{Z}/2)$, $\alpha,\beta \in \pi_0 X$. Let $\Delta(x) = \Sigma\, x' \otimes x''$, $\Delta(y) = \Sigma\, y' \otimes y''$, $\Delta(z) = \Sigma\, z' \otimes z''$. Then

(i) $[0] \# x = \varepsilon(x)[0]$ and $[1] \# x = x$

(ii) if $X = \Gamma B\mathcal{O}$, then $[-1] \# x = \chi(x)$

(iii) $(x*y) \# z = \Sigma(x \# z')*(y \# z'')$

(iv) $(x*[\alpha]) \# (y*[\beta]) = \Sigma\Sigma(x' \# y')*(x'' \# [\beta])*(y'' \# [\alpha])*[\alpha\beta$

(v) $(Q^s x) \# y = \Sigma\, Q^{s+i}(x \# Sq_*^i y)$

(vi) $\tilde{Q}^s([0]) = 0 = \tilde{Q}^s([1])$ if $s > 0$

(vii) $\tilde{Q}^s([-1]) = Q^s([1])*[-1]$

(viii) (Mixed Cartan Formula)

$$\tilde{Q}^s(x*y) = \Sigma_{a+b+c=s}\Sigma\Sigma\tilde{Q}^a(x')*Q^b(x' \# y')*\tilde{Q}^c(y'')$$

We also note that the natural map $H_*(B\mathcal{O}; \mathbf{Z}/2) \to H_*(\Gamma B\mathcal{O}; \mathbf{Z}/2)$ preserves all structure in sight. For proofs and details cf. May [27].

§3. The multiplicative structure of $H_*(B\bar{\mathcal{O}}(\mathbf{F}_q); \mathbf{Z}/2)$ and $H_*(\Gamma B\bar{\mathcal{O}}(\mathbf{F}_q); \mathbf{Z}/2)$.

In this section we investigate the structure induced by tensor product on the mod-2 homology of $B\bar{\mathcal{O}}(\mathbf{F}_q)$ and $\Gamma B\bar{\mathcal{O}}(\mathbf{F}_q)$.

3.1. Remark. In this section we will continue to use the notation $\tilde{\mathbb{N}} = \pi_0\bar{\mathcal{O}}(\mathbb{F}_q) = \{0,n,\bar{n} \mid n \geq 1\}$

$$O(n,\mathbb{F}_q) = O_+(n,\mathbb{F}_q) \qquad O(\bar{n},\mathbb{F}_q) = O_-(n,\mathbb{F}_q)$$

of II 4.7. To determine the multiplicative structure of $\tilde{\mathbb{N}}$, we note that

$$(\mathbb{F}_q,Q_\varepsilon) \otimes_{\mathbb{F}_q} (\mathbb{F}_q,Q_\delta) \cong (\mathbb{F}_q,Q_{\varepsilon\delta})$$

$\varepsilon = \pm$, $\delta = \pm$. Hence we have

$$1\cdot 1 = 1 = \bar{1}\cdot\bar{1} \qquad 1\cdot\bar{1} = \bar{1}\cdot 1 = 1$$

It follows by distributivity that $\tilde{\mathbb{N}}$ is a commutative semiring containing $\mathbb{N}$ the natural numbers as a subsemiring with

$$m\bar{n} = \begin{cases} mn \text{ if } m \text{ is even} \\ \overline{mn} \text{ if } m \text{ is odd} \end{cases} \qquad \bar{m}\bar{n} = \begin{cases} mn \text{ if } m \equiv n \bmod 2) \\ \overline{mn} \text{ if } m \not\equiv n \bmod 2) \end{cases}$$

With this notation direct sum maps $O(a,\mathbb{F}_q) \times O(b,\mathbb{F}_q)$ to $O(a + b,\mathbb{F}_q)$ while tensor product maps $O(a,\mathbb{F}_q) \times O(b,\mathbb{F}_q)$ to $O(ab,\mathbb{F}_q)$ for all $a,b \in \tilde{\mathbb{N}}$.

Passing to the group completion we see that as a ring

$$\pi_0\Gamma B\bar{\mathcal{O}}(\mathbb{F}_q) = K_0(\bar{\mathcal{O}}(\mathbb{F}_q)) = \mathbb{Z}[\gamma]/\{\gamma^2,2\gamma\}$$

where as in II 4.12 we denote $\gamma = \bar{1} - 1$.

It should also be noted that the permutative functor $\Phi\colon \bar{\mathcal{O}}(\mathbb{F}_q) \to \bar{\mathcal{O}}(\mathbb{F}_q)$ of II 4.4 is given by $\cdot\otimes(\mathbb{F}_q,Q_-)$. Hence the corresponding infinite loop map $\Phi\colon \Gamma B\bar{\mathcal{O}}(\mathbb{F}_q) \to \Gamma B\bar{\mathcal{O}}(\mathbb{F}_q)$ is given by $\#$ multiplication by the basepoint of $\Gamma_{\bar{1}}B\bar{\mathcal{O}}(\mathbb{F}_q)$.

We begin by determining the # product on the standard generators of $H_*B\overline{\mathcal{O}}(\mathbb{F}_q)$ and $H_*\Gamma B\overline{\mathcal{O}}(\mathbb{F}_q)$ of Theorems IV 2.1 and 2.3.

<u>Proposition</u> 3.2. The # product on $H_*B\overline{\mathcal{O}}(\mathbb{F}_q) = \oplus_{n\in\tilde{\mathbb{N}}} H_*BO(n,\mathbb{F}_q)$ is given on the standard generators of Theorem IV 2.1 by

(a) $v_m \# v_n = y_m \# y_n = (m,n)v_{m+n}$

(b) $v_m \# y_n = y_m \# v_n = (m,n)y_{m+n}$

<u>Proof</u>. We have the commutative diagram

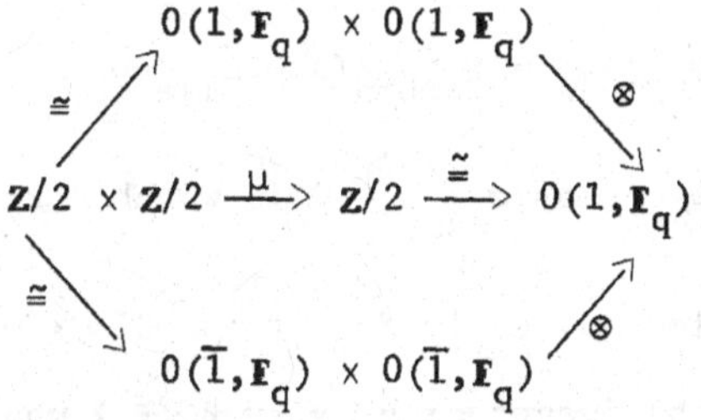

where μ denotes multiplication in $\mathbb{Z}/2$. Applying $H_*(B\cdot)$ to the diagram and using VI 3.2, we obtain (a).

We also have the commutative diagram

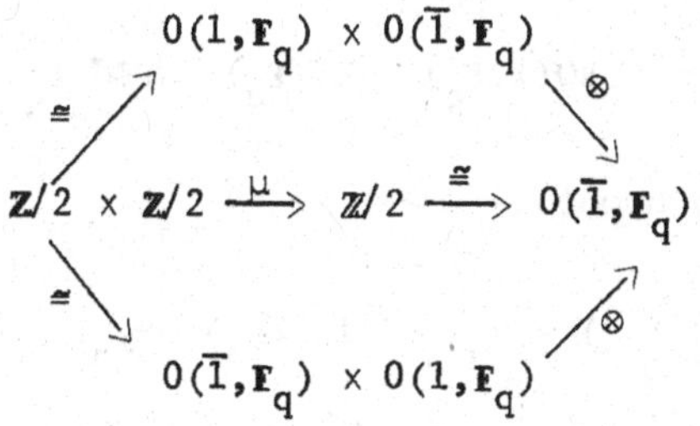

which similarly implies (b).

To compute the # product on $H_*\Gamma B\overline{\mathcal{O}}(\mathbb{F}_q)$ it turns out to be useful to use the following generators

$$v_k = \bar{v}_k * [1] \qquad k \geq 0$$

$$\tilde{u}_k = \bar{u}_k * [\gamma] \qquad k \geq 0$$

(cf. IV 2.3 regarding notation).

Proposition 3.3. The # product in $H_*\Gamma B\bar{\mathcal{O}}(\mathbf{F}_q)$ is given on generators by the formulas

(a) $v_m \# v_n = (m,n)v_{m+n} = y_m \# y_n$

(b) $\tilde{u}_m \# v_n = \tilde{u}_m \# y_n = (m,n)\tilde{u}_{m+n}$

(c) $\bar{u}_m \# v_n = \bar{u}_m \# y_n = \Sigma_{a+b=n}(a,m)\bar{u}_{m+a} * \bar{u}_b$

(d) $\tilde{u}_m \# \tilde{u}_n = 0$ unless $m = n = 0$ in which case $[\gamma] \# [\gamma] = [0$

Proof. Part (a) follows directly from Prop. 3.2 since $v_m, y_m \in H_*\Gamma B\bar{\mathcal{O}}(\mathbf{F}_q)$ are the images of $v_m, y_m \in H_*B\bar{\mathcal{O}}(\mathbf{F}_q)$ under the natural map $B\bar{\mathcal{O}}(\mathbf{F}_q) \to \Gamma B\bar{\mathcal{O}}(\mathbf{F}_q)$.

To prove part (b) we denote

$$i_1 : BZ/2 \xrightarrow{\cong} BO(1,\mathbf{F}_q) \to B\bar{\mathcal{O}}(\mathbf{F}_q) \to \Gamma B\mathcal{O}(\mathbf{F}_q)$$

$$i_2 : BZ/2 \xrightarrow{\cong} BO(\bar{1},\mathbf{F}_q) \to B\bar{\mathcal{O}}(\mathbf{F}_q) \to \Gamma B\mathcal{O}(\mathbf{F}_q)$$

Then $\Phi \circ i_1 = i_2$ and in homology

$$i_{1*}(x_n) = v_n \qquad i_{2*}(x_n) = y_n$$

while by IV 3.2

$$(i_2 - i_1)_*(x_n) = (\Phi - 1)_* i_{1*}(x_n) = (\Phi - 1)_*(v_n) = \tilde{u}_n$$

Let $\pi_1,\pi_2\colon B\mathbb{Z}/2 \times B\mathbb{Z}/2 \to B\mathbb{Z}/2$ denote the two projections. We first note that

$$i_2\circ\pi_1 \# i_1\circ\pi_2 = \Phi(i_1\circ\pi_1 \# i_1\circ\pi_2)$$

since evaluating on a generator $x_m \otimes x_n \in H_*(B\mathbb{Z}/2 \times B\mathbb{Z}/2)$ gives

$$y_m \# v_n = \Phi(v_m \# v_n)$$

which follows from Prop. 3.2 and IV 3.1. Similar reasoning shows

$$i_1\circ\pi_1 \# i_2\circ\pi_2 = \Phi(i_2\circ\pi_1 \# i_2\circ\pi_2)$$

$$i_1\circ\pi_1 \# i_1\circ\pi_2 = i_2\circ\pi_1 \# i_2\circ\pi_2$$

$$i_1\circ\pi_1 \# i_2\circ\pi_2 = i_2\circ\pi_1 \# i_1\circ\pi_2$$

Using distributivity and the above formulas we obtain

$$\begin{aligned}(i_2-i_1)\circ\pi_1 \# i_1\circ\pi_2 &= i_2\circ\pi_1 \# i_1\circ\pi_2 - i_1\circ\pi_1 \# i_1\circ\pi_2\\
&= \Phi\circ(i_1\circ\pi_1 \# i_1\circ\pi_2) - i_1\circ\pi_1 \# i_1\circ\pi_2\\
&= (\Phi - 1)\circ(i_1\circ\pi_1 \# i_2\circ\pi_2)\\
&= (1 - \Phi)\circ(i_2\circ\pi_1 \# i_2\circ\pi_2)\\
&= i_2\circ\pi_1 \# i_2\circ\pi_2 - \Phi(i_2\circ\pi_1 \# i_2\circ\pi_2)\\
&= i_2\circ\pi_1 \# i_2\circ\pi_2 - i_1\circ\pi_1 \# i_2\circ\pi_2\\
&= (i_2 - i_1)\circ\pi_1 \# i_2\circ\pi_2\end{aligned}$$

Applying these maps in homology to the element

$x_m \otimes x_n \in H_*(B\mathbb{Z}/2 \times B\mathbb{Z}/2)$ we get (b).

To prove (c) we use the distributivity formula 2.5(iii)

$$\bar{u}_m \# v_n = (\tilde{u}_m * \tilde{u}_0) \# v_n = \Sigma_{a+b=n}(\tilde{u}_m \# v_a)*(\tilde{u}_0 \# v_b)$$

$$= \Sigma_{a+b=n}(a,m)\tilde{u}_{m+a}*\tilde{u}_b = \Sigma_{a+b=n}(a,m)\bar{u}_{m+a}*\bar{u}_b$$

$$= \bar{u}_m \# y_n.$$

To prove (d) we argue as in (b):

$$(i_2 - i_1)\circ\pi_1 \# (i_2 - i_1)\circ\pi_2$$

$$= i_2\circ\pi_1 \# i_2\circ\pi_2 + i_1\circ\pi_1 \# i_1\circ\pi_2 - i_1\circ\pi_1 \# i_2\circ\pi_2 - i_2\circ\pi_1 \# i_1\circ\pi_2$$

$$= 2i_1\circ\pi_1 \# i_1\circ\pi_2 - 2i_2\circ\pi_1 \# i_1\circ\pi_2$$

$$= (2i_1 - 2i_2)\circ\pi_1 \# i_1\circ\pi_2$$

$$= 0 \# i_1\circ\pi_2$$

$$= 0$$

since $2(i_1 - i_2) = 2(\Phi - 1)i_2 = 0$.

Having computed the $\#$ product for a set of generators of $H_*(\Gamma B\bar{\mathcal{O}}(\mathbf{F}_q))$ we can extend our computations to arbitrary elements in the homology by using the distributive formula 2.5(iii). We do this explicitly for the exterior part of $H_*(\Gamma B\bar{\mathcal{O}}(\mathbf{F}_q))$.

<u>Proposition</u> 3.4. If $\bar{x},\bar{y}$ are in the image of

$$\tau_*: H_*(0) \to H_*(\Gamma_0^+ B\bar{\mathcal{O}}(\mathbf{F}_q))$$

(cf. I 2.1, III 3.1(e)) then

(a) $\bar{x} \# \bar{y} = 0$

(b) $(\bar{x}*[1]) \# (\bar{y}*[1]) = \bar{x}*\bar{y}*[1]$

unless both $\bar{x}$ and $\bar{y}$ have degree 0.

Proof. To prove (a) use Prop. 3.3(d) and induction on the algebraic degree (modulo the $*$ product) of $\bar{x}$ and $\bar{y}$ with respect to the generators $\{\tilde{u}_k | k \geq 0\}$, together with the distributive property 2.5(iii):

$$\Sigma(\bar{x}*\bar{z}) \# \bar{y} = \Sigma(\bar{x} \# \bar{y}')*(\bar{z} \# \bar{y}'') = 0$$

To prove (b), we can assume that $x \in H_*(\Gamma_i B\bar{\mathcal{O}}(\mathbf{F}_q))$, $y \in H_*(\Gamma_j B\bar{\mathcal{O}}(\mathbf{F}_q))$ where i,j is either 0 or γ. Then by 2.5(iv)

$$\begin{aligned}(\bar{x}*[1]) \# (\bar{y}*[1]) &= \Sigma\Sigma(\bar{x}' \# \bar{y}')*(\bar{x}'' \# [1])*(\bar{y}'' \# [1])*[1] \\ &= ([i] \# [j])*\bar{x}*\bar{y}*[1] \\ &= [0]*\bar{x}*\bar{y}*[1] \\ &= \bar{x}*\bar{y}*[1]\end{aligned}$$

since all the other terms in the summation are zero by (a).

We now use the fact that multiplicative homology operations are induced from the tensorial wreath product to compute them on the generators $\{v_i, y_i\}$ of $H_*(\Gamma B\bar{\mathcal{O}}(\mathbf{F}_q))$.

Proposition 3.5. $\tilde{Q}^n(v_k) = \tilde{Q}^n(y_k) = 0$ unless $k = n = 0$.

Proof. The maps

$$\mathbb{Z}_2 \times \mathbb{Z}_2 \xrightarrow{1\times\Delta} \mathcal{S}_2 \wr O(1,\mathbb{F}_q) \xrightarrow{\otimes} O(1,\mathbb{F}_q)$$

$$\mathbb{Z}_2 \times \mathbb{Z}_2 \xrightarrow{1\times\Delta} \mathcal{S}_2 \wr O_-(1,\mathbb{F}_q) \xrightarrow{\otimes} O(1,\mathbb{F}_q)$$

are both clearly trivial. Using Adem's evaluation of these maps on homology (cf. IX 2.3) we obtain

$$\Sigma_{s=0}^{[k/2]}(s,k-2s)\tilde{Q}^{n+s}(v_{k-s}) = 0$$

$$\Sigma_{s=0}^{[k/2]}(s,k-2s)\tilde{Q}^{n+s}(y_{k-s}) = 0$$

unless both $n = 0$ and $k = 0$. By induction it follows that $\tilde{Q}^n(y_k) = \tilde{Q}^n(v_k) = 0$ unless $k = n = 0$.

We can now use this result and the mixed Cartan formula (cf. 2.5(viii)) to compute the multiplicative homology operations for arbitrary elements in $H_*(\Gamma B\overline{\mathcal{O}}(\mathbb{F}_q))$. We first compute these homology operations on the image of

$$\tau_*: H_*(0) \to H_*(\Gamma B\overline{\mathcal{O}}(\mathbb{F}_q))$$

<u>Proposition 3.6</u>. $\tilde{Q}^k(\tilde{u}_n) = \begin{cases} 0 & \text{if } n \neq 0 \\ \bar{u}_k = Q^k([\gamma]) & \text{if } n = 0 \end{cases}$

<u>Proof</u>. We have by IV 2.8 and IV 3.1

$$y_n = \Phi_*(v_n) = \Phi_*(\bar{v}_n * [1]) = \Sigma_{a+b=n} v_a * \tilde{u}_b$$

Hence by the mixed Cartan formula

$$\tilde{Q}^k(y_n) = \Sigma_{a+b+c+d=n}\Sigma_{r+s+t=k}\tilde{Q}^r(v_a)*Q^s(v_b \# \tilde{u}_c)*\tilde{Q}^t(\tilde{u}_d)$$

$$= \Sigma_{b+c+d=n}\Sigma_{s+t=k}[1]*Q^s((b,c)\tilde{u}_{b+c})*\tilde{Q}^t(\tilde{u}_d)$$

$$= \Sigma_{a+d=n}\Sigma_{s+t=k}[1]*(\Sigma_{b=0}^{a}(b,a-b))Q^s(\tilde{u}_a) * \tilde{Q}^t(\tilde{u}_d)$$

$$= \Sigma_{a+d=n}\Sigma_{s+t=k}2^a[1]*Q^s(\tilde{u}_a)*\tilde{Q}^t(\tilde{u}_d)$$

$$= \Sigma_{s+t=k}Q^s([\gamma])*\tilde{Q}^t(\tilde{u}_n)*[1]$$

$$= \Sigma_{s+t=k}\overline{u}_s*\tilde{Q}^t(\tilde{u}_n)*[1]$$

If $n \neq 0$, then by Prop. 3.5

$$\Sigma_{s+t=k}\overline{u}_s*\tilde{Q}^t(\tilde{u}_n)*[1] = 0$$

for all k so by induction it follows that

$$\tilde{Q}^k(\tilde{u}_n) = 0.$$

If $n = 0$, then by Prop. 3.5

$$\Sigma_{s+t=k}\overline{u}_s*\tilde{Q}^t([\gamma])*[1] = \begin{cases} [1] \text{ if } n = 0 \\ 0 \text{ if } n \neq 0 \end{cases} = \Sigma_{s+t=k}\overline{u}_s*\overline{u}_t*[1$$

for all k. Again by induction it follows that

$$\tilde{Q}^k([\gamma]) = \overline{u}_k.$$

<u>Proposition</u> 3.7. $\tilde{Q}^k(\overline{u}_n*[1]) = Q^k(\overline{u}_n)*[1]$

<u>Proof</u>. We have by the mixed Cartan formula

$$\tilde{Q}^k(\overline{u}_n*[1]) = \tilde{Q}^k(\tilde{u}_n*[\overline{1}])$$

$$= \Sigma_{a+b=n}\Sigma_{r+s+t=k}\tilde{Q}^r(\tilde{u}_a)*Q^s(\tilde{u}_b \# [\overline{1}])*\tilde{Q}^t([\overline{1}])$$

$$= \Sigma_{a+b=n}\Sigma_{r+s=k}\tilde{Q}^r(\tilde{u}_a)*Q^s(\tilde{u}_b)*[1]$$

$$= \Sigma_{r+s=k}\tilde{Q}^r([\gamma])*Q^s(\tilde{u}_n)*[1]$$

$$= \Sigma_{r+s=k}Q^r([\gamma])*Q^s(\tilde{u}_n)*[1]$$

$$= Q^k([\gamma]*\tilde{u}_n)*[1]$$

$$= Q^k(\bar{u}_n)*[1]$$

<u>Theorem</u> 3.8. If $\bar{x}$ is in the image of

$$\tau_*: H_*(SO) \to H_*(\Gamma_0 B\bar{\mathcal{O}}(\mathbf{F}_q))$$

then

$$\tilde{Q}^k(\bar{x}*[1]) = \tilde{Q}^k((\bar{x}*[\gamma])*[1]) = Q^k(\bar{x})*[1]$$

<u>Proof</u>. By Proposition 3.4 the map

$$t_*\circ\tau_*: H_*(SO) \to H_*(\Gamma_1 B\bar{\mathcal{O}}(\mathbf{F}_q))$$

(where t: $\Gamma_0 B\bar{\mathcal{O}}(\mathbf{F}_q) \to \Gamma_1 B\bar{\mathcal{O}}(\mathbf{F}_q)$ denotes translation) is a map of Hopf algebras, the left side having its additive product $*$ and the right side the product $\#$. By Proposition 3.7

$$t_*\circ\tau_* Q^k(u_n) = \tilde{Q}^k t_*\circ\tau_*(u_n)$$

Since the u_n's generate $H_*(SO)$ it follows that

$$t_*\circ\tau_* Q^k(x) = \tilde{Q}^k t_*\circ\tau_*(x)$$

for all $x\in H_*(SO)$. Hence for any $\bar{x}$ in im τ_*

$$Q^k(\bar{x})*[1] = \tilde{Q}^k(\bar{x}*[1])$$

Next we observe that

$$\bar{x} \# [\bar{1}] = \bar{x} \# ([\gamma]*[1]) = \Sigma(\bar{x}' \# [\gamma])*\bar{x}'' = [0]*\bar{x} = \bar{x}$$

since all the other terms of the summation vanish by Proposition 3.4. Hence by the mixed Cartan formula

$$\tilde{Q}^k(\bar{x}*[\gamma]*[1]) = \tilde{Q}^k(\bar{x}*[\bar{1}]) = \Sigma_{r+s=k}\Sigma\tilde{Q}^r(\bar{x}')*Q^s(\bar{x}'' \# [\bar{1}])*\tilde{Q}^0([\bar{1}])$$

$$= \Sigma_{r+s=k}\Sigma\tilde{Q}^r(\bar{x}')*Q^s(\bar{x}'')*[1]$$

$$= \tilde{Q}^k(\bar{x}*[1])$$

$$= Q^k(\bar{x})*[1]$$

We now turn to formulating a general algorithm for computing multiplicative homology operations. We begin by looking at the generators $\{\bar{v}_i\}$.

<u>Proposition</u> 3.8. $\tilde{Q}^k(\bar{v}_n) = Q^k(\chi(\bar{v}_n))$.

<u>Proof</u>. We proceed by induction on n. For $n = 0$ this is obvious. Assume it for $i < n$. Then by the mixed Cartan relations

$$0 = \tilde{Q}^k(v_n) = \tilde{Q}^k(\bar{v}_n*[1]) = \Sigma^n_{a=0}\Sigma_{s+t=k}\tilde{Q}^s(\bar{v}_a)*Q^t(\bar{v}_{n-a})*[1]$$

$$= [\Sigma_{s+t=k}\tilde{Q}^s(\bar{v}_n)*Q^t([0]) + \Sigma^{n-1}_{a=0}\Sigma_{s+t=k}Q^s(\chi(\bar{v}_a))*Q^t(\bar{v}_{n-a})]*[1]$$

$$= [\tilde{Q}^k(\bar{v}_n) + \Sigma^{n-1}_{a=0}Q^k(\chi(\bar{v}_a)*\bar{v}_{n-a})]*[1]$$

$$= [\tilde{Q}^k(\bar{v}_n) + Q^k(\Sigma^{n-1}_{a=0}\chi(\bar{v}_a)*\bar{v}_{n-a})]*[1]$$

$$= [\tilde{Q}^k(\bar{v}_n) + Q^k(\chi(\bar{v}_n))]*[1]$$

Hence $\tilde{Q}^k(\bar{v}_n) = Q^k(\chi(\bar{v}_n))$. This completes the induction and proof.

Proposition 3.9. There is a map $\zeta: H_*(\Gamma_0^{\pm}B\bar{\mathcal{O}}(\mathbb{F}_q)) \to H_*(\Gamma_0^{\pm}B\bar{\mathcal{O}}(\mathbb{F}_q))$ such that if $\bar{x} \in H_*(\Gamma_0^{+}B\bar{\mathcal{O}}(\mathbb{F}_q))$ then

$$\tilde{Q}^k(\bar{x}) = Q^k(\zeta(\bar{x}))$$

Proof. In Proposition 3.8 we showed

$$\tilde{Q}^k(\bar{v}_n) = Q^k(\chi(\bar{v}_n))$$

while from Proposition 3.6 it follows that

$$\tilde{Q}^k(\tilde{u}_n) = 0 = Q^k(0) \text{ if } n > 0$$

$$\tilde{Q}^k([\gamma]) = Q^k([\gamma])$$

Hence we can define $\zeta(\bar{v}_n) = \chi(\bar{v}_n)$, $\zeta([\gamma]) = [\gamma]$, and $\zeta(\tilde{u}_n) = 0$ if $n > 0$. Then we proceed to define ζ inductively on the degrees of elements in the generators $\{\bar{v}_i, \tilde{u}_i\}$ with respect to $*$ using the mixed Cartan formula: thus assuming we have defined $\zeta(\bar{z})$ for $\bar{z}$ having degree $\leq$ degree $\bar{x}$ and degree $\bar{y}$ we get

$$\begin{aligned}\tilde{Q}^k(\bar{x}*\bar{y}) &= \Sigma\Sigma\Sigma_{a+b+c=k}\tilde{Q}^a(\bar{x}')*Q^b(\bar{x}'' \# \bar{y}')*\tilde{Q}^c(\bar{y}'') \\ &= \Sigma\Sigma\Sigma_{a+b+c=k}Q^a(\zeta(\bar{x}'))*Q^b(\bar{x}'' \# \bar{y}')*Q^c(\zeta(\bar{y}'')) \\ &= \Sigma\Sigma Q^k(\zeta(\bar{x}')*[\bar{x}'' \# \bar{y}']*\zeta(\bar{y}'')) \\ &= Q^k(\Sigma\Sigma\zeta(\bar{x}')*[\bar{x}'' \# \bar{y}']*\zeta(\bar{y}''))\end{aligned}$$

so we are forced to define

$$\zeta(\bar{x}*\bar{y}) = \Sigma\Sigma\zeta(\bar{x}')*[\bar{x}'' \# \bar{y}']*\zeta(\bar{y}'')$$

This completes the induction and proof.

Theorem 3.10. There is a map ρ: $H_*(\overset{+}{\Gamma_1} B\bar{\mathcal{O}}(\mathbf{F}_q)) \to H_*(\overset{+}{\Gamma_0} B\bar{\mathcal{O}}(\mathbf{F}_q))$ such that for $x \in H_*(\overset{+}{\Gamma_1} B\bar{\mathcal{O}}(\mathbf{F}_q))$

$$\tilde{Q}^k(x) = Q^k(\rho(x)) * [1]$$

Proof. We have $x = \bar{x} * [1]$ so

$$\begin{aligned}\tilde{Q}^k(x) &= \tilde{Q}^k(\bar{x} * [1]) = \Sigma\Sigma_{a+b=k} \tilde{Q}^a(\bar{x}') * Q^b(\bar{x}'') * [1] \\ &= \Sigma\Sigma_{a+b=k} Q^a(\zeta(\bar{x}')) * Q^b(\bar{x}'') * [1] \\ &= \Sigma Q^k(\zeta(\bar{x}') * \bar{x}'') * [1] \\ &= Q^k(\Sigma\zeta(\bar{x}') * \bar{x}'') * [1]\end{aligned}$$

Thus if we define

$$\rho(x) = \Sigma\zeta(\bar{x}') * \bar{x}''$$

we get

$$\tilde{Q}^k(x) = Q^k(\rho(x)) * [1]$$

This completes the proof.

Remark. If we work in the space $\Gamma B\mathcal{O}(\mathbf{R})$, we obtain an identical algorithm for computing homology operations in $BO_\otimes = \Gamma_1 B\mathcal{O}(\mathbf{R})$. Now if we denote by δ_*: $H_*(\Gamma B\mathcal{O}(\mathbf{R})) \to H_*(\Gamma B\bar{\mathcal{O}}(\mathbf{F}_q))$ the map of Hopf algebras given by $\zeta_*(\bar{e}_i) = \bar{v}_i$ and $\zeta_*([\pm 1]) = [\pm 1]$ then it follows that for $x \in H_*(BO_\otimes)$ we have by Theorem IX 2.2

$$\tilde{Q}^s(\delta_*(x)) = Q^s(\rho\delta_*(x))*[1] = Q^s(\delta_*\rho(x))*[1]$$

$$= \begin{cases} (\Phi_*\delta_*Q^s(\rho(x)))*[1] = (\Phi_*\delta_*(Q^s(\rho(x))*[1]))*[\gamma] & \text{if } q \equiv \pm 3 \pmod 8 \\ \delta_*(Q^s(\rho(x))*[1]) & \text{if } q \equiv \pm 1 \pmod 8 \end{cases}$$

$$= \begin{cases} (\Phi_*\delta_*\tilde{Q}^s(x))*[\gamma] & \text{if } q \equiv \pm 3 \pmod 8 \\ \delta_*\tilde{Q}^s(x) & \text{if } q \equiv \pm 1 \pmod 8 \end{cases}$$

Thus the formulas for the multiplicative homology operations for the polynomial part of the algebra $H_*(\Gamma_1 B\bar{\mathcal{O}}(\mathbf{F}_q))$ are formally identical to the corresponding formulas in $H_*(BO_\otimes)$ if $q \equiv \pm 1 \pmod 8$ and differ from the latter by a "twist" if $q \equiv \pm 3 \pmod 8$.

We conclude with a result relating the automorphism $\Phi\colon \Gamma B\bar{\mathcal{O}}(\mathbf{F}_q) \to \Gamma B\bar{\mathcal{O}}(\mathbf{F}_q)$ with the # product and the multiplicative homology operations.

<u>Theorem</u> 3.11. The following relations hold

(a) $\# \circ (\Phi \times \Phi) = \#$

(b) $\# \circ (\Phi \times 1) = \# \circ (1 \times \Phi) = \Phi \circ \#$

Consequently if $x,y \in H_*(\Gamma B\bar{\mathcal{O}}(\mathbf{F}_q))$

$$\Phi_*(x) \# \Phi_*(y) = x \# y,\ \Phi_*(x) \# y = x \# \Phi_*(y) = \Phi_*(x \# y)$$

Also for any $x \in H_*(\Gamma B\bar{\mathcal{O}}(\mathbf{F}_q))$

$$\tilde{Q}^s\Phi_*(x) = \tilde{Q}^s(x)$$

<u>Proof.</u> To prove (a) we note that we have the following commutative diagram of categories and functors

$$\begin{array}{ccc} \bar{\mathcal{O}}(\mathbb{F}_q) \times \bar{\mathcal{O}}(\mathbb{F}_q) & \xrightarrow{\otimes} & \bar{\mathcal{O}}(\mathbb{F}_q) \\ \downarrow{\scriptstyle \Phi\times\Phi} & & \downarrow{\scriptstyle \Phi^2} \\ \bar{\mathcal{O}}(\mathbb{F}_q) \times \bar{\mathcal{O}}(\mathbb{F}_q) & \xrightarrow{\otimes} & \bar{\mathcal{O}}(\mathbb{F}_q) \end{array}$$

Applying the functor ΓB- to the diagram and using the fact that $\Phi^2 = 1$, we obtain (a).

To prove (b) we argue similarly using the commutative diagram

$$\begin{array}{ccc} \bar{\mathcal{O}}(\mathbb{F}_q) \times \bar{\mathcal{O}}(\mathbb{F}_q) & \xrightarrow{\otimes} & \bar{\mathcal{O}}(\mathbb{F}_q) \\ \uparrow{\scriptstyle \Phi\times 1} & & \uparrow{\scriptstyle \Phi} \\ \bar{\mathcal{O}}(\mathbb{F}_q) \times \bar{\mathcal{O}}(\mathbb{F}_q) & \xrightarrow{\otimes} & \bar{\mathcal{O}}(\mathbb{F}_q) \\ \downarrow{\scriptstyle 1\times\Phi} & & \downarrow{\scriptstyle \Phi} \\ \bar{\mathcal{O}}(\mathbb{F}_q) \times \bar{\mathcal{O}}(\mathbb{F}_q) & \xrightarrow{\otimes} & \bar{\mathcal{O}}(\mathbb{F}_q) \end{array}$$

To prove the last statement we observe that we have the following commutative diagram for tensorial wreath products

$$\begin{array}{ccc} \mathcal{S}_2 \wr O(n,\mathbb{F}_q) & \xrightarrow{1\wr\Phi} & \mathcal{S}_2 \wr O_-(n,\mathbb{F}_q) \\ & {\scriptstyle\otimes}\searrow \quad \swarrow{\scriptstyle\otimes} & \\ & O(n^2,\mathbb{F}_q) & \end{array}$$

Applying $H_*(B-)$ to the above diagram we get

$$\tilde{Q}^s \Phi_*(x) = \tilde{Q}^s(x)$$

for any $x \in H_*(B^-(\mathbf{F}_q))$. We also observe that

$$\tilde{Q}^s(\Phi_*[-1]) = \tilde{Q}^s([-1]*[\gamma]) = \Sigma_{a+b+c=s}\tilde{Q}^a([-1])*Q^b([-1] \# [\gamma])*\tilde{Q}^c([\gamma])$$

$$= \Sigma_{a+b+c=s}\tilde{Q}^a([-1])*Q^b([\gamma])*Q^c([\gamma])$$

$$= \Sigma_{a+d=s}\tilde{Q}^a([-1])*\Sigma_{b+c=d}Q^b([\gamma])*Q^c([\gamma])$$

$$= \Sigma_{a+d=s}\tilde{Q}^a([-1])*Q^d([\gamma]*[\gamma])$$

$$= \Sigma_{a+d=s}\tilde{Q}^a([-1])*Q^d([0]) = \tilde{Q}^s([-1])$$

Since $H_*(\Gamma B\bar{\mathcal{O}}(\mathbf{F}_q))$ is generated by $[-1]$ and $H_*(B\bar{\mathcal{O}}(\mathbf{F}_q))$, the following inductive argument

$$\tilde{Q}^s\Phi_*(y*z) = \tilde{Q}^s(\Phi_*(y)*\Phi_*(z))$$

$$= \Sigma_{a+b+c=s}\Sigma\Sigma\tilde{Q}^a\Phi_*(y')*Q^b(\Phi_*(y'') \# \Phi_*(z'))*\tilde{Q}^c(\Phi_*(z''))$$

$$= \Sigma_{a+b+c=s}\Sigma\Sigma\tilde{Q}^a(y')*Q^b(y''\#z')*\tilde{Q}^c(z'')$$

$$= \tilde{Q}^s(y*z)$$

shows that $\tilde{Q}^s\Phi_*(x) = \tilde{Q}^s(x)$ is valid for any $x \in H_*(\Gamma B\bar{\mathcal{O}}(\mathbf{F}_q))$.

Bibliography

[1] J. F. Adams, On the groups J(X) - II, Topology 3 (1965), 137-171.

[2] __________, Vector fields on spheres, Ann. of Math. (2) 75 (1962), 603-632.

[3] __________, and S. B. Priddy, Uniqueness of BSO, Math. Proc. Camb. Phil. Soc. (to appear).

[4] M. F. Atiyah, Characters and cohomology of finite groups, Inst. Hautes Etudes Sci. Publ. Math., 9 (1961), 23-64.

[5] __________, R. Bott, and A. Shapiro, Clifford Modules, Topology 3 (Suppl. 1) (1964), 3-38.

[6] __________, and G. B. Segal, Equivariant K-theory and completion, J. Diff. Geom. 3 (1969), 1-18.

[7] A. Borel and J. P. Serre, Groupes de Lie et puissances réduites de Steenrod, Amer. J. of Math., 75 (1953), 409-448.

[8] A. K. Bousfield and D. M. Kan, Homology limits, completions and localizations, Lecture Notes in Math. 304, Springer (1972).

[9] R. Brauer, A characterization of the characters of groups of finite order, Ann. of Math. 57 (1953), 357-377.

[10] W. Browder, Torsion in H-spaces, Annals of Math. 74 (1961), 24-51.

[11] H. Cartan, Périodicité des groupes d'homotopie stables des groupes classiques, d'apres Bott, Séminaire H. Cartan (1959/60), exposé 17.

[12] __________, and S. Eilenberg, Homological Algebra, Princeton Univ. Press, Princeton, N. J., 1956.

[13] F. Cohen, T. Lada, J. P. May, The homology of iterated loop spaces, Lecture Notes in Math. no. 533, Springer-Verlag, 1976.

[14] R. R. Clough, The $\mathbb{Z}_2$ cohomology of a candidate for BImJ, Ill. J. of Math. 14 (1970), 424-433.

[15] J. Dieudonné, Le Géometrie des Groupes Classiques, Ergebnisse der Mathematik und ihrer Grenzgebiete, 5 (1955), Springer.

[16] L. E. Dickson, Linear Groups, Dover Publications (1958).

[17] Z. Fiedorowicz and S. B. Priddy, An automorphism of JO(q), (to appear).

[18] E. M. Friedlander, Computations of K-theories of finite fields, Topology 15 (1976), 87-109.

[19] J. A. Green, The characters of the finite general linear groups, Trans. Amer. Math. Soc. 80 (1955), 402-447.

[20] D. M. Kan and W. P. Thurston, Every connected space has the homology of a $K(\pi,1)$, Topology 15 (1976), 253-258.

[21] J. R. Isbell, On coherent algebras and strict algebras, J. of Algebra, 13 (1969), 299-307.

[22] S. Kochman, The homology of the classical groups over the Dyer-Lashof algebra, Trans. Amer. Math. Soc. 185 (1973), 83-136.

[23] J. Kaplansky, Linear Algebra and Geometry, Allyn-Bacon, Boston, 1969.

[24] I. Madsen, V. Snaith, and J. Tornehave, Infinite loop maps in geometric topology, Math. Proc. Camb. Phil. Soc. 81 (1977), 399-430.

[25] J. P. May, E_∞ ring spaces and E_∞ ring spectra, Lecture Notes in Math. no. 577, Springer-Verlag, 1977.

[26] ________, E_∞ spaces, group completions, and permutative categories, London Math. Soc., Lecture Notes Series 11 (1974), Cambridge University Press.

[27] ________, The homology of E_∞ ring spaces, (part of [13]).

[28] ________, A general algebraic approach to Steenrod operations, Lecture Notes in Mathematics, no. 168, Springer-Verlag, 1970.

[29] ________, The geometry of iterated loop spaces, Lecture Notes in Mathematics, no. 271, Springer-Verlag, 1972.

[30] O. F. O'Meara, Introduction to Quadratic Forms, Springer-Verlag, 1971.

[31] S. MacLane, Categories for the Working Mathematician, Springer-Verlag, 1971.

[32] J. Milnor and J. C. Moore, On the structure of Hopf Algebras Annals of Math. , 81 (1965), 211-164.

[33] S. B. Priddy, Dyer-Lashof operations for the classifying spaces of certain matrix groups, Quart. J. Math. (3), 26 (1975), 179-193.

[34] D. G. Quillen, The Adams conjecture, Topology, 19 (1971), 67-80.

[35] ____________, On the cohomology and K-theory of the general linear groups over a finite field, Ann. of Math. 96 (1972), 552-586.

[36] ____________, Higher algebraic K-theory: I, Lecture Notes in Mathematics, no. 341, Springer-Verlag, 1973.

[37] G. B. Segal, Classifying spaces and spectral sequences, Publ. Math. I.H.E.S., 34 (1968), 105-112.

[38] J. P. Serre, Linear Representations of Finite Groups, Springer-Verlag, 1976.

[39] N. E. Steenrod, Cohomology Operations, Princeton Univ. Press, Princeton, N. J., 1962.

[40] R. G. Swan, The p-period of a finite group, Ill. J. of Math. 4 (1960), 341-346.

[41] J. B. Wagoner, Delooping classifying spaces in algebraic K-theory, Topology 11 (1972), 349-371.

[42] C. DeConcini, The mod 2 cohomology of the orthogonal groups over a finite field, Advances in Math. 27 (1978), 191-229.

[43] J. M. Shapiro, On the cohomology of the orthogonal and symplectic groups over a finite field of odd characteristic, Lecture Notes in Math. no. 551, Springer-Verlag, 1976.

Index